Chemoreception in the Carotid Body

Edited by
H. Acker   S. Fidone   D. Pallot   C. Eyzaguirre
D.W. Lübbers   R.W. Torrance

With 101 Figures and 21 Tables

Springer-Verlag
Berlin Heidelberg New York 1977

International Workshop at
Max-Planck-Institut für Systemphysiologie
Dortmund

Sponsored by Max-Planck-Gesellschaft
zur Förderung der Wissenschaften e. V.
and Deutsche Forschungsgemeinschaft

Cover: Computer drawing of a glomoid.
After D. W. Lübbers, L. Teckhaus, and E. Seidl

ISBN 3-540-08455-X Springer-Verlag Berlin Heidelberg New York
ISBN 0-387-08455-X Springer-Verlag New York Heidelberg Berlin

Library of Congress Cataloging in Publication Data Main entry under title: Chemoreception in the carotid body. Papers from an international workshop held at the Max-Planck-Institut für Systemphysiologie, Dortmund, Germany, and sponsored by Max-Plack-Gesellschaft zur Förderung der Wissenschaften and Deutsche Forschungsgemeinschaft. Bibliography: p. Includes index. 1. Chemoreceptors-Congresses. 2. Carotid gland-Congresses. I. Acker, Helmut. II. Max-Planck-Gesellschaft zur Förderung der Wissenschaften. III. Deutsche Forschungsgemeinschaft (Founded 1949) QP455.C53   599'.01'826   77-13691

This work is subject to copyright. All rights are reserved, whether the whole or part of the material is concerned, specially those of translation, reprinting, re-use of illustrations, broadcasting, reproduction by photocopying machine or similar means, and storage in data banks.
Under § 54 of the German Copyright Law, where copies are made for other than private use, a fee is payable to the publisher, the amount of the fee to be determined by agreement with the publisher.

© by Springer-Verlag Berlin · Heidelberg 1977
Printed in Germany

The use of registered names, trademarks, etc. in this publication does not imply, even in the absence of a specific statement, that such names are exempt from the relevant protective laws and regulations and therefore free for general use.

Offsetprinting and Binding: Oscar Brandstetter Druckerei KG, Wiesbaden
2127/3140-543210

Preface

It is becoming traditional that periodically those of us interested in the carotid body hold an international meeting to discuss their results. In 1966 a meeting was organized by R.W. Torrance in Oxford and in 1973 by M.J. Purves in Bristol; in 1974 A.S. Paintal organized a satellite symposium of the Physiological Congress in Kashmir. The organizers of these meetings are to be commended for their efforts in publishing both the papers and discussions.

At these meetings it has become apparent that the direction of research is becoming more sharpely focused on the cellular mechanisms of chemoreception. During the meeting in Dortmund the papers dealt mostly with the different cell types in the carotid body and their environment, i.e., local PO_2 and local flow. These included light and electron microscopic studies of the morphometric and histochemical properties of the different cells, microelectrode studies of the glomoid tissue to understand the conversion of a chemical stimulation into nervous activity, as well as strictly biochemical and physiologic investigations concerning the dependence of the chemoreceptive process on O_2 consumption and the turnover of catecholamines. In spite of the variety in methodology, it was apparent that all contributors had a common interest: to understand the mechanisms of chemoreception. Although at the meeting itself there was ample time to fully discuss the various papers, it has become necessary here to shorten the papers and discussions; otherwise the cost of publication would have been prohibitive.

The organization of such a workshop and the ensuing editorial work involves a great deal of effort. Here I wish to thank Mrs. Helga Durst for her magnificant organization of the program and subsequently programing the editors. For assistance during the symposium I am grateful to Mrs. G. Blümel, Miss E. Dufau, Mr. D. Sylvester, Mr. M. Grote, Mr. W. Thutewohl and Dr. R. Heinrich. In addition, I would like to thank Mrs. Colleen Peterson (University of Utah) for typing some of the manuscripts and the ensuing discussions. For generous financial support I thank Deutsche Forschungsgemeinschaft and Max-Planck-Gesellschaft. Finally, and not least, the editors wish to thank their secretaries for their painstaking work on the manuscript. Springer-Verlag is to be congratulated on its rapid publication of this book.

At the end of this workshop it was the unanimous opinion of the participants that these periodic meetings should continue. We all look forward with interest to the next opportunity for meeting to talk about the carotid body.

Max-Planck-Institut für Systemphysiologie, Dortmund/GFR   Helmut Acker

Contents

Session I: Morphometric and Histologic Analysis of the Cell Elements in the Carotid Body Tissue

Light-Microscopic and Electron-Microscopic Studies on the Morphology of Cat Carotid Body
(E. Seidl, D. Schäfer, K. Zierold, H. Acker, and D.W. Lübbers) 1

Origin of Nerve Terminals on Glomus Cells in Cat Carotid Body: A Study of Axoplasmic Movement of Labeled Material Along Sensory Neurons of the Petrosal Ganglion
(S. Fidone, P. Zapata, and L. J. Stensaas) 9

Identification of Sensory Axon Terminations in the Carotid Body by Autoradiography
(P. G. Smith and E. Mills). 17

Degenerative Changes in Rabbit Carotid Body Following Systematic Denervation and Preliminary Results about the Morphology of Sinus Nerve Neuromas
(H. Knoche and E.-W. Kienecker). 25

Regeneration of Nerves and Nerve Terminals in Rabbit Carotid Body Following Carotid Nerve Sectioning and Suturing
(E.-W. Kienecker and H. Knoche). 30

Chemoreceptor Activity in the Rabbit Carotid Sinus Nerve During Regeneration
(D. Bingmann, E.-W. Kienecker, and H. Knoche) 36

Recovery of Chemosensory Function of Regenerating Carotid Nerve Fibers
(P. Zapata, L.J. Stensaas, and C. Eyzaguirre) 44

Studies of Normal and Wobbler Mutant Carotid Bodies
(D.J. Pallot and T.J. Biscoe) 51

Fine Structure of Pressoreceptor Terminals in the Carotid Body (Mouse, Cat, Rat)
(K. Gorgas and P. Böck). 55

Capillary Distances and Oxygen Supply to the Specific Tissue of the Carotid Body
(D.W. Lübbers, L. Teckhaus, and E. Seidl) 62

Session II: Electrophysiological Characteristics of the Cell Elements in the Carotid Body

Effects of Temperature and Stimulating Agents on Carotid Body Cells
(C. Eyzaguirre, M. Baron, and R. Gallego) 71

Are the Conventional Electrophysiological Criteria Sufficient to Give Evidence of an Electrogenic Sodium Transport Across Neuronal Membranes?
(W. Zidek, E.-J. Speckmann, H. Caspers, and A. Lehmenkühler) 79

Reaction of Cultured Carotid Body Cells to Different Concentrations of Oxygen and Carbon Dioxide
(F. Pietruschka, D. Schäfer, and D.W. Lübbers) 86

Meaning of the Type I Cell for the Chemoreceptive Process – An Electrophysiological Study on Cultured Type I Cells of the Carotid Body
(H. Acker and F. Pietruschka) 92

Session III: Histochemical and Biochemical Investigation of the Transmitters in the Carotid Body

Dopamine Beta-Hydroxylase Activity in the Cat Carotid Body
(C. Belmonte, C. González, and A.G. García) 99

Endogenous Acetylcholine Levels in Cat Carotid Body and the Autoradiographic Localization of a High Affinity Component of Choline Uptake
(S. Fidone, S. Weintraub, W. Stavinoha, C. Stirling, and L. Jones) 106

Molecular Biology of Chemoreceptor Function: Induction of Tyrosine Hydroxylase in the Rat Carotid Body Elicited by Hypoxia
(I. Hanbauer) 114

Effects of Hypoxia on Carotid Body Type I Cells and Their Catecholamines. A Biochemical and Morphologic Study
(S. Hellström) 122

Loss of Histochemically Demonstrable Catecholamines in the Glomus Cells of the Carotid Body After α-Methylparatyrosine Treatment
(M. Grönblad and O. Korkala) 130

Enzymes and Inhibitors of the Catecholamine Metabolism in the Cat Carotid Body
(H. Starlinger) 136

Session IV: The Afferent and Efferent Chemoreceptive Pathway of the Carotid Body

A Pharmacologic Study on a Possible Inhibitory Role of Dopamine in the Cat Carotid Body Chemoreceptor
(K. Nishi) 145

Blockade of Carotid Body Chemosensory Inhibition
(P. Zapata and F. Llados) 152

Variable Influences of the Sympathetic Nervous System Upon Carotid Body Chemoreceptor Activity
(R. G. O'Regan) 160

Mechanism of Inhibition of Chemoreceptor Activity by Sinus Nerve Efferents
(P. Willshaw) 168

Further Studies on the Fluctuation of Chemoreceptor Discharge in the Cat
(M.J. Purves and J. Ponte) 175

Carotid Body Chemoreceptor Afferent Neurons in the Solitary Tract Nucleus Area of the Cat
(J. Lipski, R.M. McAllen, and A. Trzebski) 182

Multifactor Influences on the Functional Relationship Between Ventilation and Arterial Oxygen Pressure
(H. Kiwull-Schöne and P. Kiwull) 190

Session V: Morphometric Analysis of Ultrastructural Changes in the Carotid Body Tissue

Histofluorescent and Ultrastructural Studies on the Effects of Reserpine and Calcium on Dense-Cored Vesicles in Glomus Cells of the Rat Carotid Body
 (A. Hess) 201

Ultrastructural Changes in Sensory Nerve Endings Accompanying Increased Chemoreceptor Activity: A Morphometric Study of the Rat Carotid Body
 (D. McDonald) 207

Dense-Cored Vesicles and Cell Types in the Rabbit Carotid Body
 (A. Verna) 216

The Carotid Body Chief Cell as a Paraneuron
 (S. Kobayashi) 221

Session VI: Environmental Conditions for the Chemoreceptive Process in the Carotid Body

Factors Affecting O_2 Consumption of the Cat Carotid Body
 (W.J. Whalen and P. Nair) 233

Mathematical Analysis of Oxygen Partial Pressure Distribution of the Carotid Body Tissue
 (U. Grossmann, R. Wodick, H. Acker, and D.W. Lübbers) 240

Comparative Measurements of Tissue PO_2 in the Carotid Body
 (H. Weigelt and H. Acker) 244

A Functional Estimate of the Local PO_2 at Aortic Chemoreceptors
 (A.S. Paintal) 250

Role of Calcium Ions in the Mechanism of Arterial Chemoreceptor Excitation
 (M. Roumy and L.M. Leitner) 257

Tissue PO_2 in the Cat Carotid Body During Respiratory Arrest After Breathing Pure Oxygen
 (D. Bingmann, H. Schulze, H. Caspers, H. Acker, H.P. Keller, and D.W. Lübbers) 264

Relationship Between Local Flow, Tissue PO_2, and Total Flow of the Cat Carotid Body
 (H. Acker and D.W. Lübbers) 271

Effects of Temperature on Steady-State Activity and Dynamic Responses of Carotid Baro- and Chemoreceptors
 (D. Bingmann) 277

Manipulation of Bicarbonate in the Carotid Body
 (R.W. Torrance) 286

Subject Index 295

List of Participants

PD Dr. H. Acker
Max-Planck-Institut
für Systemphysiologie
Rheinlanddamm 201
4600 Dortmund/GFR

Prof. Dr. C. Belmonte
Facultad de Medicina
Departamento de Fisiologia y Bioquimica
Valladolid/Spain

Dr. D. Bingmann
Physiologisches Institut
der Universität Münster
Westring 6
4400 Münster/GFR

Prof. Dr. M. H. Blessing
Pathologisches Institut der
Städt. Krankenanstalten Köln
Ost-Merheimer Straße
5000 Köln/GFR

Prof. Dr. P. Böck
Anatomisches Institut
Techn. Universität München
Biedersteiner Straße 29
8000 München 40/GFR

Prof. Dr. H. Caspers
Physiologisches Institut
der Universität Münster
Westring 6
4400 Münster/GFR

Prof. Dr. C. Eyzaguirre
The University of Utah
Medical Center
College of Medicine
Department of Physiology
50, North Medical Drive
Salt Lake City, UT 84 132/USA

Dr. S. J. Fidone
The University of Utah
Medical Center
College of Medicine
Department of Physiology
50, North Medical Drive
Salt Lake City, UT 84 132/USA

Dr. K. Gorgas
Anatomisches Institut der
Universität Köln
Josef-Stelzmann-Straße 9
5000 Köln-Lindenthal/GFR

U. Grossmann
Max-Planck-Institut für
Systemphysiologie
Rheinlanddamm 201
4600 Dortmund/GFR

M. Grönblad
Yliopiston Anatomian Laitos
00170 Helsinki/Finland

Dr. I. Hanbauer
Department of Health
Education and Welfare
Section of Biochemical Pharmacology
Hypertension-Endocrine Branch
NIH, Building 10
Bethesda, MD 20014/USA

Dr. St. Hellström
c/o Dr. E. Costa's Laboratory
of Preclinical Pharmacology
NIMH
Washington, DC/USA

Prof. Dr. A. Hess
College of Medicine and Dentistry
of New Jersey Rutgers Medical School
Department of Anatomy
University Heights
Piscataway, NJ 08 854/USA

Dr. S. Ji
Max-Planck-Institut
für Systemphysiologie
Rheinlanddamm 201
4600 Dortmund/GFR

Dr. J.O. Jost
Anatomisches Institut
der Universität Münster
Vesaliusweg 2–4
4400 Münster/GFR

Dr. E.W. Kienecker
Anatomisches Institut
der Universität Münster
Vesaliusweg 2–4
4400 Münster/GFR

PD Dr. P. Kiwull
Institut für Physiologie
Ruhr-Universität Bochum
Postfach 2148
4630 Bochum/GFR

Dr. H. Kiwull-Schöne
Institut für Physiologie
Ruhr-Universität Bochum
Postfach 2148
4630 Bochum/GFR

Prof. Dr. H. Knoche
Anatomisches Institut
der Universität Münster
Vesaliusweg 2–4
4400 Münster/GFR

Dr. S. Kobayashi
Department of Anatomy
Niigata University
School of Medicine
951 Niigata/Japan

Dr. O. Korkala
Yliopiston Anatomian Laitos
Institutum Anatomicum Universitatis
Siltavuorenpenger 20
00170 Helsinki 17/Finland

Dr. L.-M. Leitner
Laboratoire de Cytologie
Université de Bordeaux II
Avenue des Facultés
3300 Talence/France

J. Lipski
Department of Physiology
Institute of Physiological Sciences
Medical Faculty
Warsaw 00927/Poland

Prof. Dr. H.H. Loeschcke
Physiologisches Institut
der Universität Bochum
Friederikastraße 11
4630 Bochum/GFR

Prof. Dr. D.W. Lübbers
Max-Planck-Institut
für Systemphysiologie
Rheinlanddamm 201
4600 Dortmund/GFR

Dr. D. McDonald
University of California
Department of Anatomy
School of Medicine
The Cardiovascular Research Institute
San Francisco, CA 94 143/USA

Dr. K. Nishi
Department of Pharmacology
Kumamoto University, Medical School
2-2-1 Honjo
Kumamoto/Japan

Dr. R.G. O'Regan
Department of Physiology
University College
Earlsfort Terrace
Dublin 2/Ireland

Prof. Dr. A.S. Paintal
The V. Patel Chest Institute
Delphi University
P.O. Box 2101
Delhi – 7/India

Dr. D. Pallot
University of Leicester
Department of Anatomy
6, University Road
Leicester LEI 7RH/England

Dr. F. Pietruschka
Max-Planck-Institut
für Systemphysiologie
Rheinlanddamm 201
4600 Dortmund /GFR

Prof. Dr. M.J. Purves
Department of Physiology
The Medical School
University of Bristol
University Walk
Bristol BS8 1TD/England

Dr. Michel Roumy
St. Georges Hospital
Medical School
Department of Physiology
Tooting
London SW17 OQT/England

Dr. D. Schäfer
Max-Planck-Institut
für Systemphysiologie
Rheinlanddamm 201
4600 Dortmund/GFR

Dr. H. Schulze
Physiologisches Institut
der Universität Münster
Westring 6
4400 Münster/GFR

Dr. E. Seidl
Max-Planck-Institut
für Systemphysiologie
Rheinlanddamm 201
4600 Dortmund/GFR

Dr. P.G. Smith
Department of Physiology
and Pharmacology
Duke University, Medical Center
Durham, NC 27 710/USA

Prof. Dr. E.-J. Speckmann
Physiologisches Institut
der Universität Münster
Westring 6
4400 Münster/GFR

Dr. H. Starlinger
Max-Planck-Institut
für Systemphysiologie
Rheinlanddamm 201
4600 Dortmund/GFR

Prof. Dr. W. Thorn
Institut für Organische Chemie und Biochemie
Dr. Martin-Luther-King-Platz 6
2000 Hamburg 13/GFR

Prof. Dr. R.W. Torrance
St. John's College
Oxford OXI 3 JP/England

Prof. Dr. med. A. Trzebski
Institute of Physiological Sciences
Medical Faculty
Krakowskie Przedmiescie 26/28
Warsaw 00927/Poland

Dr. A. Verna
Laboratoire de Cytology
Université de Bordeaux II
Avenue des Facultés
3300 Talence/France

G. Walter
Anatomisches Institut
der Universität Münster
Vesaliusweg 2–4
4400 Münster/GFR

Dr. H. Weigelt
Max-Planck-Institut
für Systemphysiologie
Rheinlanddamm 201
4600 Dortmund/GFR

Prof. Dr. W.J. Whalen
St. Vincent Charity Hospital
2351 East 22nd Street
Cleveland, OH 44 115/USA

Prof. Dr. W. Wiemer
Universitätsklinikum Essen
Physiologisches Institut
Hufelandstraße 55
4300 Essen/GFR

Dr. P. Willshaw
Sherrington School of Physiology
St. Thoma's Hospital
Medical School
Lambeth Palace Road
London SE1 7EH/England

Prof. Dr. P. Zapata
Departamento de Neurobiologia
Universidad Catolica de Chile
Casilla 114-D
Santiago 1/Chile

W. Zidek
Physiologisches Institut
der Universität Münster
Westring 6
4400 Münster/GFR

Dr. K. Zierold
Max-Planck-Institut
für Systemphysiologie
Rheinlanddamm 201
4600 Dortmund/GFR

Session I

Morphometric and Histologic Analysis of the Cell Elements in the Carotid Body Tissue

Light-Microscopic and Electron-Microscopic Studies on the Morphology of Cat Carotid Body

E. Seidl, D. Schäfer, K. Zierold, H. Acker, and D. W. Lübbers*

The purpose of this work is to examine the qualitative morphologic findings on cat carotid body by means of quantitative data obtained from histologic and electron-microscopic studies. Combined with physiological and biochemical values, these data could increase understanding of the structure and function of the carotid body, especially as regards its oxygen consumption.

Light-Microscopic Findings

The carotid body of an adult cat (1) was fixed by perfusion with buffered glutaraldehyde (pH 7.4, 420 mosmol, 4°C (13)), embedded in Epon, and cut into 1.5 µm semithin serial sections. Two Sects. 30 µm apart were stained with toluidine blue (Sects. 1 and 2 in Table 1) and used to make single light micrographs with the Leitz interference contrast device (objective/oil, magnification 1250_x; (8)). The color micrographs were montaged so that 300 × 500 µm^2 areas were available for planimetric evaluation. To facilitate this evaluation the contours of vessels, specific tissue, and nerves were marked with different colors on transparent foil (Fig. 1). The clusters of type I and type II cells, which are called glomoids (9) and together with an adjacent vessel form a functional unit, are surrounded by the stratum nervosum. It was not difficult to mark the vessels and the boundary between vessels and specific tissue. The delimitation of the specific tissue from the stratum nervosum was fixed in the region where the connective tissue was already dispersed. The parallel filaments were classed with the glomoid (specific tissue; (8)), and the other tissue was considered stratum nervosum (2). Additionally, a 7-µm thick Sect. from the Bouin fixed carotid body of another adult cat from anoxia experiment was analyzed. Sects. 1 and 2 were from the periphery, and Sect. 3, from the central region of the carotid body. In Sect. 3 the glomoids were cut lengthwise, and in Sects. 1 and 2, crosswise.

Table 1 shows the area occupied by large nerves, specific tissue, and stratum nervosum. These area measurements correspond to volume measurements (3,11). The quality of specific tissue is unexpectedly small compared with the quantity of stratum nervosum (more than 60%) and the vessels (slightly more than 20%). So far it cannot be determined whether the condition of anoxia, the fixation technique, the central or peripheral region, the direction of cutting through the glomoids, or differences in the construction of the carotid body (10) are responsible for the differing proportions of specific tissue. From the number and total area of the dissected vessels the mean cross-sectional

---

*The authors are indebted to Mrs. T. Brand and Mrs. M. Seiffert for skilful technical assistance.

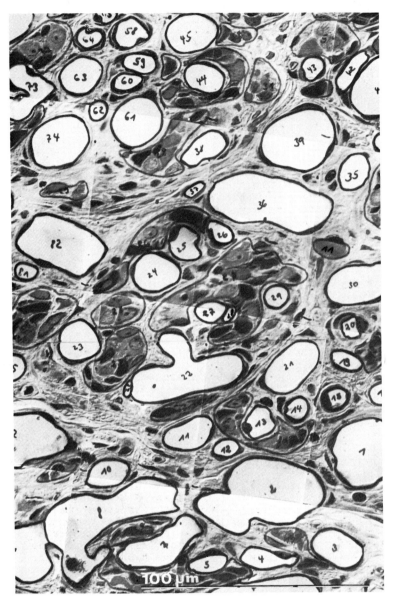

Fig. 1. Composite of several light micrographs showing semithin Sect. from glutaraldehyde-perfused cat carotid body. *Numbered areas*: white, vessels; gray, specific tissue

area of the vessels and the mean diameter were calculated (Table 2). The mean diameter was between 15 and 20-µm. Since some vessels were cut at an angle, however, the true diameter would certainly be smaller.

The median vascular diameter was determined from the frequency distribution, which indicated that half the vessels had a diameter less than 12 µm. It is possible that the vessels were dilated by perfusion fixation (Sects. 1 and 2) or by anoxic congestion (Sect. 3).

Table 1. Area occupied by tissue components from three Sects. of cat carotid body

Components	Sect. 1 (%)	Sect. 2 (%)	Sect. 3 (%)
Vessels	21.6	22.2	23.0
Large nerves	2.2	2.5	–
Specific tissue	15.8	13.3	21.7
Stratum nervosum	60.4	62.0	55.3

Table 2. Vascular characteristics from three Sects. of cat carotid body

Sects.	Vessel area (%)	No. of vessels	Mean vessel diam. (μm)	Median vessel diam. (μm)
1	21.6	175	15.4	11.4
2	22.2	141	17.3	12.6
3	23.0	96	19.3	12.3

Electron-Microscopic Findings

The electron-microscopic studies were restricted to the morphometric analysis of the glomoids. The electron micrographs were of the material used to provide Sects. 1 and 2; the utrathin Sect. was cut immediately after the semithin Sect. 2. Two glomoids were montaged at a magnification of 20,000× (Fig. 2). Type I cells, type II cells, collagen, mitochondria, and nuclei were traced on transparent foil or marked with colored pencils on the picture. The areas to be investigated were counted with the aid of the semiautomatic evaluation device MOP KM II of Fa. Kontron (square lattice of test lines of 2.5 mm spacing). We classed with the glomoid the type II cells, the small processes adjacent to type I cells, and separate, immediately adjoining parts of pseudopods. This classification allowed a clear delimitation of the glomoid, which corresponded roughly to that defined for the light micrograph. In contrast to the light microscope the electron microscope makes visible the spaces between the glomoid cells, and in these spaces we saw collagen and nerves. Consequently, only with the combined technique of light-microscopic and electron-microscopic analyses can the true portion of specific parenchyma (type I cells plus type II cells) be determined. Table 3 shows the percent distribution of type I cells, type II cells, nerves, and collagen of two glomoids (samples 1 and 2).

Table 3. Distribution of components of two glomoids

Components of glomoids (= 100%)	Sample 1 (%)	Sample 2 (%)
Type I cells	45.3	75.3
Type II cells	20.8	10.7
Nerves	5.9	2.5
Collagen	28.0	11.5

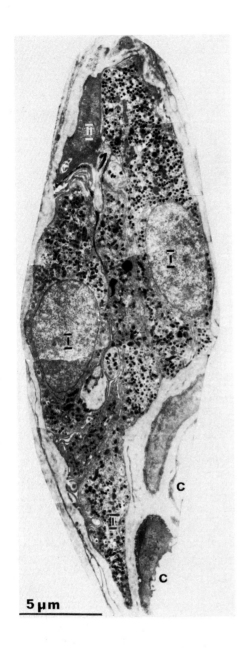

Fig. 2. Electron micrograph of glomoid sample 2. I, type I cells; II, type II cells; C, vessels

The proportions of the four components differed considerably in the two samples: the first sample contained 66% and the second 86% specific tissue.

We assume that about 75% of glomoid parenchymal tissue determined with the light microscope consisted of type I plus type II cells. Our findings show that the quantity of specific tissue in the cat carotid body is a quarter of that found by Laidler and Kay (4) for the young Wistar rat.

Table 4 shows the percent of the whole cell volume that the cytoplasm volume occupies. The cytoplasm volume was 50-75% of the whole cell volume.

Table 4. Cytoplasm volume in relation to whole cell volume

Components	Sample 1 (%)	Sample 2 (%)
Type I cells	49.2	74.0
Type II cells	47.6	70.9
Nerves	100	100

The results show that the nuclei of the carotid body parenchymal cells are considerably larger than the nuclei of liver cells, which occupy only 6% of the cell volume (5,6,12). However, the measured ratio of nuclei to cytoplasm was influenced to a large degree by the position of the cutting line across the cell (central or peripheral).

We also studied the mitochondrial concentration in the carotid body parenchyma (7). Table 5 shows the percent of the cytoplasm volume occupied by the mitochondria. This percent is smaller than in nerve cells and considerably smaller than in liver cells, where it amounts to 16%, according to unpublished results by Schäfer and Zierold. Based on the proportion of mitochondria, which is small in comparison to the rat liver, a strong respiratory intensity cannot be assumed.

Table 5. Mitochondrial volume in relation to cytoplasm volume

Components	Sample 1 (%)	Sample 2 (%)
Type I cells	6.9	2.5
Type II cells	2.6	2.1
Nerves	5.3	13.0

Summary

Morphometric studies performed in three histologic Sects. (magnification 1250×) and two glomoids (secondary magnification 20,000×) provided quantitative data on the morphologic structure and composition of the carotid body. Noteworthy are the rather constant proportions of vessels (about 20%) and stratum nervosum (about 60%) as well as the small proportion of specific tissue (20%) (the latter percent would be reduced to 15% if the volume of type I and type II cells were considered). These preliminary results will be followed up by further studies.

References

1. Acker, H.: Der lokale Sauerstoffdruck im Glomus caroticum der Katze und seine Bedeutung für die nervöse chemoreceptive Aktivität. Habilitationsschrift, Ruhr-Universität Bochum 1975
2. Böck, P., Stockinger, L., Vyslonzil, E.: Z. Zellforsch. 105, 543-568 (1970)
3. Henning, A.: in: Proc. Symp. Quantitative Methods in Morphology. Weibel, E.R., Elias, H. (eds.). Berlin-Heidelberg-New York: Springer-Verlag 1967
4. Laidler, P., Kay, J.M.: J. Pathol. 117, 183-191 (1975)
5. Reith, A.: Lab. Invest. 29, 216-228 (1973)
6. Rohr, H.P., Riede, U.N.: in: Current Topics in Pathology. Grundmann, E., Kirsten, W.H. (eds.). Berlin-Heidelberg-New York: Springer-Verlag 1973, pp 1-48
7. Schäfer, D., Schäfer, S., Lübbers, D.W.: Arzneim. Forsch. 23, 1611-1612 (1973)
8. Seidl, E.: Leitz Mitt. Wiss. Techn. 69, 57-58 (1973)
9. Seidl, E.: in: The Peripheral Arterial Chemoreceptors. Purves, M.J. (ed.). New York: Cambridge University Press 1975, pp 293-299
10. Seidl, E.: Anat. Embryol. (Berl.) 149, 79-86 (1976)
11. Weibel, E.R., Kistler, G.S., Scherle, W.F.: J. Cell Biol. 30, 23-38 (1966)
12. Weibel, E.R., Stäuble, W., Gnägi, H.R., Hess, F.A.: J. Cell Biol. 42, 68-91 (1967)
13. Weissenfels, N., Schäfer, D., Hündgen, M.: Cytobiologie 3, 188-2o1 (1971)

DISCUSSION

Pallot: Have you divided the nerve component into those fibers passing through a complex and those that have some terminal specialization associated with type I cells?

Zierold: No, only the nerve tissue recognized in the glomoid was classified as nerve tissue.

Paintal: Your second slide of the light-microscopic analysis indicated that the large nerves occupied 2.2%. By large nerves I presume you mean the myelinated fibers.

Zierold: This was a reference to the nerves outside the glomoids. The nerves in the glomoids were not distinguished.

Paintal: Your work would have given you an opportunity to determine the diameter of these nerve fibers just before they become the sensitive portions of the apparatus. Do you have any measurements of the diameters of the nerve fibers?

Zierold: No.

Belmonte: How do you remove the carotid bodies? If the sympathetic tone changes, perhaps the diameter of the vessels also changes.

Lübbers: That is a very important question. It is always difficult to standardize the volume of the vessels, but we took just the glomus caroticum to get general information about the different compartments. It may be change somewhat, but basically the amount of tissue relating to the vasculature should not differ by more than 10 or 15% maximum.

Our main concern in this experiment was to examine the hypothesis that the glomus caroticum has a rather large respiratory activity, and we were very surprised that the actual amount of mitochondria found in the specific cells is so small. This is the opposite of what we had expected.

McDonald: Did you find a difference in the volume or percent area occupied by the blood vessels in carotid bodies fixed by perfusion and those fixed by immersion? Also, what were the conditions under which you induced hypoxia and how did those differ from your control conditions?

Zierold: In our special case there was no difference between the material fixed by perfusion and by immersion.

Lübbers: In hypoxia we normally find a dilated vasculature with red cells in the vessels that are normally empty.

McDonald: What was the PO_2 you achieved in the hypoxia?

Acker: This was a special case of anoxia. A general problem we have is finding open capillaries in the carotid body. The animal was ventilated with nitrogen and died. In this special case all the vessels were open and filled with red cells. If we perfuse with glutaraldehyde, we also find open vessels under these special conditions.

Eyzaguirre: Going back to the question asked by Lübbers, I wonder if you measured or counted the number of mitochondria in nerve terminals.

Zierold: No.

Paintal: Following up on what Eyzaguirre has asked, your percent of mitochondria in the cytoplasm is 5%. Do you think this is somewhat further away from the nerve terminal, which in many other receptors has a much higher ratio of mitochondria to cytoplasm?

Torrance: When you counted the mitochondria in nerve tissue, did you include the nerve terminals in your count?

Zierold: No, we did not distinguish terminals from fibers.

Lübbers: We have to be careful with these numbers because there are only three samples. It takes a lot of time to do all the counting, and it would be necessary to have 20 or 50 samples because we are testing only a few cells from the glomus caroticum, not the whole. In this situation it is not clear whether these cells are representative.

Pallot: If you count the mitochondria in the unmyelinated nerve fibers within the actual glomoid using the electron microscope the problem is that, depending upon the level of Sect. of the nerve fiber, the mitochondrial content will vary because the mitochondria are grouped over small areas of the nerve fiber. This may explain your range between 5.3 and 13%.

Torrance: Can one assume that most of the oxygen uptake in the carotid body goes through mitochondria? If so, what is the fraction of mitochondria that exists in nerve tissue? In the Biscoe hypothesis this is crucial because he argues that the oxygen consumption of the nerve fibers must be very great. Do your observations support this?

Lübbers: Our data do not support Biscoe's hypothesis. We have compared numbers mainly with liver cells and have found that the number of mitochondria in the nerves is the same as in liver cells.

Torrance: You estimate the percent of the carotid body that is nerve and the percent of nerve that is mitochondrial. You have also done this for the other tissues in the carotid body; hence, if the oxygen consumption per mitochondrion is constant, you should be able to say what fraction of total oxygen consumption of the carotid body occurs in nerve tissue.

Lübbers: I am sorry that we do not have good data about the oxygen consumption.

Torrance: I am talking about a proportion of the oxygen consumption.

Lübbers: Yes, but how can we deduce from the consumption of oxygen something about the fraction? We do not know about the oxygen consumption of the stratum nervosum.

Torrance: I am asking if you can assume that the maximum oxygen consumption that the tissue can achieve is proportional to the concentration of mitochondria in it. If you can, then you can estimate how much of the consumption of oxygen is located in the nerve fibers. Also, from your measurements of the capillaries and their fraction of the volume, you can make some calculation of the maximum distance likely for diffusion of oxygen from the vessels.

Lübbers: We shall discuss this in a later paper.

Origin of Nerve Terminals on Glomus Cells in Cat Carotid Body: A Study of Axoplasmic Movement of Labeled Material Along Sensory Neurons of the Petrosal Ganglion*

S. Fidone, P. Zapata and L. J. Stensaas

The site of termination of chemosensory fibers in the carotid body, once considered to be well established by the pioneering studies of de Castro (3), has been the object of renewed interest and speculation. De Castro concluded that sensory fibers in the cat carotid nerve terminate on glomus (type I) cells in the carotid body, because he observed that such nerve endings were unaltered following intracranial Sect. of the glossopharyngeal (IXth) nerve root central to its sensory ganglion (deefferentation). These findings have been challenged, however, by the more recent ultrastructure studies of Biscoe and co-workers (1). They repeated de Castro's original experiment and claimed that the synapses between carotid nerve terminals and glomus cells are *efferent*, not afferent; it appeared to them that the nerve endings did degenerate with deefferentation of the carotid nerve. This interpretation was supported by the concurrent discovery by Neil and O'Regan (11) of an efferent control of chemoreceptor discharge by fibers passing along the carotid nerve. This issue was complicated by the curious finding that the time course of nerve ending degeneration following decentralization was sometimes as much as 1 year, a finding that Biscoe et al. (1) compared to similarly long degeneration times for neurons in certain areas of the central nervous system.

Hess and Zapata (8) and Nishi and Stensaas (12) repeated the experiments involving intracranial Sect. of the IXth nerve, found that nerve endings on glomus cells were unaltered, and confirmed de Castro's original findings in direct contradiction to the results from Biscoe's group. Hess and Zapata (8) suggested an alternative explanation for the long degeneration times reported by Biscoe et al., namely, that trauma to the ganglion during the operative approach to the nerve roots produced slow necrosis of the sensory ganglion cells and led to the observed axonal changes. They drew attention to the fact that two sensory ganglia are commonly located along the IXth nerve, the petrosal ganglion and the smaller superior, or Ehrenritter's, ganglion. The latter is located intracranially and might have been damaged when Biscoe et al. (1) removed a portion of the nerve roots. Thus a series of arguments and counterarguments have developed sustaining the controversy.

In an attempt to resolve this issue, we decided to utilize the normal physiological phenomenon of axoplasmic flow to demonstrate the distribution of nerve terminals from neurons in the sensory ganglion (petrosal ganglion). Perikaryal uptake, incorporation, and axonal transport of $\{^3H\}$ amino acid to the axon terminals have been shown to provide a specific and sensitive method for identifying neuronal

---

[+]This work was supported by U.S. Public Health Service grants NS-1a636, NS-07938, NS-10864, and NS-05666.

projections (4). In this study {³H}proline was delivered to cells of the cat petrosal ganglion and the peripheral distribution of the labeled material determined by means of electron-microscopic autoradiography and liquid scintillation spectrometry. The details of the methods have been reported elsewhere (6,7,13).

The time course of net accumulation of radioactivity in the carotid body following administration of {³H}proline to the petrosal ganglion is shown in Fig. 1, expressed both as total carotid body counts per minute (cpm) (open bars) and as percent of radioactivity remaining in the petrosal ganglion (closed bars). The latter may more accurately relect the time course of accumulated radioactivity, since the number of ganglion cells labeled by the application of {³H}proline might be expected to vary in each experiment. The upward trend in total carotid body cpm in Fig. 1 suggests that migration of radioactivity out of the ganglion cells with time results in increased accumulation of radioactivity in the carotid body.

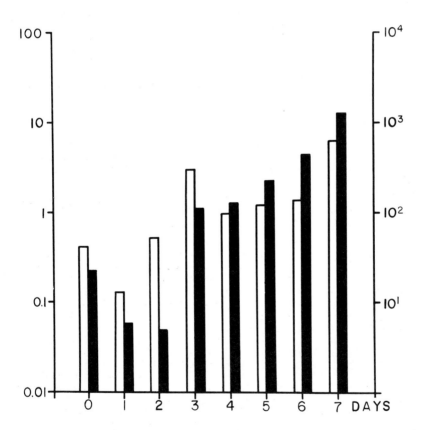

Fig. 1. Time course of accumulation of radioactivity in cat carotid body following administration of {³H}proline into petrosal ganglion. *Open bars*: counts per minute (cpm)/carotid body, expressed logarithmically. *Closed bars*: cpm/carotid body as percent of cpm remaining in petrosal ganglion, expressed logarithmically
▬▬▬▬ Carotid body/petrosal ganglion (%)
▭▭▭▭ Carotid body cpm

The short lengths of nerve (20-25 mm) involved in these experiments precluded accurate measurements of the rates of exoplasmic transport in the carotid or glossopharyngeal nerves, but the data suggested the presence of fast, intermediate, and slow components. The velocity of the fast component was estimated to be *in excess* of 200 mm/day, whereas the slow and intermediate components had velocities ranging from 1 to 70 mm/day. The trough in the curve of Fig. 1 at 1 and 2 days probably represents the interim between the arrival at the carotid body of the fast and the slower components.

The distribution of label in the carotid body was examined autoradiographically at 1-day intervals from 1 to 7 days after ganglionic injection of {^{3}H}proline. At all intervals the autoradiographic tracks were localized almost exclusively over nerve fibers and nerve terminals, with very few tracks over glomus cells, sustentacular cells, or other constituents of the carotid body. Both myelinated (Fig. 2) and unmyelinated axons in the carotid body were found to be labeled. Nerve terminals on glomus cells were identified by the presence of clear-cored vesicles and mitochondria. In addition to these organelles, electron-dense material, or junctional densities, were frequently observed at sites of membrane apposition between nerve terminal and glomus cell. Labeled axon profiles in carotid body autoradiographs included bouton endings (Fig. 4), calyciform endings, and terminal enlargements of intermediate shape and size (Fig. 3). Labeling of neural elements in the carotid body was sometimes quite effective. As much as 60-90% of the nerve endings in a given ultrathin Sect. autoradiograph was labeled. The incidence of labeling in glomus and sustentacular cells was extremely slow; only an occasional cell was found to be labeled and then only with a single overlying track.

Selective labeling of sensory nerve terminals in cat carotid body is dependent on several conditions. These are {1} that fibers of passage through the ganglion do not incorporate and tranport labeled amino acid to the carotid body, {2} that passive movement of label from the injection site is negligible, and {3} that no appreciable number of autonomic (efferent) ganglion cells are present in the petrosal ganglion whose axons enter the carotid body.

1. Experiments were performed in which {^{3}H}proline was administered to a desheathed region of the IXth nerve immediately distal to the petrosal ganglion. After 3 and 24 h no significant migration of labeled material occurred along the nerve. This finding argues against the possibility that fibers of passage through the petrosal ganglion can incorporate and transport labeled amino acid and agrees with observations by other investigators that axons lack or have limited capacity for incorporation and transport of amino acids (4).

2. Application of {^{3}H}proline to the nerve, as described above, resulted in movement of the label only 4-6 mm distally along the nerve after 3 h or after 24 h. Passive transfer of labeled material following ganglionic injection should then be no more than 4-6 mm after 24 h. The carotid body was always farther than 6 mm from the ganglion, minimally 13-15 mm and usually farther, and at 24 h labeled nerve endings were already identifiable in carotid body autoradiographs. Crushing the ganglion prior to the injection of {^{3}H}proline resulted in radioactivity only 1-2% of what would normally be present along the nerve after 3 days in an uninjured preparation. The limited passive movement of label observed in these experiments agrees with reports by other investigators who attribute this to uptake of amino acids by Schwann cells and connective tissue elements surrounding the nerve (9). Finally, it should be noted that the prominent localization of label to nerve endings in the carotid body, as shown autoradiographically,

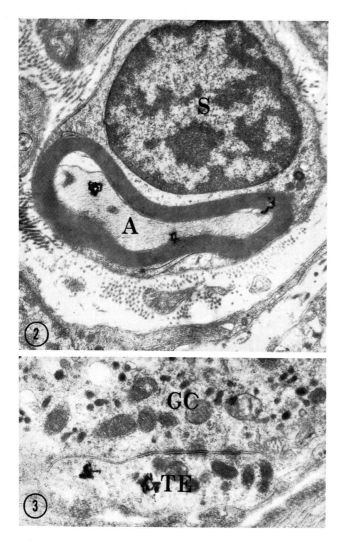

Fig. 2. Labeled myelinated axon (A) and Schwann cell (S) in stroma of carotid body 2 days after exposure of petrosal ganglion to {^{3}H}proline. 18,000×

Fig. 3. Labeled axon terminal enlargement (TE) of intermediate size and shape apposed to surface of glomus cell (GC) 1 day after labeling petrosal ganglion. 25,000×

cannot be reconciled with the passive spread of significant amounts of radioactivity from the ganglion to the carotid body. If the latter were to occur, one would expect to find nearly uniform labeling of the cellular constituents of the carotid body, including glomus cells. One would certainly not predict nerve endings to be the exclusive site of protein synthesis in the carotid body. Although synaptosomal fractions are capable of some degree of polypeptide synthesis, it is well established that the cell body is the principal site for amino acid incorporation into proteins.

3. Early anatomical studies of the glossopharyngeal and carotid nerves revealed the presence of a small number of autonomic ganglion cells

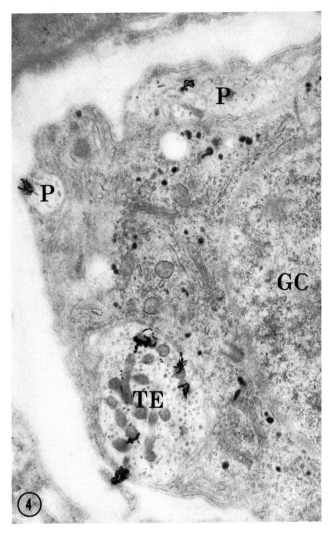

Fig. 4. Preterminal (P) and terminal axon enlargement (TE) 6 days after labeling petrosal ganglion. Terminal axon profile apposed to glomus cell (GC) is typical of bouton-type enlargements. 18,700×

scattered along these nerves in the cat (2). As indicated by Torrance (14), these ganglion cells might be strays from a large population of autonomic neurons present in the petrosal ganglion, an arrangement sometimes seen in nerve trunks near autonomic ganglia. Although there are no reports of efferent neurons in the petrosal ganglion, anatomical and physiological studies of the nodose ganglion have indicated the possibility of a synapse between preganglionic fibers from the brain stem and autonomic efferent neurons (5). There can be little doubt that the petrosal ganglion is predominantly a sensory ganglion, but the existence of a small population of autonomic ganglion cells in the petrosal ganglion might constitute part of an efferent pathway from brain stem to carotid body. Transport of labeled material into the terminals of visceromotor neurons following ganglionic injection of

{^{3}H}proline would thus nullify the conclusion that nerve endings on glomus cells arise exclusively from sensory neurons.

In consideration of this problem, the petrosal ganglion was carefully examined for the presence of autonomic ganglion cells. Ultrastucture analysis of 902 neurons at 17 levels of two cat petrosal ganglia provided no evidence for the presence of synapses on cell somata or on their processes (13). The results of this study are consistent with the concept that the petrosal ganglion is strictly a sensory ganglion, containing unipolar nerve cells exclusively. It is reasonable to conclude, therefore, that labeled nerve endings observed on glomus cells in carotid body autoradiographs following injection of {^{3}H}proline in the petrosal ganglion are distal processes of sensory neurons whose cell bodies are located in the ganglion. These observations, coupled with the high percentage of labeled endings in some autoradiographs, suggest that most nerve endings on glomus cells in the cat carotid body arise from sensory neurons in the petrosal ganglion. The possibility cannot be ruled out, however, that the remaining small percentage of unlabeled endings in an autoradiograph might be efferent terminals arising from neuronal perikarya located elsewhere (10).

References

1. Biscoe, T.J., Lall, A., Sampson, S.R.: J. Physiol. *208*, 133-152 (1970)
2. Castro, F. de: Trab. Lab. Invest. Biol. (Univ. Madr.) *24*, 365-432 (1926)
3. Castro, F. de: Trab. Lab. Invest. Biol. (Univ. Madr.) *25*, 331-380 (1928)
4. Cowan, W.M., Gottlieb, D.I., Hendrickson, A.E., Price, J.L., Woolsey, T.A.: Brain Res. *37*, 21-51 (1972)
5. Daly, I., Hebb, C., in: Pulmonary and Bronchial Vascular Systems. Bancroft, H. et al., (eds.). London: Arnold 1966
6. Fidone, S.J., Zapata, P., Stensaas, L.J.: Brain Res. (1977) (in press)
7. Fidone, S.J., Zapata, P., Stensaas, L.J.: Brain Res. *124*, 9-28 (1977)
8. Hess, A., Zapata, P.: Fed. Proc. *31*, 1365-1382 (1972)
9. Lasek, R.: Brain Res. *7*, 360-377 (1968)
10. McDonald, D.M., Mitchell, R.A.: J. Neurocytol. *4*, 177-230 (1975)
11. Neil, E., O'Regan, R.G.: J. Physiol. *200*, 69-71 (1969)
12. Nishi, K., Stensaas, L.J.: Cell Tissue Res. *154*, 303-319 (1974)
13. Stensaas, L.J., Fidone, S.J.: Brain Res. 124, 29-39 (1977)
14. Torrance, R.W.: in: The Pulmonary Circulation and Interstitial Space. Fishman, A.P., Hecht, H.H. (eds.). Chicago: Univ. Press 1969, pp. 223-237

DISCUSSION

Nishi: There are many small nerve fibers of about 0.1 μm diameter in the carotid body: many people think these nerve fibers have a chemoreceptor function. You show that the large nerve endings have their cell bodies in the petrosal ganglion so they are presumably afferent, but could you identify these small nerve fibers as afferent of efferent?

Fidone: We observed labeling over both large and small diameter nerve fibers. Unfortunately, for a labeled unmyelinated nerve fiber it is not known whether its parent fiber is myelinated or unmyelinated.

Kobayashi: Why did you use tritiated proline? Second, where was the peptide located in the nerve endings?

Fidone: Tritiated proline has been demonstrated to be marker of choice for tracing neuronal projections. It appears to have a much lower autoradiographic background than leucine, for example. With regard to the second question, this type of autoradiography does not permit localization in the nerve terminal because of limited resolution. You would have to resort to the use of physical developers and punctate grain size in order to answer that question.

McDonald: What happens to the labeled protein when it begins to disappear from the carotid body? Do you have any evidence of transfer from the nerve to the glomus cell?

Fidone: The half-life of the fast component appears to be about 18-24 h which fits with that described in the ciliary ganglion. The material does not appear to move transsynaptically. We see very little labeling in glomus cells. For example, of 178 neural elements among 76 glomus cells, the average track concentration was about four tracks per neural element, but only two tracks were found overlying all 76 glomus cells. The petrosal ganglion innervates the taste buds so we have also looked at the termination of nerve fibers in the taste buds. There, interestingly enough, we seem to find a greater degree of movement from the terminal into the receptor cells but not into the supporting cells. We are presently studying this. We want to cross-innervate the nerves going to the taste buds and the carotid body to see whether this transsynaptic movement is a function of a specific donor neuron or specific sensory cell or both.

McDonald: What is your conclusion, based on finding this difference between the taste bud and the glomus cell of the carotid body, about their comparative roles?

Fidone: The obvious trophic interdependence. When you denervate the taste bud, it disappears; when you denervate the carotid body, nothing happens. The implication is intriguing. It may be something is moving from afferent nerve fiber into the sensory receptor cell because not all afferent neurons reinnervate the taste buds. It has to be a potential gustatory neuron.

Pallot: If you look around the periphery of a type II/type I complex, what percentage of the nerve fibers are labeled?

Fidone: It is often quite high, although you can find groups of fibers that are all labeled and other groups that are all unlabeled.

O'Regan: The functional unit in the carotid body is basically 10-20 cells, which, according to McDonald and Mitchell's paper, is innervated by one nerve fiber that branches extensively. In these particular cells with tight junctions between them, could a small number of efferents, say 2-5%, which McDonald and Mitchell showed, by affecting one cell not affect the whole glomus complex? You thus would have a very small number of efferent and a large number of afferent endings so that you would be biased in favor of an afferent interpretation.

Fidone: That remaining 5% may have a very important function, but this study does not address itself to that problem.

O'Regan: But could we clarify what the functional unit in the carotid body is?

Fidone: I cannot at the moment. Perhaps it is the glomus cell, perhaps not.

Böck: Have you seen any fine structure specialization in the post-synaptic region such as accumulation of dense-cored vesicles or anything else?

Fidone: In looking at the autoradiograph we do see occasional conglomerations of vesicles, but the truth is we simply did not study that point carefully.

Nishi: In the cat it is very difficult to see specializations of the sensory nerve terminal, so I think Fidone is right. If you want to see some kind of specialization at the sensory synapse, you have to change your material, e.g., to the rat, but in the cat you cannot see any special kind of apparatus.

Fidone: It is certainly more difficult to find. I wonder if McDonald would agree with that.

McDonald: In our analysis comparing the rat and cat carotid bodies we have found that reciprocal synapses are five times more common in the rat than in the cat carotid body.

Torrance: You are measuring something that is quantal in these fibers, and some of the fibers were labeled on the basis of one track. Have you done any statistics of the type done by Katz on the quantal composition of very small end-plate potentials and could you thereby increase the number labeled?

Fidone: No. We have not tried that because the data indicated that a high percentage of the label was localized in the nerve endings and the fibers.

Pallot: Without getting into the argument about whether there are true efferent fibers in the sinus nerve, it has been shown repeatedly by Biscoe and Sampson, and by O'Regan and others that if you record from the central cut end of the sinus nerve, you can record centrifugal activity. If you have this high proportion of label, where do these nerve fibers go in the carotid body? Have you any thoughts on this?

Fidone: No, except we do see 5% unlabeled endings. Where they are coming from is uncertain.

Identification of Sensory Axon Terminations in the Carotid Body by Autoradiography

P. G. Smith and E. Mills

The innervation of the carotid body has been the subject of several studies employing nerve degeneration methods and electron microscopy. There is agreement that sinus nerve transection results in degeneration of the nerve endings on type I cells (1). It has been reported in the cat and rat that decentralization of the glossopharyngeal ganglia is without effect on these nerve endings (7,12,15,19). In constrast, Biscoe et al. (2) reported reductions in the number of nerve endings on type I cells and degenerative changes in remaining nerve endings after cutting the glossopharyngeal nerve rootlets and allowing longer survival periods than those of other studies. They were also able to record chemoreceptor discharge from the sinus nerve of carotid bodies that displayed degenerate nerve endings.

In attempting to resolve this conflict, we have investigated the innervation of the carotid body using a nondegenerative technique. We have employed electron-microscopic autoradiography in conjunction with axoplasmic transport in order to identify the terminations of petrosal ganglion neurons in the cat carotid body. This method allows direct visualization of the intact nerve terminals making it possible to describe both their morphology and relationships with other cell types.

Fig. 1 shows the relevant gross anatomy of the cranial portion of the glossopharyngeal nerve, which has been exposed by removing the tympanic bulla and a portion of the petrous temporal bone. The rootlets emerge from the medulla as a bundle and enter the jugular foramen where the superior or Ehrenritter ganglion is formed. The inferior or petrosal ganglion lies flattened against the surface of the bulla. The nerve of Jacobsen emerges from the upper extent of this ganglion. The brackets denote the portion of the nerve that may be considered to be encased in bone. Examination of the ganglia with light microscopy reveals cell bodies distributed about the periphery of the nerve rootlets in the superior ganglion and cell groups interspersed with fibers in the inferior ganglion. Perikarya are essentially similar in both, having diameters of 20-60 µm in cell profiles containing the nucleus. In electron microscopy the cells appear to be similar to dorsal root ganglion cells and are found to be void of synaptic contact within the ganglia.

We injected the petrosal ganglion of cats with 20 µCi of L-$\{3,4,5-^3H(N)\}$ leucine (81.3 Ci/mM) through a glass micropipette. Cycloheximide was perfused through the carotid bifurcation to prevent incorporation of stray amino acid by carotid body cells. Animals were maintained for 8-48 h to allow labeling of nerve terminals by fast axoplasmic transport (11). At the end of this period the head was perfused with fixative, and carotid bodies and ganglia were removed. The ganglia were embedded in Paraplast, sectioned serially, and coated with Kodak NTB-2 emulsion. Exposures from 1 to 2 weeks revealed labeling of petrosal ganglion cells with few grains overlying other elements within

Fig. 1. Gross anatomy of cranial portion of glossopharyngeal nerve. Typanic bulla and petrous temporal bone were removed to expose the nerve. Black background was inserted beneath the nerve to enhance contrast. Superior ganglion, SG; inferior or petrosal ganglion, IG; nerve of Jacobsen, NJ; vagal rootlets, VR; medulla, M. Portion of nerve surrounded by bone

the ganglion (Fig. 2a). An autoradiograph of the superior ganglion (Fig. 2b) taken from the same Sect. as Fig. 2a shows no activity, indicating that the injection was localized to the petrosal ganglion.

Carotid bodies were prepared for electron microscopy, and semithin Sects. were placed on parlodian-coated slides and dipped in Ilford L-4 emulsion. Autoradiographs were developed at intervals of up to 6 months. Examination of carotid body autoradiographs revealed deposition of silver grains over nerve terminals and, less frequently, over both myelinated and unmyelinated nerve fibers. The distribution of labeled structures in the body was not uniform with some glomera showing intense labeling, whereas others showed little activity. We attributed this to uneven distribution of the isotope within the ganglion, since

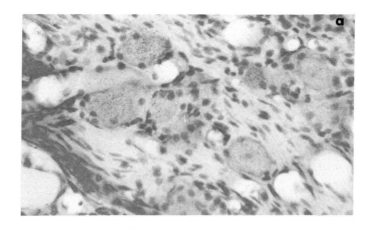

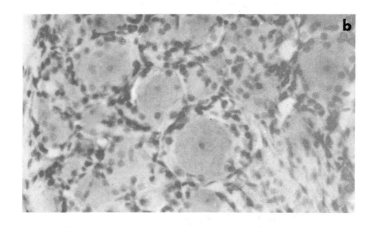

Fig. 2. (a) Light-microscopic autoradiograph of petrosal ganglion injected with {^{3}H}leucine 48 h prior to fixation. Ganglion cells display numerous silver grains indicating uptake of labeled amino acid. Other structures within ganglion display little activity. 15 days exposure. 600×; (b) Light-microscopic autoradiograph of superior ganglion from same Sect. as 2a. Ganglion cells show no labeling, indicating that injection was restricted to petrosal ganglion. 15 days exposure. 600×

perikarya are found over a large area of nerve and display variability in the number of grains associated with them. In addition, it is likely that the bulk of the labeled protein may not have reached the carotid body in the relatively short transport times employed in this study.

Labeled nerve endings were found to be in synaptic contact with type I cells. Such endings contained mitochondria, spherical clear-cored vesicles, and occasional dense-cored vesicles. In single-section electron micrographs, labeled endings appeared to be both calyciform and bouton-shaped. The synaptic morphology of labeled petrosal ganglion nerve terminals was examined, and polarity was ascribed to them on the basis of generally accepted criteria for chemical synapses in the nervous system (4,20). Fig. 3a is an electron-microscopic autoradiograph of a labeled nerve terminal in contact with a type I cell. Two types of membrane specialization may be identified. Regions of puncta

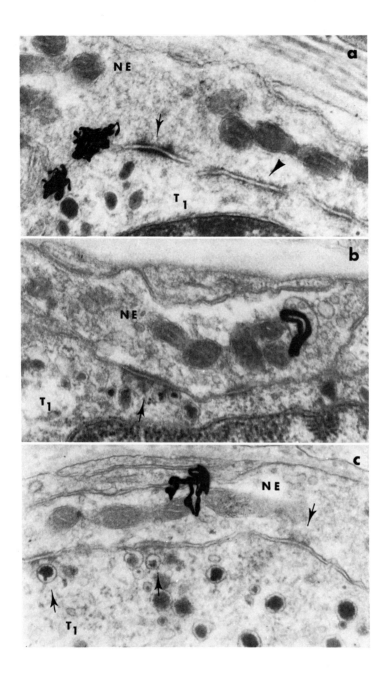

Fig. 3. (a) Labeled nerve ending (NE) on type I cell (T_1). Puncta adhaerentia (*arrowhead*) occur along nerve type I cell membranes. In addition, asymmetrical region of synaptic contact is present with cone-shaped presynaptic densification along nerve membrane (*arrow*) and postsynaptic band along type II cell membrane. 60,000×.
(b) Labeled nerve ending (NE) on type I cell (T_1). Presynaptic projections occur along type I cell membrane (*arrow*), and postsynaptic band is present on nerve ending. 60,000×. (c) Labeled nerve ending (NE) in reciprocal synaptic contact with type I cell

adhaerentia, characterized by symmetrical densification of the nerve and type I cell membranes, are present. In addition, there is an asymmetrical region with electron-dense cones along the type I cell membrane. This is interpreted as a synaptic complex in which the nerve ending is presynaptic to the type I cell. In Fig. 3b another nerve terminal is seen in contact with a type I cell. In this case the postsynaptic band occurs along the nerve terminal membrane, and the presynaptic projections are on the type I cell membrane, indicating a polarity opposite to that in the previous. In favorable Sects. it is possible to identify both pre- and postsynaptic specializations on the same nerve terminal, which is shown in Fig. 3c. Two synaptic regions are present where the nerve terminal appears to be postsynaptic to the type I cell. The occurrence of both types of synaptic complex on the same nerve terminal, known as a reciprocal synapse, has been described in the carotid bodies of the rat (15,16), the duck (5), and, most recently, the cat (21).

By applying horseradish peroxidase to the cut central end of the sinus nerve, Kalia and Davies (13) demonstrated labeling of the ipsilateral petrosal ganglion. Since petrosal ganglion cells show no characteristics that might suggest they are other than sensory in function, it seems safe to conclude that the petrosal ganglion contains cell bodies of sensory nerve fibers to the carotid body. Since en passant fibers lack the ability to take up and transport radioactive amino acid in quantities sufficient for detection with autoradiography (8,10,14), we can conclude that we have selectively labeled sensory nerve terminals in the carotid body. Our results are in agreement with those of Fidone et al. (9,10) that sensory nerve terminals are in synaptic contact with the type I cell. In addition, the sensory nerve endings form afferent, efferent and reciprocal synapses with the type I cell.

Results from these autoradiographic studies do not appear to be compatible with those of Biscoe et al. (2). One possible explanation that might reconcile our different conclusions is that, as a consequence of intracranial transsection, a retrograde degenerative response was initiated in the sensory neurons. Although it is generally believed to be unlikely, such responses to central axotomy may occur in some instances (6,17). Since Biscoe et al. did not examine the petrosal ganglion, this possibility cannot be ruled out. If the ganglion cells degenerate in a retrograde fashion, involvement of the peripheral processes would probably require a longer time course than that of Wallerian degeneration.

Assuming this to be the case, it is still necessary to account for the presence of chemoreceptor activity in a nerve fiber with degenerate endings. Although the electrophysiological properties of degenerating nerves have not been studied, it may be pointed out that Nishi (18) has recorded chemoreceptor activity in a sinus nerve cut 20 h previously. But in structural studies Hess and Zapata (12) found degenerative changes in the nerve terminals when the sinus nerve had been

---

60,000×. (c) Labeled nerve ending (NE) in reciprocal synaptic contact with type I cell (T_1). Two synaptic regions are present where type I cell appears to be presynaptic to nerve ending. A third region is apparent where synaptic morphology indicates that the nerve ending is presynaptic to the type I cell. (*arrows*) indicate presumed direction of transmission. 60,000× (From Smith and Mills, 1976)

cut for 24 h. Therefore, the possibility cannot be excluded that nerve fibers whose terminations are undergoing degeneration may be capable of continued chemoreceptor activity. Furthermore, fibers giving rise to nerve terminals on type I cells are known to display considerable branching (3). Thus, a reduction in the number of nerve endings on type I cells, as seen in the degeneration experiment, need not necessarily reflect a decrease in the number of fibers giving rise to these terminations. This may account for the ease with which chemoreceptor activity could be recorded in carotid bodies showing degenerative changes.

Finally, it is relevant to point out that on theoretical grounds chemoreceptor activity may not require the integrity of the nerve ending. Biscoe (1) provides a numerical argument that suggests that small unmyelinated fibers may in anoxic conditions display sufficient depolarization to account for chemoreceptor discharge. The argument is used to support the contention that free nerve endings arising from unmyelinated fibers are the chemoreceptors. It is known that unmyelinated fibers of the proper dimension also give rise to calyciform endings on type I cells (19). In the degeneration study the authors systematically evaluated the status of the nerve terminals but not the condition of the axons giving rise to the terminations. Therefore, if one assumes that it is the small diameter fiber that is essential for chemoreception, then it is possible that such a fiber might be functional at a time the nerve ending shows degeneration.

References

1. Biscoe, T.J.: Physiol. Rev. *51*, 437-495 (1971)
2. Biscoe, T.J., Lall, A., Sampson, S.R.: J. Physiol. *208*, 133-152 (1970)
3. Biscoe, T.J., Pallot, D.J.: Experentia *28*, 222 (1972)
4. Bloom, F.E., Aghajanian, G.K.: J. Ultrastruct. Res. *22*, 361-375 (1968)
5. Butler, P.J., Osborne, M.P.: Cell Tissue Res. *163*, 491-502 (1975)
6. Carmel, P.W., Stein, B.M.: J. Comp. Neurol. *135*, 145-166 (1969)
7. Castro, F. de, Rubio, M.: in: Arterial Chemoreceptors. Torrance, R.W. (ed.). Oxford: Blackwell, 1968, pp. 267-270
8. Cowan, W.M., Gottlieb, D.I., Hendrickson, A.E., Price, I.L., Woolsey, T.A.: Brain Res. *37*, 21-54 (1972)
9. Fidone, S.J., Stensaas, L.J., Zapata, P.: J. Neurobiol. *6*, 423-427 (1975)
10. Fidone, S.J., Stensaas, L.J., Zapata, P.: This symposium
11. Hendrickson, A.E.: J. Comp. Neurol. *144*, 381-397 (1972)
12. Hess, A., Zapata, P.: Fed. Proc. *31*, 1365-1385 (1972)
13. Kalia, M., Davies, R.O.: Physiologist *19*, 246 (1976)
14. Lasek, R.J.: Brain Res. *7*, 360-377 (1968)
15. McDonald, D.M., Mitchell, R.A.: J. Neurocytol. *4*, 177-230 (1975)
16. Morgan, M., Pack, R.J., Howe, A.: Cell Tissue Res. *157*, 255-272 (1975)
17. Nathaniel, E.J.H., Nathaniel, D.R.: J. Ultrastruct. Res. *45*, 168-182 (1973)
18. Nishi, K.: Br. J. Pharmacol. *55*, 27-40 (1975)
19. Nishi, K., Stensaas, L.J.: Cell Tissue Res. *154*, 303-319 (1974)
20. Peters, A., Palay, S.L., Webster, H. de F.: in: The Fine Structure of the Nervous System. New York: Evanston. Hoeber Medical Division. London: Harper and Row 1970
21. Smith, P.G., Mills, E.: Brain Res. *113*, 174-178 (1976)

DISCUSSION

Weigelt: Which amino acid did you use?

Smith: Tritiated leucine.

Torrance: Dr. Fidone, are there any differences between your study and Smith's?

Fidone: His conclusions are similar to ours; the only differences are very minor. We obtained rather intense labeling of almost everything including the connective tissue elements in the petrosal ganglion, but other than that his results are identical to ours.

McDonald: One point that has been mentioned is the branching of the chemoreceptor axons. In the rat there are fewer than 500 nerve fibers in the sinus nerve and in excess of 10^4 type cells. There is thus a tremendous amount of branching; at least 10 glomus cells are innervated by each nerve fiber, because not all fibers in the sinus nerve go to the carotid body. Do you have any information about whether a group of glomus cells is innervated by a single or several axons? If you assume that when a sensory ganglion cell takes up the label and all of its peripheral processes are labeled, then the presence of unlabeled nerve endings in a group would suggest that the group is innervated by more than one axon.

Smith: I have no data on branching. However, we found that the labeled endings tended to occur in groups, which might suggest derivation from a common labeled source, e.g., one neuron or a group of neurons. To resolve this you would have to resort to intracellular injection techniques, which we have not done.

Torrance: Do you agree with that, Dr. Fidone?

Fidone: Yes. You would have to label just one ganglion cell and then find its terminals in the carotid body. We considered that and then thought better of it! The leakage out of the cell to neighboring cells would be very difficult to control.

Eyzaguirre: In regard to branching, Kondo in our laboratory has conducted an extensive serial electron-microscopic reconstruction of the carotid body and its innervation. He found that the so-called sensory unit is 10-15 cells to one nerve fiber. Very often one glomus cell is innervated by more than one fiber.

Trzebski: My question is to McDonald and Eyzaguirre. Is there any special difference between rats and cats as regards the number of type I cells and nerve terminals and nerve fibers?

Eyzaguirre: Kondo did his studies in the rat, and the only information we have about the cat was obtained by examining some of de Castro's slides, work presented a few years ago in Bristol. Although it was very difficult to follow the slides, the ratio of cells to fibers appeared to be about ten to one in the cat. We have not done quantitative studies in the cat in our laboratory.

McDonald: In the rat there are about 500 axons and 5000 glomus cells. In the carotid body of the cat it seems to be ten times that number of glomus cells. We have not counted the number of axons in the carotid sinus nerve of the cat although it certainly is more than 500.

Paintal: Fidone mentioned that the label moved from the nerve terminal to the receptor cell in the taste bud, and he also mentioned that there is no movement of the label into the type I cell. I want to ask both Smith and Fidone whether any label was found in the type II cell.

Smith: No, I did not find any labeling in type II cells. My data differ somewhat from Fidone's in that, in some cases, there were type I cells that showed some labeling in the region of the nerve ending. This was not a frequent occurrence, and the difference might be due to the fact that we used different amino acids and might have been labeling different proteins.

Fidone: We found that the type II cells were even less frequently labeled than the type I cells. I do agree that when glomus cells were labeled, it was near the nerve terminal.

O'Regan: Is there any possibility that when you apply your radioactive substances they spread to the brain stem to be taken up by cells that could be efferents?

Smith: I do not think so. The superior ganglion showed very little label, usually restricted to a few fibers that seemed to pass up to the brain stem. Of course cells do transport in both directions so we would be labeling brain stem and also carotid body. But this would be restricted to the axons, and I do not think there would be much trans-synaptic flow and therefore flow back down through efferents.

Acker: Did you see any labeled nerve fibers on the capillaries?

Smith: We looked but we did not see any. We concluded that the afferent nerve fibers are restricted to the glomera.

McDonald: What did you define as an interstitial enlargement? Could some of these have been next to blood vessels?

Fidone: An interstitial enlargement was something that contained an accumulation of clear vesicles and mitochondria but was not apposed to a glomus cell. Now if there was a blood vessel with some labeled nerve fibers passing nearby, we would describe this just by saying that those nerve fibers were passing nearby. We could not see any accumulation of label circumferentially orientated around the vessels. If such an innervation were at all extensive, we would probably have seen that.

O'Regan: Did you try injecting into the superior cervical ganglion?

Smith: No, we did not.

Degenerative Changes in Rabbit Carotid Body Following Systematic Denervation and Preliminary Results about the Morphology of Sinus Nerve Neuromas

H. Knoche and E.-W. Kienecker

We wanted to investigate the regenerative ability of the carotid sinus nerve by iso- and heteromorphic regeneration. First we had to clarify which nerves take part in the innervation of the carotid body and where their terminals end. To do this we performed the following experimental studies on the carotid body of the rabbit (number of operated rabbits in brackets): {a} vagotomy, by removal of the nerve between the brain stem and its peripheral ganglion (2); {b} sympathectomy (6); {c} sectioning (17); and {d} coagulation of the carotid nerve (9). The tissue was fixed by perfusion from the left ventricle with 2% phosphate-buffered glutaraldehyde and embedded in Epon after dehydration in ethanol.

After vagotomy no structural changes occurred in the carotid body tissue. The peripheral nerve tissue and the periglandular, interstitial, and intraglandular plexus are unaffected as are the glomoids and the nerve terminals in type I cells (8).

To demonstrate sympathetic nerve fibers the fluorescence method (4) was applied to the carotid body (Fig. 1). Fine fluorescent adrenergic nerve fibers were seen in the normal carotid body situated near blood vessels between the intensively fluorescent glomoids. Shortly after sympathectomy the specific fluorescence of the fine adrenergic nerve fibers disappeared whereas the glomoids still fluoresced. These findings suggest alterations in the ultrastructure of the nerve tissue in the carotid body. After sympathectomy peripheral nerve fibers within the periglandular plexus show with electron-microscopy a light axoplasm and sintering of their cytoplasmic structures such as mitochondria, neurotubules, and neurofilaments as a consequence of degeneration. Degeneration signs also occurred in some nerves and nerve endings situated in apposition to sinusoids and glomoids (7). Typical osmiophilic bodies, fat droplets and disorganized cytoplasm and mitochondria, could be seen in these degenerating axons.

The sectioning of the carotid sinus nerve was carried out after accurate definition of all nerves lying in the area of the carotid artery bifurcation to prevent sectioning or injury of other nerves (Fig. 2). After sectioning the carotid sinus nerve, its proximal end was resected. Several changes could then be seen in the ultrastructure of the carotid body. In the nerve plexus degenerating nerve fibers containing lysosomelike-destroyed mitochondria and brightened axoplasm were found. These axons were encapsulated by reactive Schwann cells with much rough and smooth endoplasmic reticulum, ribosomes, and microtubules. After breakdown of the nerve fibers resulting from degeneration, the axon debris was phagocytosed by Schwann cells, which later involuted. Similar degenerative signs were also observed in the glomoids.

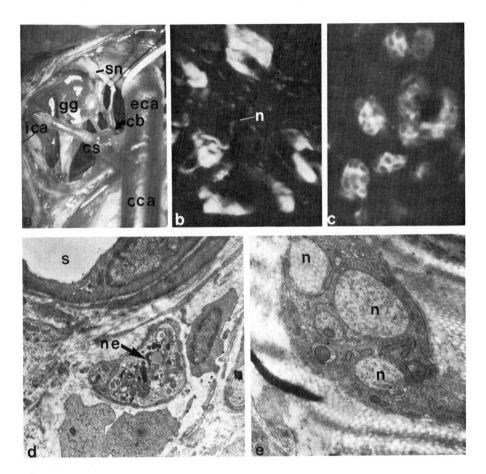

Fig. 1. (a) Operation situs of rabbit carotid bifurcation area. cca, common carotid artery; ica, internal carotid artery; eca, external carotid artery; cg, superior cervical ganglion; sn, carotid sinus nerve; cb, carotid body; cs, carotid sinus. Note two sympathetic branches from superior cervical ganglion reaching carotid body area, one in apposition to sinus nerve. 25×. (b and c) Fluorescence of carotid body after formaldehyde treatment. (b) Normal carotid body with weak fluorescing glomoids and some sympathetic nerves (n). 250×. (c) 8 days after superior cervical ganglion resection there are no more fluorescing sympathetic nerve fibers. 250×. (d) Degenerating sympathetic nerve terminal (ne) near a sinusoid (s) with shrinking mitochondria and sintering of its axoplasm 12 h after superior cervical ganglion resection. 12,640×. (e) Degenerating sympathetic nerves (n) in periglandular plexus of carotid body 12 h after resection of superior cervical ganglion. 15,000×

Partially destroyed and shrinking mitochondria appeared in axon terminals that had a light axoplasm. After this stage of degeneration a total breakdown of the axon terminals occurred, as previously described by a number of investigators (1,2,3,5,6). This breakdown left lacunae within the tissue that were not filled by cytoplasmic extensions of the surrounding cells; some lacunae contained osmiophilic material.

The region of type I cell associated with degenerating nerve endings may also show degenerative changes; such regions were devoid of membranes. While the lacunae produced by degeneration of nerve endings persisted even after a period of 3 weeks, the type I cells were ap-

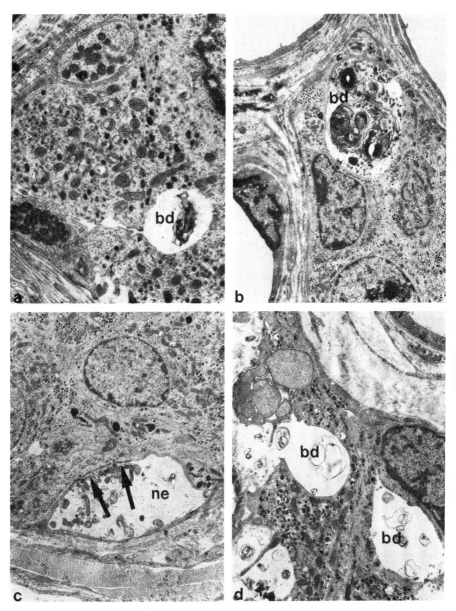

Fig. 2. Ultrastructure of typical breakdown (bd) of nerve endings (ne) in carotid body following sinus nerve sectioning. (a) 4 days after sectioning accompanied by breakdown of adjacent type I cell cytoplasm. 15,000×. (b) 8 days after sectioning. Note huge axonal debris. 5100×. (c) 2 days after sectioning. Note intact synaptic structure (*arrows*), whereas axoplasm has degenerated. 6890×. (d) 3 weeks after sectioning. 11,830×

parently able to repair their affected regions. Large lacunae containing axonal debris occurred in some parts of the glomoids; their large size suggests that they were the remains of the large nerve terminals that occur in normal carotid body. Eventually all nerve endings were disturbed by degeneration. Many lacunae were then found in

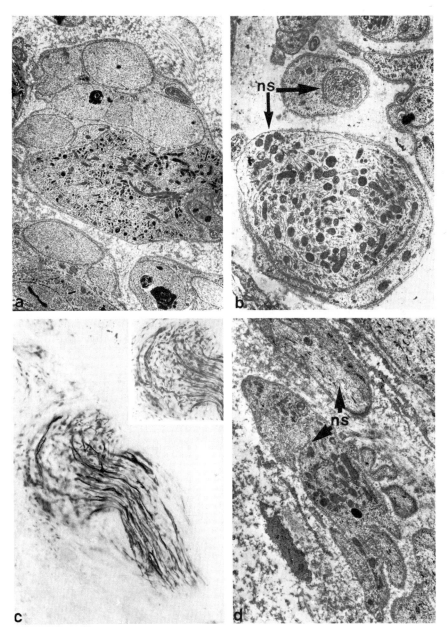

Fig. 3. a and b. Ultrastructure of regenerating nerve sprouts (ns) in carotid body following sinus nerve coagulation. (a) 2 days after coagulation. 6826×. (b) 4 days after coagulation. 2592×. (c and d) Sinus nerve neuroma 8 days old. (c) silver impregnation. 125×. (d) Nerve sprouts (ns) in neuroma

the glomoids. Not only the nerve terminals were affected by degeneration, but also the type I cell cytoplasm showed some additional changes. Lysosomelike, osmiophilic bodies and fat droplets could be seen accompanied by a rich rough and smooth endoplasmic reticulum, microtubules, and phagocytized osmiophilic material within the cytoplasm of type II cells. Three weeks after sectioning the carotid nerve, no nerve terminals in apposition to type I cells were seen.

More impressive degenerative signs can be produced by coagulation, a method that also permits the study of the first stages of regeneration (Fig. 3). At first the ultrastructure of the carotid body was typified by giant, multilocular nerve and nerve terminal destructions. Within the periglandular, interstitial, and intraglandular plexus the same degenerative processes described above occurred. Four days after nerve coagulation the first signs of regeneration could be seen. Because of the distribution of nerves, regeneration proceeds from the periglandular to the interstitial and to the intraglandular plexus. Huge bulb-like axon sprouts filled with mitochondria appeared in the periglandular plexus after 3 days, and after 4 days they could be found within the interstitial plexus as well. There were also the typical signs of hyperneurotization, which accompanies all nerve regenerations (9).

Electrophysiological recordings of chemo- and barosensory activity 8 days after coagulation gave identical results to experiments conducted 4 weeks after anastomosis of the sectioned carotid sinus nerve. These results will be discussed by Bingmann et al. in this volume (p. 36).

Recently we have performed structural and functional studies of the neuromas produced by implantation of the distal cut end of the carotid sinus nerve into the connective tissue of the mylohyoid muscle. Eight days after operation there were numerous axon sprouts, which spread into the loose connective tissue forming the neuroma. There is no chemoreceptive tissue as is normally found in the carotid body. The axons in this neuroma end in a clublike way.

References

1. Abbot, C.P., De Burgh Daly, M., Howe, A.: Acta Anat. (Basel) *83*, 161-185 (1972)
2. Biscoe, T.J., Stehbens, E.W.: J. Exp. Physiol. *52*, 31-36 (1967)
3. Biscoe, T.J., Lall, A., Sampson, S.R.: J. Physiol. *208*, 133-152 (1970)
4. Falck, B., Hillarp, N.A., Thieme, G., Torp, A.: J. Histochem. Cytochem. *10*, 348-352 (1962)
5. Hess, A., Zapata, P.: Fed. Proc. *31*, 1365-1384 (1972)
6. McDonald, M.D., Mitchell, R.A.: J. Neurocytol. *4*, 177-230 (1975)
7. Knoche, H., Kienecker, E.W.: (to be published) (1976)
8. Knoche, H., Kienecker, E.W., Schmitt, G.: Z. Zellforsch. *112-4*, 494-515 (1971)
9. Schröder, J.M.: Melsunger Medizinische Mitteilungen *46*, 37-52 (1972)

DISCUSSION

For the discussion of this paper, see page 40

Regeneration of Nerves and Nerve Terminals in Rabbit Carotid Body Following Carotid Nerve Sectioning and Suturing

E.-W. Kienecker and H. Knoche

We have previously presented details of the innervation of rabbit carotid body. Here we discuss if a regeneration of carotid sinus nerve fibers and terminals may take place after certain kinds of lesion and if under optimal conditions there is a recovery of chemo- and barosensory activities.

There are three basic ways to produce a nerve fiber lesion for degeneration followed by regeneration:

1. Nerve compression followed by primary degeneration and regeneration. The extent of the compression and primary destruction cannot be established exactly. There may occur a locally reversible axon lesion or, more usually, a segmental fiber change, which is characterized by demyelination and subsequent remyelination of nerve fibers. During nerve restitution the axons are in a state of neurolytic swelling and axon retraction (7).

2. Nerve coagulation followed by secondary degeneration, also called Wallerian degeneration, and then by isomorphic regeneration. The extent of coagulation necrosis varies greatly from experiment to experiment. Typical Wallerian degeneration occurs after such treatment. The possibility of regeneration depends in large part on the extent of the coagulation necrosis. If there is regeneration, it is isomorphic, which means all recovering axons are situated in their Büngner's cords (3,5).

3. The third method consists of nerve suturing after sectioning in order to achieve Wallerian degeneration by heteromorphic regeneration (3,5,7,8). Nerve compression is a very good method to test the recovery of function of nerves (9), but it does not give evidence of the regenerative faculty of the nerve. The processes of nerve coagulation and suture involve ideal conditions for studying specific nerve regeneration, both iso- and heteromorphic. The ultrastructure of the carotid body following carotid nerve Sect. and coagulation has been demonstrated by Knoche and Kienecker (1). We present here the results of heteromorphic regeneration experiments.

Experiments were performed on 9 adult rabbits. After dissecting the carotid artery bifurcation, the carotid sinus nerve was dissected, and the adjacent connective tissue and the epineurium of the nerve were carefully removed to prevent the secondary formation of a neuroma. After this procedure a suture was applied with atraumatic material 10-0 before cutting the nerve. The sectioning of the nerve was performed within the loop of this suture, and the two ends of the suture were tied to give accurate apposition of the end of the nerve without tension (4,6). Six to 8 days are required for the nerve to grow through the anastomosis (7); subsequently the nerve grows at a rate of 0.5-2.0 mm/day. The animals were therefore fixed for electron microscopy at

1, 2, 3, 4, 5, 7, and 8 weeks after the operation by perfusion via the left ventricle with 2% phosphate-buffered glutaraldehyde, and the carotid bodies were processed for electron microscopy by standard techniques (Fig. 1).

Because the first nerve sprouts reached type I cells 4 weeks after operation, we used this time for physiological studies. Characteristic of the carotid body ultrastructure 3 weeks after the nerve suture are huge, degenerative debris due to Wallerian degeneration of nerves and their terminals. There are breakdowns of nerve terminals with the formation of lacunae some of which contain osmiophilic material. Reactive changes of type I cells can also be detected. Lysosomelike vesicles appear in the cytoplasm, and adjacent to degenerated nerve endings a disorganization of the cytoplasm occurs, which may lead to cytoplasmic breakdown. These alterations of type I cells decrease during regeneration . A few nerve sprouts in apposition to type I cells may be found in the carotid body's periphery resulting from a shorter regeneration distance, but there are no junctional connections. The main feature at this stage is the hyperneurotization of the periglandular plexus.

After a typical Wallerian degeneration has taken place, as shown by carotid nerve Sect., the first signs of reinnervation in the carotid body occurred 3 weeks after the operation with the appearance of the first regenerating axon bundles. Within the periglandular plexus the axon sprouts entered the interstitial plexus and then ramified to the glomoids via type II cells, following their Büngner's cords. After much time elapsed restoration of nerve terminals on type I cells occurred. The terminals formed enlargements with junctional densities, which contained differentiated structures such as mitochondria and clear-cored vesicles. Nonmyelinated nerve fibers regenerated in the typical manner of hyperneurotization and were characterized by a light cytoplasmic matrix, including neurofilaments, mitochondria, some dense vesicles, and microtubules. Some regenerating myelinated axons were found, but they were very rare. The very first few axon terminals without typical junctional connections were detected 3 weeks after operation at the periphery of the organ. Thereafter there was a slow increase in the number of new terminal enlargements. Junctional densities contacting type I cells in the usual way were formed in the beginning of the 5th week.

The Büngner's cords induce the new nerve endings to sprout into their former positions. The axons reached these positions by conduction via reactive Schwann cells whose cytoplasm contained axonal and myelin debris, ribosomes, rough and smooth endoplasmic reticulum, and microtubules. The nerve terminal enlargements, which could be very small at first, did not fill the lacunae left by the degenerated terminals, but there was continuous growth until all empty spaces disappeared. During the whole time the nerve endings were in direct apposition to type I cells. One week later the regeneration had proceeded to the interstitial plexus where hyperneurotization could then be found. There was a slight increase in the number of nerve sprouts that contact type I cells. In general, this stage of regeneration resembles the one 3 weeks after nerve suture, except for a distinct decrease in degeneration.

The essential regenerative changes occur 5 weeks after nerve anastomosis (Fig. 2). By then most of the type I cell alterations have receded. Repair of cytoplasmic breakdowns is observed and osmiophilic lysosomelike vesicles disappear. The hyperneurotization reaches the interstitial connective tissue between the glomoids. Numerous nerve terminals can also be found in contact with type I cells. For the first

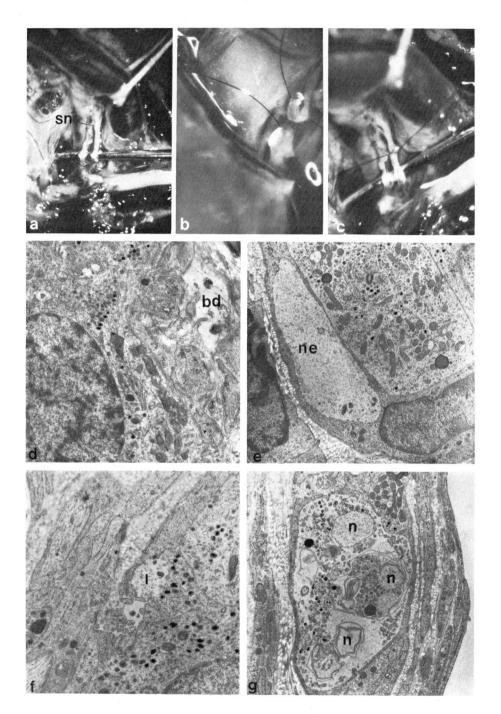

Fig. 1. a–c. Operation microphotographs of sinus nerve (sn) cutting and suturing. (a) Preparation of sinus nerve. 25×. (b) Cutting of sinus nerve within sutures loop. 45×. (c) Suturing of cut sinus nerve. 25×. (d–g) Degeneration of nerves (n) and nerve endings (ne) in carotid body 3 and 4 weeks after sinus nerve suture. (d) Typical breakdown (bd) after 3 weeks. 10,131×. (e) 3 weeks. 9519×. (f) Empty lacunae (l) after 4 weeks. 12,000×. (g) Axonal debris of nerves after 4 weeks. 10,200×

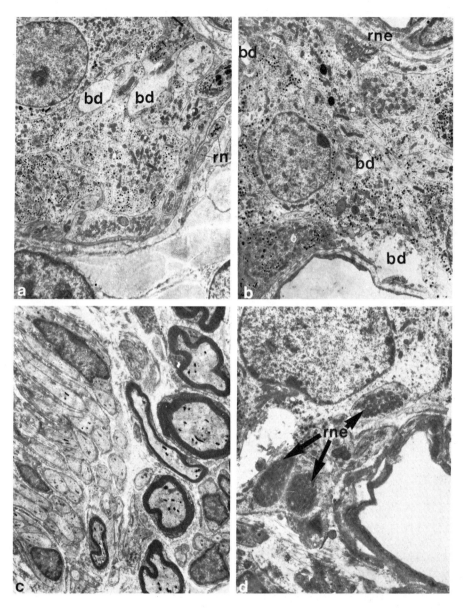

Fig. 2. a-d. Ultrastructure of carotid body 5 weeks after sinus nerve suturing. First regenerating nerves (rn) and nerve endings (rne) in carotid body. bd, breakdown. (a) 3519×. (b) 5402×. (c) 2600×. (d) 7540×. Note hyperneurotization in periglandular plexus

time there occur junctional densities on the membranes of type I cells and appositional nerve terminals. The number of regenerated nerve endings is about 30-40% of that for normal carotid body. The size of the terminal enlargements is relatively small, so there are still some empty spaces in the lacunae. Seven and 8 weeks after nerve anastomosis by suture, the hyperneurotization has reached type II cells, and about

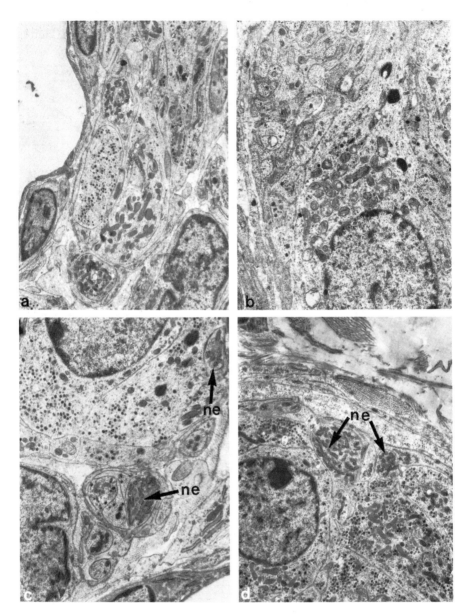

Fig. 3. a-d. Ultrastructure of carotid body 7 and 8 weeks after sinus nerve suturing. Hyperneurotization within intraglandular plexus and type II cells. (c and d) mature nerve endings (ne) in carotid body. (a) 9129×. (b) 10,500×. (c) 14,580×. (d) 8400×

two-thirds of all degenerated nerve terminals are replaced by new terminal enlargements that fill the lacunae in the usual way (Fig. 3). A differentiation into terminal types is possible. One type is compactly filled with mitochondria, others are bulblike, enpassant invaginated, as is seen in the normal carotid body (2).

Finally, there is no possibility for a total regeneration resulting from heteromorphic restitution. Only a certain number of axon sprouts reach Büngner's cords in the distal end of the carotid sinus nerve, and only these can regenerate in the ordinary way.

References

1. Kienecker, E.-W., Knoche, H.: Glomus caroticum. Arbeitsgespräch, März 1976, Physiol. Inst. der Universität Münster
2. Knoche, H., Kienecker, E.-W., Schmitt, G.: Z. Zellforsch. *112/4*, 494-515 (1971)
3. Knoche, H., Addicks, K., Kienecker, E.-W.: Versammlung der Anatomischen Gesellschaft (1976)
4. Millesi, H.: Melsungen Medizinische Mitteilungen *46* (Heft 116), 181-188 (1973)
5. Mumenthaler, M., Schliak, H.: Lasionen peripherer Nerven, 2nd. ed. Stuttgart: Georg Thieme Verlag
6. Samii, M., Kahl, R.-J.: Melsungen Medizinische Mitteilungen *46* (Heft 116), 197-202 (1973)
7. Schröder, J.M.: Melsungen Medizinische Mitteilungen *46* (Heft 116), 37-58 (1972)
8. Thomas, P.K., Jones, D.G.: Regeneration of the perineurium after nerve section. J. Anat. *101*, 45-55 (1967)
9. Zapata, P., Stensaas, C.J., Eyzaguirre, C.: Brain Res. *113*, 235-253 (1976)

DISCUSSION

For the discussion of this paper, see page 40

Chemoreceptor Activity in the Rabbit Carotid Sinus Nerve During Regeneration

D. Bingmann, E.-W. Kienecker, and H. Knoche

In other papers in this volume Knoche and Kienecker have demonstrated that after suture of the cut carotid sinus nerve regeneration of mature nerve terminals within the carotid body occurred after 7-8 weeks. After 4 weeks the first axon sprouts reappeared in the vicinity of the type I cells, and 1 week later the first contacts were formed between nerve endings and the so-called specific cells. This slow rate of regeneration has enabled us to study whether receptor fiber units respond to hypoxia and hypercapnia before nerve terminals are apposed to the glomus cells.

In anesthetized, relaxed, and artificially ventilated rabbits the carotid sinus nerves that had been sutured 4 weeks earlier were prepared. At the entry into the nervus glossopharyngeus the carotid sinus nerves were cut. After removing the connective tissue, fiber potentials were recorded bipolarly from the whole sinus nerve. In the neurogram thus obtained the typical, synchronized baroreceptor activity was missing. Even with stepwise changes in blood pressure baroreceptor discharges could not be detected either visually or in an impulse frequency signal obtained by electronically counting spikes in 1-s periods (I/s). Blood pressure was recorded continuously in the femoral artery. P_aO_2 and P_aCO_2 were monitored continuously in a femoral arteriovenous (A-V) loop. Using this arrangement the chemoreceptive properties of the active fibers in the regenerating sinus nerve were tested.

Fig. 1 shows the response of these fibers to a ventilatory arrest. When the ventilation pump was switched off, chemoreceptor discharge increased continuously from 50 to 300 I/s. At the same time P_aCO_2 rose from 40 to more than 60 mm Hg, whereas P_aO_2 decreased below 50 mm Hg. Changes in both gas tensions were recorded with a considerable delay, because the circulation time in the A-V loop was about 20 s. After reventilation the receptor discharge rate decreased abruptly as is usually observed in receptor units of intact carotid bodies. During ventilatory arrest P_aO_2 and P_aCO_2 changed simultaneously. Therefore, responses of these regenerating fiber units to hypercapnic stimuli were tested with P_aO_2 values ranging above normal levels using the so-called apnea technique (3). A detailed description of this experimental procedure is given in another paper by Bingmann et al. in this volume (p. 30).

Fig. 2 illustrates such an experiment. In the course of apnea P_aCO_2 rose continuously from 45 to more than 100 mmHg, while blood pressure remained stable on a slightly elevated level. Paralleling the rise in P_aCO_2 the P_aO_2 decayed. However, at the end of this ventilatory arrest P_aO_2 still ranged above hypoxic values. During this period receptor activity rose significantly, and activity bursts became prominent in the fiber. With the rising PCO_2 the number of spikes in each burst increased considerably, whereas the number of bursts per unit time was reduced. After reventilating the animal, receptor activity returned to its initial level. The groupings of discharges persisted and were

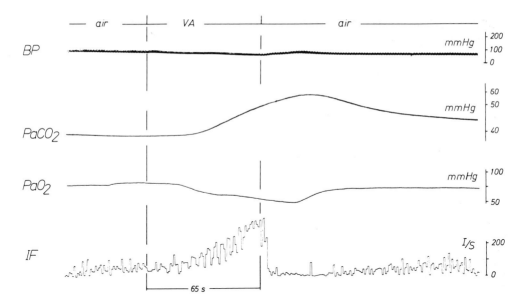

Fig. 1. Responses of regenerating fiber units of carotid sinus nerve to ventilatory arrest (VA) after artificially ventilating the rabbit with air. Activity of fiber units is shown as impulse frequency curve (I/F), which was obtained by counting electronically action potentials in 1-s periods. Gas tensions P_aO_2 and P_aCO_2 were recorded in bypass system with a considerable delay resulting from a long circulation time of blood in arteriovenous loop. BP, arterial blood pressure

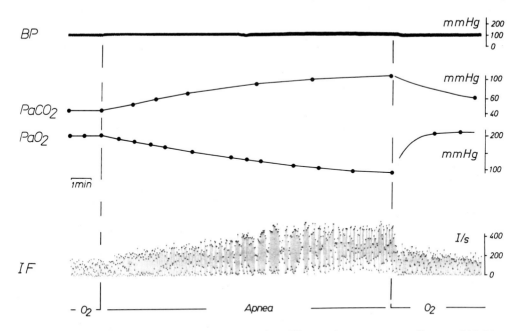

Fig. 2. Reactions in activity of regenerating fiber units to apnea after ventilating the animal 25 min with pure oxygen. Receptor activity, shown as impulse frequency curve, increased when P_aCO_2 rose while P_aO_2 ranged above 100 mm Hg. BP, blood pressure

running in parallel with the cycles of respiration. Up to a respiratory frequency of 28/min the bursts could be triggered, but at higher frequencies the fluctuations in the discharge rate returned to a rhythm of 12-15/min. After an inflation of the lungs and after the administration of norepinephrine the groupings disappeared. This might indicate that these fluctuations in the rate of discharge in the fiber units reflect repetitive changes in the microcirculation of the carotid body. Aside from these questions, however, Fig. 1 and 2 demonstrate that these regenerating nerve receptor units encode hypoxia and hypercapnia in a way to similar intact carotid bodies. In contrast to the findings of Zapata et al. (5) these chemoreceptive properties were reinitiated before nerve terminals were apposed to the type I cells.

These findings point to the hypothesis of Biscoe (2) and Mitchell et al. (4) that nerve terminals in the carotid body are the receptor elements of this organ. Mitchell et al. came to this conclusion when they observed chemoreceptor activity in neuromas of the carotid sinus nerve. These neuromas were 12-18 months old. The nerve endings in the sutured carotid sinus nerve, however, encored hypoxia and hypercapnia after a few weeks. Thus the question, do nerve endings in a neuroma of the carotid sinus nerve exhibit chemoreceptive properties already a few days after the formation of this neuroma?

To answer this question electrical activity was recorded bipolarly from a cut carotid sinus nerve that ended in a 8-day-old neuroma. In the whole nerve fiber potentials appeared only rarely. The frequency of these potentials did not change significantly when the animal breathed pure oxygen and subsequently hypoxic gas mixtures. Ventilatory arrest and apnea as well did not affect the discharge rate. In an early experiment the fiber activity increased markedly after the animal had died. The same rise in the discharge rate was obtained in another animal when the neuroma was vascularly isolated. In this situation the frequency of discharge in the sinus nerve could be widely affected by blowing nitrogen, oxygen, and air over the surface of the neuroma. This observation is shown in Fig. 3A. The first neurogram was recorded after death of the animal. The neurograms labeled 2 and 3 were recorded while nitrogen was being blown over the neuroma. From the frequency curve in Fig. 3B it can be seen that the discharge rate increased each time nitrogen was blown over the nerve neuroma preparation, whereas air decreased the activity in the nerve. Similar reactions were found in cat carotid bodies about 15 min after death of the animal when nitrogen or air was blown over the surface of the glomus (1).

As a whole, the experiments showed the following:

1. After suture of the cut carotid sinus nerve chemoreceptor responses to hypoxic and hypercapnic stimuli were recorded from regenerating fibers at a time when synapses or appositions between nerves and specific cells in the carotid bodies were still missing, and only seldomly axon sprouts were detected in the vicinity of type I cells.

2. Naturally perfused neuroma preparations of the carotid sinus nerve did not respond to hypercapnic and hypoxic stimuli, which are known to excite chemoreceptors in the intact carotid body. Only after stopping microcirculation did the discharge rate of nerve action potentials in the neuroma preparations increase when nitrogen was blown onto the neural node. These observations support the hypothesis of Biscoe (2) and Mitchell et al. (4) that nerve endings are chemoreceptive transducers in the carotid body. The unresponsiveness of the 8-day-old neuroma to hypoxia and hypercapnia might indicate that the specific structures of the carotid body have to form those conditions that enable

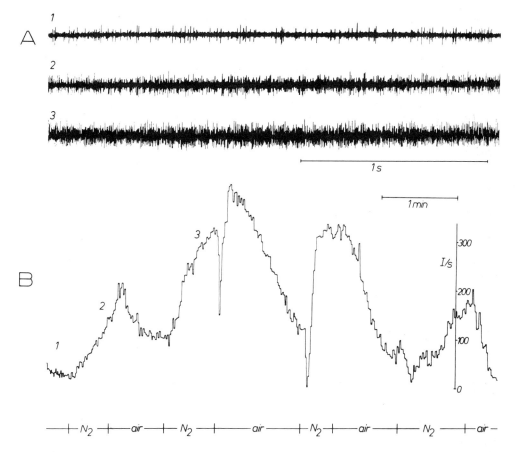

Fig. 3. (A) Neurograms of a sinus nerve that ended in an 8-day-old neuroma: (1) recorded immediately after death of the animal; (2 and 3) nerve activities when nitrogen was blown onto neuroma. (B) Changes in discharge rate of fibers led from while nitrogen and air were blown onto neuroma. Impulse frequency curve was obtained by counting electronically in 1-s periods the action potentials displayed in (A)

the transducers to sense changes in P_aO_2, P_aCO_2, and pH within physiological ranges.

References

1. Acker, H., Lübbers, D.W., Weigelt, H., Bingmann, D., Schäfer, D., Seidl, E.: in: Oxygen Transport to Tissue. Bicher, H.J., Bruley, D.F. (eds.). New York-London: Plenum Press, 1973, Vol. XXXVIIA, pp. 609-615
2. Biscoe, T.J.: Physiol. Rev. *51*, 437-495 (1971)
3. Caspers, H., Speckmann, E.-J.: Ärztl. Forsch. *25*, 241-255 (1971)
4. Mitchell, R.A., Sinha, A.K., McDonald, D.M.: Brain Res. *43*, 681-685 (1972)
5. Zapata, P., Stensaas, L.J., Eyzaguirre, C.: Brain Res. *113*, 235-253 (1976)

DISCUSSION

Joint Discussion on Discourses

H. Knoche, E.-W. Kienecker:
Degenerative Changes in Rabbit Carotid Body Following Systematic Denervation and Preliminary Results About the Morphology of Sinus Nerve Neuromas

E.-W. Kienecker, H. Knoche:
Regeneration of Nerves Terminals in Rabbit Carotid Body Following Carotid Nerve Sectioning and Suturing

D. Bingmann, E.-W. Kienecker, H. Knoche:
Chemoreceptor Activity in the Rabbit Carotid Sinus Nerve During Regeneration

Wiemer: What happened to the baroreceptors in these sectioning experiments?

Kienecker: We do not know. Bingmann is examining this.

Bingmann: I cannot make any comment on this point except that Eyzaguirre has also been studying this.

Eyzaguirre: I think Zapata will discuss this in the next paper.

Willshaw: Have you made any neuromas of nerves that were not originally chemoreceptor nerves?

Kienecker: We did perform one experiment using the superior laryngeal nerve. There was no activity.

Bingmann: But again results are preliminary.

Willshaw: Did you examine the two types of neuroma histologically?

Kienecker: You have seen the neuroma of the sinus nerve. The other neuroma is being examined now; we do not think there are any differences.

Bingmann: I want to stress that the superior laryngeal neuroma results are preliminary. It is only speculation, but I think that the nerve endings are the receptors and the specific structure forms the internal conditions that enable the receptors to do their job in the physiological range.

Pallot: Have you tried the neuroma experiment after removing the carotid body so that you can, under those circumstances, be absolutely certain that no nerve fibers have grown back to the carotid body?

Kienecker: No, we did not remove the carotid body, but there was no nerve connection to the carotid body.

Leitner: Have you used cyanide on the neuroma?

Kienecker: No.

O'Regan: Were there any nerve endings left following Sect. of the sinus nerve? Second, what is the function of the glomus cells if they are not necessary for chemoreception? They have nerve terminals so presumably they have some function.

Kienecker: About 2 or 3 weeks after sectioning the sinus nerve there are no nerve endings left on the type I cells.

Belmonte: I would like to come back to that point about baroreceptor activity. We have done some experiments on regeneration of carotid sinus nerves and have observed that most of the fibers respond to tapping of the area; it was very difficult to see chemoreceptor activity over the mechanical activity. The second point is that when we removed the carotid body, which was theoretically far away from the neuroma, all chemoreceptor activity disappeared in those regenerating fibers; our results are contrary to yours. Have you any explanation for the lack of mechanoreceptor activity in the regenerating fibers?

Bingmann: How many weeks did you wait before making the electrophysiological recording?

Belmonte: We waited 4-8 weeks, and after 4 weeks the mechanoreceptor activity could be recorded.

Bingmann: I cannot comment on this because in our preparation we have not seen baroreceptor activity, and we have tested neuroma only at 4 weeks. Perhaps if we look at neuroma at 6 or 8 weeks we would also be able to find baroreceptor activity.

Belmonte: Did you prod the neuroma?

Bingmann: Yes, we did. We changed blood pressure by injections of epinephrine, also by inflation of the lung, and by application of rapid stimuli.

Trzebski: I wonder if the criteria to identify a response as baroreceptor are exact enough. We know that the baroreceptors are rate sensitive. Perhaps rapid changes of pressure within the carotid sinus would be a more effective procedure to evoke activity than just raising mean systemic blood pressure. Looking at the histology of the carotid sinus, did you see any histologic changes within the carotid sinus as well as you did in the carotid body?

Bingmann: First we tested to see if baroreceptor activity was present with rapid changes; the slow changes were another test.

Kienecker: We are going to investigate the carotid sinus regeneration after nerve Sect.. We do not know yet whether there is any regeneration.

Purves: We have measured sinus nerve activity in the muskrat during a dive and have observed a very remarkable grouping of the discharge that occurs with each heartbeat and is accentuated by the accompanying bradycardia. What was the period of the grouping in the apnoic test?

Bingmann: The frequency was about 12-15/min, and these frequencies could be triggered after reventilation by the pump. I think these groupings reflect changes in microcirculation that result from changes in the autonomic nerves driven by neurons in the respiratory centers of the brain stem.

Purves: You did not demonstrate a bradycardia?

Bingmann: No.

Pallot: Have you looked at the histology of the carotid sinus during regeneration? An explanation of the failure to record baroreceptor activity is that the nerve fiber, before it makes any specific junction in the carotid body, can act as a chemoreceptor. But in order to get baroreceptor activity you have to have reinnervation of the sinus to get the transmural pressure changes.

Kienecker: We are just to investigating that point.

Loeschcke: Have you ever tried to find out whether there is an interaction of CO_2 and O_2 effects on your neuroma?

Bingmann: It seems to be so.

Acker: Can I conclude from your experiments that the carotid body is able to respond to hypoxia only when the flow is stopped or when the flow is disminished?

Bingmann: Why is that?

Acker: Because with the perfused neuroma you cannot get any response to hypoxia.

Bingmann: I am convinced that the nerve endings in the neuroma do not react to changes in gas tension because the PO_2 in the neuroma is too high under the test conditions. Perhaps the low PO_2 values in the normal carotid body demonstrate a part of the function of the specific tissue so that even at a high rate of circulation within the carotid body and at high P_aO_2 values in the circulating blood the tissue PO_2 is low.

Acker: You should compare your preparation with that of Eyzaguirre, because the superfused carotid body has the highest responses when the flow over it is stopped. How would you explain this?

Bingmann: I think we need to know the changes in PO_2 in the neuroma after blowing nitrogen over its surface. We need PO_2 values in the neuroma at a level where in the normal carotid body chemoreceptors would react to changing P_aO_2.

Eyzaguirre: Would it be possible to remove the neuroma and put it in vitro? If so, then the sensitivity of the neuroma could be compared with the carotid body and nerve in vitro in terms of the amount of hypoxia and hypercapnia needed to stimulate. It may be that we are dealing here with a difference in threshold. Perhaps this method would provide better data than just blowing nitrogen over the carotid body after killing the cat.

Bingmann: That is a good suggestion. We have just begun constructing a chamber for this task!

Paintal: Could you tell us what kind of cells surround the neuroma?

Kienecker: Loose connective tissue, fibroblasts, reactive Schwann cells, and large numbers of blood vessels, which involute from the third month.

Torrance: We have the very simple diagram of the carotid body as a type I cell with a big nerve ending against it. Can you give us a rough diagram of the branching of a nerve fiber within a neuroma?

Kienecker: There are the usual bulklike endings that are found in any other neuroma.

Torrance: Are these enclosed by cells like Schwann cells or are they free?

Kienecker: I cannot answer that because we have to look at the ultrastructure.

Torrance: It has already been remarked that there are large numbers of unmyelinated nerve fibers within the carotid bodies as though the nerve fibers go around and around in circles. Did you get the impression in your regeneration that the nerve fibers disappeared and then gradually reappeared?

Kienecker: There is a typical hyperneurotization resulting from the involution of many Schwann cells after the degeneration. In some Schwann cells new axons build up so there seems to be a very large number of unmyelinated nerve fibers. We know that after heteromorphic regeneration only a part of the nerve fibers reach their old terminals. Not all nerve sprouts find their way into the old terminals.

McDonald: What about the lack of response of the neuroma to hypercapnia? You say that after circulation is stopped the neuroma responds to hypoxia. What about the response to hypercapnia?

Bingmann: We did not observe changes in the activity during circulatory arrest with changing PCO_2. It might mean these nerve endings at this early stage cannot respond to PCO_2. Perhaps it is a mechanism that differs widely from that of hypoxia. Mitchell did see a response to hypercapnia, but his neuroma were old in comparison to this one.

McDonald: I did the ultrastructure on these neuromas, so perhaps I can comment. It was confirmed that there were no glomus cells; there were a lot of nerve fibers, some myelinated and some unmyelinated. It is difficult to define a terminal since we did not do serial Sect. to actually find the end of the nerve. But there were unmyelinated portions of the nerve that contained a variety of organelles found in the normal carotid body nerve endings. The main issue we did resolve was the absence of glomus cells, which supported our findings by fluorescence microscropy that there were no catecholamine-containing cells.

Trzebski: Are there any data on the carbonic anhydrase activity within the neuromas?

Kienecker: I do not know.

O'Regan: You say the fluctuations in discharge are due to autonomic effects. Are they abolished when you cut the sympathetic supply to the carotid body?

Bingmann: No. We tried it but the fluctuations still persisted. But there are many small fibers that may go into the carotid body via the sinus nerve.

O'Regan: Did you try catecholamines on the neuroma or the regenerated carotid body?

Bingmann: No. We did not try any chemical agents.

Eyzaguirre: My suggestion would be to get these neuromas out of the animal. First, we have carotid body tissue in the neck, which has been described by de Castro. The danger is great of contamination of those small fibers. Second, removal would allow the possibility of studying the threshold properties of the neuroma, which I think is an interesting problem.

Recovery of Chemosensory Function of Regenerating Carotid Nerve Fibers*

P. Zapata, L. J. Stensaas, and C. Eyzaguirre

The site at which physiological stimuli elicit carotid nerve activation is still disputed. To evoke an increase in discharge in slowly adapting receptors (such as carotid body chemoreceptors), stimuli must depolarize the sensory endings. Although hypoxia can depolarize nerve membranes (8), hypercapnia has the opposite effect (9). Furthermore, high acidity is required to alter nerve membrane potentials (15) or the generation of baroreceptor impulses (6). Thus to postulate a direct action of small changes in PO_2, PCO_2, and pH on carotid body sensory endings, one has to assume that these terminals are quite different from others. The idea that carotid body nerve endings do not possess specific chemosensitivity has been further sustained by successful functional reinnervation of carotid bodies with vagal (2) and superior laryngeal nerve fibers (18).

These results support the hypothesis that the carotid body parenchymal cells are essential for transduction. However, this theory has been challenged by Mitchell et al. (12) who have reported that carotid nerve neuromas - isolated from the carotid bodies - present spontaneous discharge, which are increased by hypoxia, hypercapnia, and injections of NaCN, ACh, or nicotine. Their conclusion is that the nerve endings themselves are the chemosensitive transducers. To test this hypothesis, we crushed the carotid nerves at different distances from the carotid bodies and studied the preparations several days later. After the nerves are crushed, the neuronal sheaths remain in continuity to guide the outgrowing axons toward the glomus tissue with little or no neuroma formation. The reappearance of carotid nerve sensory discharges was established by electrophysiological recordings, while regeneration of nerve fibers in the same specimens was studied by electron microscopy. A detailed report has been published elsewhere (19). Experiments were performed on anesthetized cats. Carotid nerve crushes were made with a pair of Dumont tweezers, and although a single crush induced immediate disappearance of baro- and chemosensory discharges, triple crushes were applied to ensure complete axonotmesis. Furthermore, light and electron microscopy of the nerve distal to a crush performed 3 days earlier showed only regenerating fibers and no intact axons. To exclude sympathetic fibers the ganglioglomerular nerves were excised 3 days before recording and fixation.

Electrophysiological recordings of the whole carotid nerves between 3 and 5 days after a crush showed absence of both chemo- and barosensory discharges. However, chemosensory activity reappeared at the 6th day after crushing. Regenerating carotid nerves reestablished control levels

---

*This work was supported by grants NO-05666, NO-07938, and NO-10864 from the U.S. Public Health Service. Thanks are due to Mrs. Caroline Zapata for preparation of the figures.

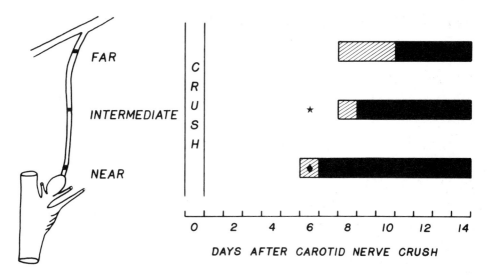

Fig. 1. Summary of electrophysiological observations on recovery of chemosensory activity in regenerating carotid nerves after crushes at different levels. (*solid areas*) Maximal activity between normal range. (*slanted areas*) Disminished activity as compared to normal controls. (*star and diamond*) Special cases described in text

of activity shortly after resumption of chemosensory discharges. Fig. 1 shows that reappearance of chemosensory discharges at normal response levels depended on the distance between the site of nerve crush and the carotid body. When the crush was made close (1-2 mm) to the glomus, chemosensory discharges reappeared by the 6th day, and were still fewer (slanted area) then those present in the normal contralateral nerve, although responses were within the normal range in one case (diamond). By the 7th day and after responses of regenerating carotid nerves were similar to those of normal nerves (solid area). When the nerve was crushed at an intermediate (5-6 mm) level between the glomus and the glossopharyngeal trunk, spontaneous chemosensory discharges reappeared at the 8th day, and normal responses were recorded from the 9th day on; however, in one experiment (star) discharges of a single or few chemosensory units were ewoked by the largest dose of NaCN (25 µg/kg, i.v.), although no spontaneous activity or responses to asphyxia were detected. When the carotid nerve was crushed far (10-12 mm) from the carotid body, no responses were detected after 6 or 7 days (even after intracarotid injection of 100 µg ACh, nicotine, or NaCN); diminished chemosensory activity occurred between 8 and 10 days, and responses attained normal values 11 days after the crush.

Our results suggest that recovery of chemosensory function does not depend on the degree of maturation of regenerating carotid nerve fibers; it is inversely related to the distance between the site of crush and the glomus. This was clear when both carotid nerves were crushed simultaneously, one close to and the other far from the carotid bodies. In one animal recordings were made 6 days after the initial operation. The mean frequency of spontaneous sensory activity was 18 Hz, and the maximal frequency after 1 min of asphyxia was 48 Hz in the nerve crushed close to the glomus; no spontaneous or induced discharges were obtained from the contralateral nerve crushed far from its glomus. In another animal recordings were performed 8 days after the crushes. Spontaneous activities were ca. 50 and 27 Hz and maximal responses elicited by asphyxia ca. 195 and 75 Hz, respectively, for the nerves crushed near to and far from the carotid bodies.

Ultrastructural observations showed that no nerve terminals or intact myelinated axons were present in the carotid body 3 days after a carotid nerve crush. Prominent empty spaces appeared between glomus (type I) and sustentacular (type II) cells in areas ordinarily occupied by nerve terminals. This finding suggests that degenerating nerve endings become detached from parenchymal (types I and II) cells and retract into nearby Schwann cells where they are broken down. A few unmyelinated axons enclosed by Schwann cells and identified as regenerating axons sprouts were present in the neural pole of the carotid body. Five days after a crush, profiles of axon sprouts ensheathed by Schwann cells, but without apposition to parenchymal cells, were commonly found.

Fig. 2 shows that unmyelinated axons apposed to the surface of glomus cells appeared 6 days after crushing close to the glomus. Some were undifferentiated, but a few appeared as mature terminals with synaptic vesicles and junctional densities. By the 8th day and after, regenerating axons were numerous in the interlobular zones. After 2 weeks undifferentiated axon sprouts came commonly into apposition with glomus cells, but some of them appeared as specialized terminals, both calyciform and bouton types. Immature myelinated axons were first observed in the glomus 3 weeks after the nerve crush, but they did not appear normal before 6 weeks. By the 7th week after nerve crushing, nonspecialized appositions between unmyelinated axons and glomus cells became less numerous, while calyciform and bouton terminals regained their normal size and appositional relations.

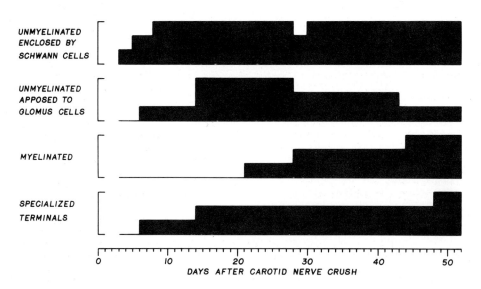

Fig. 2. Summary of time course of axon regeneration in carotid body after carotid nerve crush near glomus. Each type of element, classified as minimal, more numerous, and abundant, is represented by one of the three widths of horizontal bars.

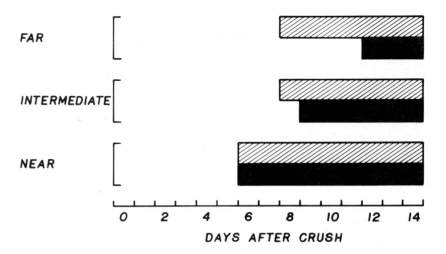

Fig. 3. Summary of time course of appearance of regenerating axons in carotid bodies after carotid nerve crushes at different distances from glomera. (*slanted areas*) Undifferentiated unmyelinated axons apposed to glomus cells. (*solid areas*) specialized terminals - calyciform or bouton type - contacting glomus cells

Fig. 3 shows that reestablishment of contacts between sensory nerves and parenchymal cells depended on the distance between the site of nerve crush and the glomus. When the crush was made at intermediate levels or far from the glomus, appearance of nonspecialized appositions between unmyelinated axons and glomus cells occurred 2 days after similar developments on the side where the nerve was crushed close to the glomus. Similarly, reappearance of specialized terminals (either calyciform or bouton types) was delayed by 3 days when the crush was made at the intermediate level and by another 3 days when the crush was made far from the glomus. The fact that unmyelinated axons are present in the carotid body only 3 days after crushing the nerve close to the glomus indicates a fast rate of regeneration. Despite the early appearance of axons in the carotid body, another 3 days were needed for recovery of chemosensory function; this coincides with establishment of contacts between axon sprouts and parenchymal cells. At equal intervals of 6 days, crushes made far from the carotid bodies induced similar degrees of nerve regeneration; but there was not enough time for the sensory fibers to reach the carotid body cells. Concomitantly, there was no chemosensory discharge even when strong natural (asphyxia) or pharmacologic (nicotine, ACh, NaCN) stimuli applied.

Comparison of Fig. 1 and 3 suggests that reapposition of nerve endings to carotid body cells is essential for the recovery of chemosensory transduction. Restoration of function resulting from the apposition of regenerating axons to parenchymal cells does not indicate which type of cell plays the essential role in chemosensory transduction. The growth cones contact, at approximately the same time, glomus and sustentacular cells. Thus sensory endings may excited by transmitters released by natural stimuli (7) from type I cells, or the ending excitability may be modulated by chemical release from these cells (17). Another view is that sensory nerve terminals are excited by mechanical deformation (13) caused by K^+ leakage from type II cells (11).

These results disagree with those of Mitchell et al. (12), who proposed that carotid nerve endings are the sole chemoreceptive transducers based on the absence in their neuromas of the yellow green fluorescence characteristic of type I cells. However, in only 1 of their 4 experiments was the carotid body excised. Therefore, several other possibilities may explain their results: {1} presence of type II or poorly fluorescent glomus cells in the neuromas; {2} establishment of nerve connections between the neuromas and the original carotid bodies, or {3} regenerating nerve fibers connecting the neuromas to aberrant glomera. They recorded from the carotid neuromas 12-18 months after the initial operation; thus there was ample time for regenerating fibers to reach remnants or accessory glomera. This suggestion is based on the variable morphology of cat carotid bodies (14), the presence of miniature glomera along the carotid artery (3,10), and our own experience that after sectioning, fibers in central stumps grow in many directions and reach distant structures (18).

Our results complement those of Verna et al. (16). They reported loss of chemoreceptive properties of carotid nerves after destruction of carotid body cells by local freezing; only when some glomus and sustentacular cells were preserved could chemosensory discharges be elicited by different stimuli. An interesting aspect of our observations is that recovery of spontaneous chemosensory activity and attainment of nearby normal values for maximal responses occur at a time when relatively few axons are apposed to parenchymal cells. One must take into consideration that normally each carotid nerve fiber branches at some distance from the glomus (5) and that each intraglomeral branch is apposed to several glomus type I cells (1,4). Thus only a few appositions may suffice to restore function in a given nerve fiber. Furthermore, sensory endings do not have to be fully differentiated and with a large apposed area to glomus cells to begin firing sensory impulses.

References

1. Biscoe, T.J., Pallot, D.: Experientia *28*, 33-34 (1972)
2. Castro, F. de: Acta Physiol. Scand. *22*, 14-43 (1951)
3. Castro, F. de: Arch. Int. Pharmacodyn. Ther. *139*, 212-234 (1962)
4. Eyzaguirre, C., Gallego, A.: in: The Peripheral Arterial Chemoreceptors. Purves, M.J. (ed.). London: Cambridge U. Pr. 1975, pp. 1-23
5. Eyzaguirre, C., Uchizono, K.: J. Physiol. *159*, 268-281 (1961)
6. Eyzaguirre, C., Zapata, P.: J. Physiol. *195*, 557-588 (1968)
7. Eyzaguirre, C., Zapata, P.: in: Arterial Chemoreceptors. Torrance, R.W. (ed.). Oxford: Blackwell 1968, pp. 213-251
8. Lorente de Nó, R.: Stud. Rockefeller Inst. Med. Res. *131*, 114-147 (1947)
9. Lorente de Nó, R.: Stud. Rockefeller Inst. Med. Res. *131*, 148-194 (1947)
10. Matsuura, S.: J. Physiol. *235*, 57-73 (1973)
11. Mills, E.: Fed. Proc. *31*, 1394-1398 (1972)
12. Mitchell, R.A., Sinha, A.K., McDonald, D.M.: Brain Res. *43*, 681-685 (1972)
13. Paintal, A.S.: J. Physiol. *189*, 63-84 (1967)
14. Seidl, E.: Anat. Embryol. *149*, 79-86 (1976)
15. Spyropoulos, C.S.: J. Neurochem. *5*, 185-194 (1960)
16. Verna, A., Roumy, M., Leitner, L.-M.: Brain Res. *100*, 13-23 (1975)
17. Zapata, P.: J. Physiol. *244*, 235-251 (1975)
18. Zapata, P., Hess, A., Eyzaguirre, C.: J. Neurophysiol. *32*, 215-228 (1969)
19. Zapata, P., Stensaas, L.J., Eyzaguirre, C.: Brain Res. *113*, 235-253 (1976)

DISCUSSION

McDonald: You showed the number of nerve endings next to type I cells at various periods after crush. On day 0 there were no nerve endings. This implies that there is immediate disappearance of all types of sensory nerve endings next to the glomus cells, but this is not so. In your quantitative studies of nerve endings next to type I cells did you find a decrease in number followed by an increase in number? Is there an overlap of degeneration and regeneration that could confuse the picture?

Zapata: We showed the appearance of nerve terminals only from the third day. For the first 2 days we did not make any type of quantitative estimation for degeneration of these terminals. This has been done before. Our present results concern the reappearance of nerve terminals close to the carotid body cells.

McDonald: At 3 days there were no nerve endings at all? At 6 days what was the proportion of normal?

Zapata: On the third day after crush close to the carotid body we did not find any intact nerve terminals apposed to the glomus cells; those terminals close to glomus cells were in process of degeneration or some kind of retraction. There was only membranous, but not cytoplasmic, material close to the glomus cells.

Pietruschka: Did you find any morphologic alterations in the type I cells during nerve degeneration, specially in the number of the dense-cored vesicles?

Zapata: We did not make a quantitative study. However, we have the impression that carotid bodies excised about 50 days after the crush had an increased number of granules. The problem is that sometimes it is very difficult to differentiate large dense-cored vesicles and lysosomal dense bodies.

Kienecker: Is this a primary degeneration or a secondary, real Wallerian degeneration? It may be possible when you crush the nerve that there is only a neurologic swelling with an axon retraction within the glomus.

Zapata: We tested that only in the animals in which the nerve was crushed far from the carotid body, where we found that all of the fibers in the sinus nerve after, say, 5 days were in a process of degeneration. The problem is that if you allow such a time for primary and secondary degeneration there is no way to differentiate between them. They have the same time delay. I do not believe that for the physiological recovery of chemosensory function it is going to make any difference.

Torrance: What about the nerve bundles within the carotid body? Were the nerve bundles absent after the crush?

Zapata: You can find that all of these nerve bundles are in a clear process of degeneration, myelin destruction, and so on.

Purves: The suggestion has been made that the fine unmyelinated fibers form a different population from the fibers that make synaptic contact with type I cells. It seems from your data that the rate of regeneration, the time course, was similar for all groups. Do you think you have looked at enough fibers to be able to say whether there are genuinely different populations or that the unmyelinated fibers might form fine preterminal parts of the fibers that terminate on the type I cells?

Zapata: The very first few days after the crush all the nerve endings contacting the glomus cells seemed to connect with unmyelinated fibers. Myelination was the most delayed of the regenerative processes. It takes usually more than 2 weeks to start, and it was completed after about $1^1/_2$ months.

O'Regan: Did you, like Biscoe and Stehbens or McDonald and Mitchell, notice any normal nerve endings after Sect. of the sinus nerve that did not degenerate?

Zapata: No. But our material does not include the endings of fibers coming from the sympathetic superior cervical ganglion because in all our animals we also cut that nerve.

Bingmann: The fact that restoration of the morphologic and functional qualities of the nerve endings has a similar time course of about 4 days is not proof that the synapse is a necessary part of the system that encodes hypoxia and hypercapnia. In one of your publications (*Brain Res*. 113: 235, 1976) you said that the number of synapses was low when you recorded activity that appeared to be normal.

Zapata: There is good correlation between the reappearance of terminals close to type I cells. There is no such correlation between the normalization of chemosensory function and the normalization of glomus nerve synapses. Perhaps it is not necessary to have all the complete apparatus for the generation of discharge in chemoreceptors. Perhaps only a few appositions are needed, and there is a large satefy factor in this preparation.

Acker: You did single-fiber preparations?

Zapata: No. Experiments were done with a whole nerve.

Kienecker: Do you believe that the type I cell is a sensory cell that can initiate chemoreceptor activity? Unlike other sensory cells type I cells can form neoplasms.

Zapata: This series of experiments does not differentiate between the role played by type I cells and type II cells. It is not absolutely necessary that the transducer process occurs in these cells. There is still the possibility that chemosensory transduction is at the level of chemosensory nerve terminals, but the terminals must be close to type I cells. Without this association we cannot record activity.

Wiemer: Does the position of your nerve crush, besides affecting the time course of regeneration, have any effect on the functional result of the regeneration as reagrds completeness of response to cyanide or other stimuli?

Zapata: No.

Speckmann: Have you observed any significant changes in the spike shape recorded after your crushes? Does the rate of rise or fall or maximum amplitude change?

Zapata: I could not differentiate for the amplitude and size, but the recordings were always done on a part of the carotid nerve central to the crush.

Studies of Normal and Wobbler Mutant Carotid Bodies

D. J. Pallot and T. J. Biscoe

Biscoe et al. (2) reported that after Sect. of the glossopharyngeal nerve central to its sensory ganglion the nerve endings on type I cells of the carotid body degenerated with a long time course. At a time when most of these nerve endings has disappeared, it was still possible to record a normal chemoreceptor discharge. These experiments have been the object of criticism. Here we want to show that in the mutant mouse wobbler, it is possible to record a normal chemoreflex when subsequent electron-microscopy shows the carotid body type I cells are effectively devoid of normal synaptic vesicle-containing nerve endings.

The mutant wobbler was originally described by Falconer (4) and its neuropathology first examined by Duchen and Struch (3). They showed that the abnormality was a degeneration of nerve cells in the motor system within the brain and spinal cord accompanied by progressive denervation of skeletal muscle. No abnormalities were seen in the sensory innervation of skin, mucosae, teeth, or pacinian corpuscles, and muscle spindles were affected only where denervation atrophy of extrafusal fibers was very pronounced. The gene responsible for the defect is recessive, and hence in any little born of heterozygous (normal) parents 25% of the young are affected. Normal and mutant animals were anesthetized with pentobarbital sodium (Nembutal, Abbott) 6 µg/g intraperitoneally. Tracheal and peritoneal cannulae were inserted and the animals were placed in a plethysmograph with the cannulae exteriorized. The animals could be given varying gas mixtures via the tracheal cannulae. Respiration was monitored with a pressure transducer.

Reflex activity originating from the carotid body was tested as follows. The animals were permitted to breathe 15% oxygen in nitrogen until respiration reached a steady state; the respiratory response to the inhalation of one or two breaths of 100% oxygen was then recorded. In three experiments the region of the carotid bifurcation was dissected free from surrounding tissues, hence denervating the carotid body, and the response to inhalation of 100% oxygen was tested once more. Following the physiological experiments, the animals were fixed by perfusion via the left ventricle with a mixture of 2.5% glutaraldehyde and 1% paraformaldehyde in a phosphate buffer at pH 7.2, the carotid bifurcation was removed, and processed by standard techniques for electron-microscopy. In a few experiments the animals were ventilated for 10 min with 100% oxygen prior to perfusion.

Sequential Sects. for counting nerve endings were cut at a separation of 8 µm and mounted on single-slot grids. One hunderd type I cells sectioned through the plane of the nucleus were examined, and the number of profiles of synaptic vesicle-containing nerve endings associated with each were counted.

Physiological experiments showed that the mutant mouse responded as does the normal mouse to a single or double breath of oxygen, namely, there is a fall in minute volume starting after 5 s or so (i.e., after another breath or two) with both tidal volume and rate typically depressed. Denervation of the carotid sinus region abolishes this response in both normal and mutant mouse where a breath or two of oxygen now typically produces an increase in minute volume.

In the structural studies of the normal mouse between 130 and 170 normal synaptic vesicle-containing nerve endings were found associated with each type I cell examined. Between 26 and 32% of type I cells in the normal mouse had no nerve endings in the nuclear plane: around individual cells from one to seven separate nerve-ending profiles could be counted.

The situation in the carotid body of the mutant animal was completely different; between 0 and 7 normal nerve endings were found per 100 type I cells. Counts of degenerating nerve endings were made and ranged from 17 to 60 per 100 type I cells. Animals were examined at ages from 6 weeks to 7 months. While the absolute figures for numbers of endings were variable it was not possible to detect any trend associating age with number of nerve endings. Indeed, the largest numbers of normal and degenerating nerve endings were found in the same 6-month old animal.

In four animals (two mutant and two controls) the number of type II cell nuclei were counted; an average of five type I cells was found per type II cell nucleus in both mutant and control animals. In two or three animals some changes in type I cells in the mutant were found. These were similar to the changes described by Knoche and Kienecker (5) in type I cells after sinus nerve section. These experiments show that the wobbler mouse is capable of responding to hypoxia and hyperxia, but that the carotid body type I cells are not innervated in the normal manner and have few if any synaptic vesicle-containing endings on the type I cells.

Whatever the cause of the degeneration found in these animals and in the experiments described by Biscoe et al. (2), the chemoreceptor function of the carotid body appears to be normal. That being the case it is difficult to ascribe a chemoreceptor function to the synaptic vesicle-containing endings and to the type I cells with which they make contact. Nevertheless, other possibilities exist. One explanation is redundancy. Thus, it might be that only a very small percentage of normal nerve endings are required for normal function. Then the remaining endings in the study of Biscoe et al. (2) might suffice for normal function, and the one or two nerve endings found here might also be adequate.

A further explanation is that in the present study the loss of nerve endings does not result from a true degeneration of the whole nerve ending and associated nerve fiber and cell body, but rather that it might be a process affecting only the nerve ending while the nerve fiber remains intact. Under these conditions it might be that the nerve fiber that previously terminated on a type I cell is now acting as the chemosensor (1).

In conclusion, the wobbler mouse reacts normally to the inhalation of 100% oxygen. In the mouse we have demonstrated that this reflex is mediated via the carotid body. Subsequent electron-microscopy shows that very few normal nerve endings are found associated with the carotid body type I cells. Even if the various degenerating structures are counted as normal there is still a deficit amounting to about 70-90% of the normal number of nerve endings. We interpret these results as showing that the nerve endings associated with type I cells are not afferent.

References

1. Biscoe, T.J.: Physiol. Rev. *51*, 437 (1971)
2. Biscoe, T.J., Lall, A., Sampson, S.R.: J. Physiol. *208*, 133 (1970)
3. Duchen, L.W., Struch, S.J.: J. Neurol. Neurosurg. Psychiat. *31*, 535 (1968)
4. Falconer, R.: Mouse News Letter (1950)
5. Knoche, H., Kienecker, E.W.: this book

DISCUSSION

McDonald: In any of the P_aO_2 mutant mice you have studied did you measure the P_aO_2 just prior to starting the perfusions? Some of the nerve endings and glomus cells contain swollen mitochondria with small numbers of vesicles in the sensory nerve endings and simulate conditions of hypoxia. Does this result from the motor lesion causing the mice to underventilate and hence making them chronically hypoxic?

Pallot: I think that is a fair conjecture. However, the mouse has proved rather difficult to fix. What we have done, because we wondered whether their hypoxia before fixation could produce the abnormal nerve endings, is to ventilate some of the animals with 100% oxygen for 10 min prior to fixation. This made no difference in the number of normal nerve endings found, or rather the number still fell within the range shown in other animals.

McDonald: But what if the condition was chronic and the animals were chronically hypoxic because of the disease?

Pallot: I cannot answer that. However, up to the age of 4 weeks the affected animals are indistinguishable from their normal littermate controls. The earliest age at which we have done a physiology experiment and subsequent electron microscopy is 6 weeks. There appears to be no difference between the 6-week-old mutant animal and one 6 months of age; the number of the nerve endings is similar.

Purves: The somatic motor deficit is presumably the most obvious and easily tested feature in these animals. Have you or Duchen looked at sensory and autonomic function in these animals?

Pallot: No one has studied physiology of other sensory systems. Duchen has studied the innervation of spindles and Pacinian corpuscles. They are apparently normal. We are looking now at the innervation of the adrenal medulla.

Hess: My question is very similar. Did you count the number of petrosal ganglion cells or look at them?

Pallot: I have not counted the number of petrosal ganglion cells, but we have looked at some ganglion cells with the electron microscope. I have not found any degenerating petrosal ganglion cells. What we have seen in the main nerve trunk as it is traversing the ganglion is some degenerating nerve fibers enclosed in Schwann cells.

Hess: But the nerve terminals on these cells are presumably formed from the processes of petrosal ganglion cells.

Pallot: That may be, but you asked if I have found any degenerating nerve cells in the petrosal ganglion, and the answer is no.

Kobayashi: In your electron micrographs two populations of membrane-bound granules in the type I cell, one small and one large, seem to be apparent. The small one could be the same as what I described in 1970 in the normal mouse carotid body. It occurs in the region immediately beneath the synaptic contact between the type I cell and nerve terminal. In your electron micrographs the small population of granules do not have any dense precipitate inside the limiting membrane. In my electron micrographs they have a precipitate, but in McDonald's electron micrographs the small granules have no precipitate. What are your ideas on this structural difference?

Pallot: The only comment I can make is that it represents a difference in fixation technique.

McDonald: I am interested in the conclusions you draw from these data. Do you believe that it takes relatively few nerve endings next to the glomus cells in order to give rise to a normal ventilatory response to this brief exposure to high oxygen tension?

Pallot: I think there are three possible conclusions. One, these nerve endings are not sensory. This is perhaps less likely in view of other work described in this volume. Two, there is such a massive redundancy of nerve terminals within the carotid body that, in fact, if 20% of the normal number remains, there is still a normal response. The third is that this lesion on closer study could be shown to be a lesion of the actual nerve terminal rather than the nerve fibers. Although the terminal is degenerate, the nerve fibers themselves are still intact, and under these conditions the nerve fibers can act as the transducer.

Torrance: You talk about the nerve endings in relation to the type I cell. How do they relate to the type II cell?

Pallot: The endings are enveloped by type II cells as in the normal animal. However, in some cases the endings are so degenerate that, in fact, they are just spaces within the tissue.

Wiemer: Do you have any indication whether the respiratory muscles are working normally in the mutant? If they are not, there could be a fourth possibility: the reaction recorded when the animal breathes pure oxygen is weaker than normal because it results from a partial respiratory failure superimposed on a stronger hypoxic drive than in normal animals.

Pallot: To answer that question you have to know what the P_aO_2 is, and this we have not yet measured. Also the response to 100% oxygen must be quantified. We have not done this because we felt that, in a reflex that is altered by the depth of anesthesia, such a quantification would be meaningless. However, our impression during the experiments has often been that the wobbler, in fact, has a greater response to 100% oxygen than the control animal.

Torrance: But you will accept the point that if your stimulus is set as the gas breathed, the resultant P_aCO_2 may well be lower. Therefore you have a bigger stimulus to remove with your 100% oxygen so you would expect a bigger change in ventilation.

Pallot: That may be, but the fact remains that the mutant animal responds to 100% oxygen by decreasing its ventilation. If you then denervate the region around the carotid artery on both sides of the animal, it no longer responds to the 100% oxygen in this way. Hence the mutant apparently has a carotid body that is capable of responding to hypoxia and hyperoxia.

Fine Structure of Pressoreceptor Terminals in the Carotid Body (Mouse, Cat, Rat)*

K. Gorgas and P. Böck

Introduction

Depressor nerve endings were identified at the supplying arteries of the carotid body in mammals (6) and birds (9) by means of silver impregnation methods. The arborized endings were found to originate from myelinated axons, forming widerspread reticulated and flattened terminations in the adventitial layers of small arteries. This paper presents electron-microscopic findings on the fine structure of these sensory endings.

Materials and Methods

Swiss mice, albino rats, and cats were anesthetized with ether and fixed by perfusion (the fixative consisted of 5% glutaraldehyde and 4% polyvinylpyrolidone buffered to pH 7.4 with phosphate buffer). Perfusion was carried out for 5 min, after which the carotid bifurcations were prepared and fixed by immersion in the same solution for 2 h. The aldehyde was washed out with buffer overnight, and the specimens were postfixed for 2 h in 1% OsO_4 buffered with sodium-cacodylate. Dehydration was performed in a series of ethanols and in popylene oxide, and embedding was done in Araldite. Serial semithin and thin Sects. were cut on a Reichert OmU2 ultramicrotome. The electron microscope was a Siemens EM 101 A.

Results and Discussion

Profiles of baroreceptor terminals were identified in single Sects. by their rich content of mitochondria (Figs. 1 and 2). This was the most striking morphologic feature, which in general (5,7), is consistent with that of baroreceptor terminals in the carotid sinus (4,10) and with sensory terminals in the arterial tunica adventitia. The individual profiles reached 3.5 µm in diameter and included abundant mitochondria, most of which showed only a single longitudinally positioned crista. Glycogen particles and myelin bodies were also regular constituents within the axoplasm. Neurotubules were rarely seen between the mitochondria, and neurofilaments appeared to be gathered near the axolemma. Vesicles were mostly of the clear type, their diameters ranging from 500 to 2000 Å. The major part of the surface of the axon terminal was covered by a Schwann cell. Where a Schwann cell envelope was absent, the free areas of the axolemma were provided with a basal lamina continuous with that of the Schwann cell. In these regions the axons formed lamellar or fingerlike protrusions, which were characterized by their rich

---

*Supported by a grant from the Deutsche Forschungsgemeinschaft Bo/525-1

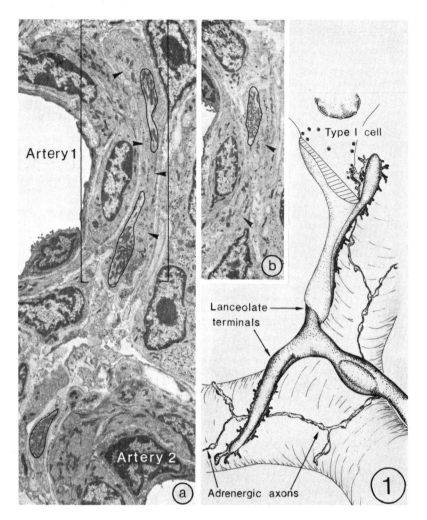

Fig. 1. Drawing shows two lanceolate baroreceptor terminals at branching site of an artery. Mechanoreceptor terminals, which are partially covered by processes of a Schwann cell, originate from same stem axon. Upper receptor is seen to make specific contact with carotid body type I cell, as seen in the mouse. Drawing is based on serial sectioning work, two views of which are shown in (a) and (b). Profiles of baroreceptor terminals are identified by their rich content of mitochondria. (*arrowheads*) Cytoplasmic processes of carotid body type I cells

content of neurofilaments and some small, clear vesicles known as receptor matrix of mechanoreceptors (1).

The entire profiles described were located in the tunica adventitia of small arteries (1-2 layers of smooth muscle cells). Over wide areas their Schwann cell cover or free axolemma was never less than 0.3 μm from the surface of the vascular smooth muscle cells. In circumscribed regions axon protrusions approached the plasmalemmata of smooth muscle cells up to 1000 Å. In these cases, however, basal lamina material remained interposed (Fig. 2). Faint collagen fibrils were often seen to reach the basal lamina of the Schwann cells or of the axon protrusions.

Serial Sects. revealed that the receptor terminals described above belonged to far-reaching, free, branched lanceolate terminals of a complex type (2). These adhered helically or circularly to the arterial wall (Fig. 1) and were regularly found at the branching sites of arteries. In mice and rats lanceolate terminals were shorter and their degree of branching less than in cats. The stem axons belonging to the branched lanceolate terminals were nonmyelinated within the carotid body. Judging from their numerical relationships to Schwann cells (10) they very probably belong to the premyelinated category of axons. This view also substantiated by the fact that lanceolate terminals, located at arteries near the hilus of the organ, were found to originate from myelinated axons. The diameters of these stem axons varied from 1 to 2 µm. Abundant neurotubules and a few mitochondria were seen in cross sections. Accumulations of mitochondria, myelin bodies, and glycogen were found at intervals along the length of the lanceolate terminals.

Slender processes of carotid body type I cells also approached the adventitia of small arteries and reached the lanceolate pressoreceptor terminals. Type I cell processes and the baroreceptor terminals were embedded within the same Schwann cell (or carotid body type II cell) and were seen in one case to make contact with each other (mouse, Fig. 2). In this peculiar situation a fingerlike process of the baroreceptor terminal was indented into the process of the type I cell, and several dense-cored vesicles of varying size were accumulated near the contact zone. The results of Kondo (8), who studied the wall of the aortic arch in chickens, are of particular interest in this connection. He found that various nerve endings and granulated cells (paraganglionic cells) were gathered in the tunica media in those regions supplied by the depressor nerve. Some of these axon terminals were characterized by a perfusion of mitochondria, and the granulated cells were shown to contain catecholamines (3). It is suggested that this intimate spatial relation between mechanoreceptor terminals and catecholamine-containing cells is of functional significance.

Contact areas between baroreceptor terminals and carotid body type I cells were observed in mice. In the cat, lanceolate terminals and processes of type I cells were found to be enveloped by the same Schwann cell but did not touch each other. This situation has not yet been observed in the rat. The number of baroreceptors at arteries in the carotid body in rats is low in comparison with cats and mice. Apart from these differences the general morphology of baroreceptor terminals is the same in all species studied.

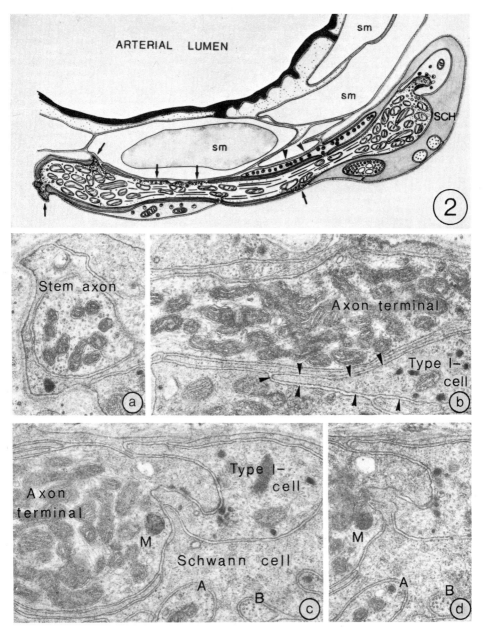

Fig. 2. Schematic drawing of ideal Sect. of baroreceptor terminal at small artery within mouse carotid body. sm, smooth muscle cells; SCH, Schwann cell or carotid body type II cell. (*dotted lines*) Basal lamina material. Pressoreceptor terminal is characterized by abundant mitochondria. Protrusions of axon terminal contain receptor matrix (*arrows*). Cytoplasmic processes of carotid body type I cells, which include dense-cored vesicles, come in close contact with vascular smooth muscle cells (*arrowheads*). Small cytoplasmic processes of enveloping Schwann cell (or carotid body type II cell) are usually interposed between mechanoreceptor terminal and process

References

1. Andres, K.H.: Z. Zellforsch. *75*, 339 (1966)
2. Andres, K.H., Düring, M.v.: in: Handbook of Sensory Physiology. Iggo, A. (ed.). Berlin: Springer-Verlag 1973, Vol. II, pp. 3-28
3. Bennett, T.: Z. Zellforsch. *114*, 117 (1971)
4. Böck, P., Gorgas, K.: Cell Tissue Res. *170*, 95 (1976)
5. Burnstock, G.: Clin. Exp. Pharmacol. Physiol. (Suppl.) *2*, 7 (1975)
6. Castro, F. de: Trav. Labor. Rech. Biol. Univ. Madrid *24*, 365 (1926) and *25*, 331 (1928)
7. Hagen, E., Wittkowski, W.: Z. Zellforsch. *95*, 429 (1968)
8. Kondo, H.: Anat. Rec. *178*, 253 (1974)
9. Nonidez, J.F.: Anat. Rec. *62*, 47 (1935)
10. Rees, P.M.: J. Comp. Neurol. *131*, 517 (1967)

DISCUSSION

<u>Nishi</u>: I am rather confused. The characteristics of the terminal are the accumulation of mitochondria, the presence of small cored vesicles, and also the accumulation of glycogen granules. These are the characteristics of the mechanoreceptor nerve endings. However, particularly in the cat carotid body there were a lot of nerve endings associated with type I cells that possess similar characteristics. I think it is rather difficult to distinguish these nerve terminals as definitive mechanoreceptor terminals.

<u>Böck</u>: These characteristics are those of a sensory terminal. Our dedinition is based on the work of Andres describing the lanceolate mechanoreceptor ending, and our observations are in good agreement with his. The size of mitochondria in the endings is relatively small, and these mitochondria are provided with a single christa, which is orientated in the long axis of the organelle. We tried to find these terminals near, or in close relation to, the adventitial layer of arteries. I agree with you that very similar profiles can be found at the surface of carotid body tissue; I feel that those axon terminals that include the small type of mitochondria may be related also to something like a mechanoreceptor. In the synaptic contacts shown elsewhere the mitochondria are larger. I think it is possible to differentiate these terminals. I was pleased to see that mitochondria-rich axon profiles can be found on the small arteries within the carotid body. In Burnstock's review this is the only criterion used to identify a pressoreceptor.

<u>Bingmann</u>: Do these receptors degenerate after cutting the sinus nerve?

<u>Böck</u>: I have no information as to where the stem axons originate.

---

of type I cell. In one case, specific contact was seen between these structures, as illustrated at right end of receptor terminal. (a-d) Essential details from electron micrographs on which schematic drawing is based. (a) Stem axon of premyelinated type, rich in neurotubules and mitochondria. (b) Cell processes of type I cells (*arrowheads*), characterized by dense-cored vesicles, are usually separated from axon terminals by lamellae of Schwann cell cytoplasm. (c and d) Consecutive Sects. showing specific contact between axon terminal and type I cell. M, myelin body; A and B, corresponding profiles

Willshaw: Gerhich and Moore showed that chemoreceptor afferent activity could be correlated with elctrocardiogram or pulse. Perhaps these mechanoreceptors may in some way influence chemoreceptor activity?

Böck: We have no hypothesis on the meaning of type I cells related to mechanoreceptors.

Wiemer: From your schematic drawings it seems that another criterion for mechanoreceptors is the relation to the small arteries. Your scheme showed that the ending in contact with the type I cell was connected to other endings that had contact with the arteries. Is this always so?

Böck: From one stem axon there originate at least two but probably more lanceolate terminals. In single Sects. it is hard to decide which axons belong to which lanceolate terminal.

Trezebski: Have you any data on the diameter of the fibers that give rise to the structures you described?

Böck: The diameter of a stem axon is usually 1-2 µm and lanceolate terminal enlargements are 3.5-4 µm in diameter.

McDonald: Have you any evidence from your three-dimensional reconstructions that nerve terminals apposed to glomus cell bodies or within a group of glomus cells can be in continuity with terminals next to arterioles?

Böck: I cannot answer this question. I think the axons that supply the lanceolate terminals are running as small nerves traversing the carotid body tissue, but whether some branching also reaches the carotid body type I cells. I do not know. In the periphery we have seen that processes of the chromaffin cells can reach the mechanoreceptor terminals.

Acker: If these mitochondria of Böck are in mechanoreceptors, which are perhaps responsible for flow regulation, is there any possibility that they regulate or affect the diameter of arterioles? It is really impossible to think that these mitochondria affect the chemoreceptor process.

Böck: To answer your question, the lanceolate terminals are only the sensor, and an effector function must be ascribed, for instance, to sympathetic vascular innervation. I do not know if the pressoreceptor terminals can influence the chemoreceptor terminals.

Acker: It is striking that the mechanoreceptors are very close to the type I cell.

Böck: This is correct.

Purves: Can you explain how you distinguish between the pressoreceptor terminals and the calyceal type of nerve terminals on type I cells? Is there the possibllility of confusion, or can you really distinguish between them?

Böck: I think we can really distinguish them. The first point is the localization of the lanceolate terminal, which is always in the adventitial layer of an artery outside the carotid body tissue, whereas the calyceal contacts involve the type I cell somata. The second point is the characteristic accumulation of peculiar mitochondria - extremely small, elongated, with a single crista - within the pressoreceptor terminals. There are also glycogen particles and various types of vesicles, while nerve endings on type I cells include fewer and larger mitochondria. These mitochondria are also show more than one crista, the amount of glycogen is very low, they include only vesicles of the synaptic variety, and besides the synaptic vesicles there are some large, dense vesicles like those in synapses.

Paintal: I have no difficulty in accepting that you have lanceolate endings at the bifurcation of your small vessels in the carotid body. I think the main problem is to establish that there is actual continuity between the lanceolate ending and the small terminal that is close to the type I cell. If this is true, there is a contradiction of Müller's law of specific nerve endings because there would be pressoreceptor and chemoreceptor activity in the same fiber. You said the axons were 1-3 μm in diameter. Was this in the cat?

Böck: That was measured in mice, but the dimensions are very similar in all animals investigated.

Capillary Distances and Oxygen Supply to the Specific Tissue of the Carotid Body

D. W. Lübbers, L. Teckhaus, and E. Seidl

If a low tissue PO_2 is the immediate stimulus for the cheomreceptive process, one has to look for a mechanism that produces a low local PO_2 in the carotid body in spite of the high blood flow through it. It has been suggested that the low PO_2 may be brought about by large diffusion distances within the carotid body (1,2).

To test this hypothesis we used a part of Seidl's histologic material to obtain information about the distances for diffusion of oxygen within the carotid body. The coordinates of the structures to be studied were digitalized and provided the material to be evaluated. In addition to a statistical evaluation, three-dimensional reconstruction of the tissue allowed the quantification of individual structures that may be especially important in the function of the carotid body (see Fig. 3).

Statistical evaluation proceeds as follows:

1. The organ, embedded in paraffin, in serially sectioned. (Linear shrinkage 15-20%)

2. The structures in the tissue Sect. of interest - the border lines of the vasculature, specific tissue (cells of type I and II), and the stratum nervosum - are marked with different colors on the micrographs or transparent film.

3. These drawings are digitalized using a data tablet digitizer (Summagraphics), which has a spatial resolution of 0.254 mm. In our micrographs 100 μm correspond to 13.7 cm, so we can obtain rather more than 5 points per micrometer.

4. The digitalized points of the border lines are connected so that closed areas of the different tissue structures are obtained. Then we compute which points are within the closed areas using a point distance of 1 μm.

5. This grid of points is analyzed every 1 μm in the x direction, measuring the distances in this direction to the border line of the nearest vessel. Then the kind of tissue (specific tissue or stratum nervosum) in the space between the vessels is tested, and the measurements are used to make histograms, one for points in the grid that are in specific tissue and the other for points in stratum nervosum.

6. Sine the vessels are not randomly distributed in every part of the carotid body, the procedure is repeated in directions at various angles (0°, 45°, 90°, 135°) to the x direction.

Fig. 1 shows the histograms obtained. In Fig. 1A for specific tissue the maximum distances of the vessels are between 10 and about 50 μm, with the median 34.06μm; 81% of the distances are less than 60μm and 96% are less than 90 μm. Fig. 1B shows the distances to the vessels from

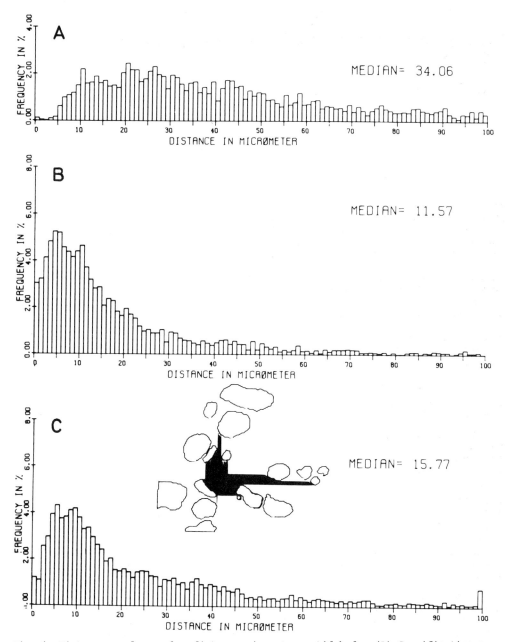

Fig. 1. Histograms of vascular distances in cat carotid body. (A) Specific tissue; (B) stratum nervosum; (C) centerline of specific tissue

points within the stratum nervosum. Here small distances are more frequent than for specific tissue, and the median is less than one-third as great.

This statistical analysis uses the oversimplification that the oxygen reaching a point in a tissue from any direction can be adequately analyzed by measuring the distance to the nearest capillary in that direction. In reality several capillaries surround any small tissue

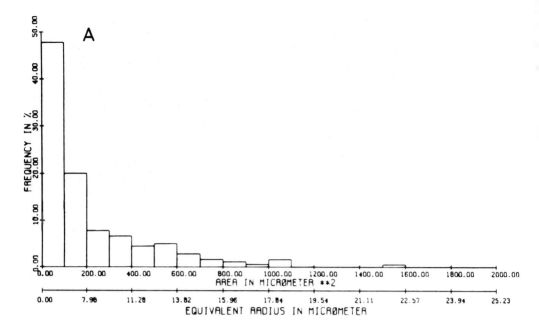

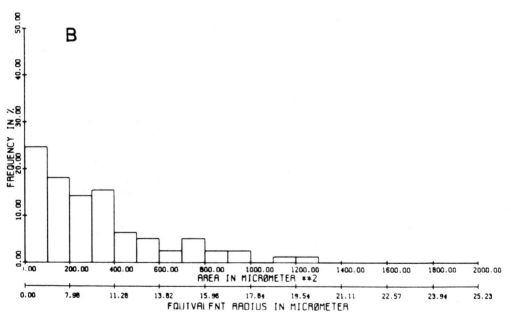

Fig. 2. Histograms of areas of vessels (A) and specific tissue (B) in cat carotid body

volume and influence each other in supplying it with oxygen. To test the influence of adjacent capillaries that supply the same element of specific tissue with oxygen, we measured the distances from the centerline of an area of specific tissue to the border of the nearest vessel,

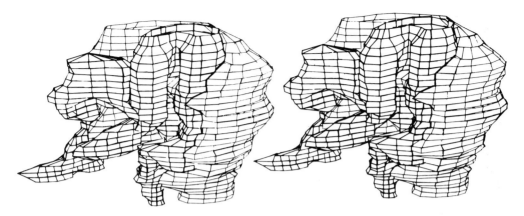

Fig. 3. Reconstruction of a capillary and adjacent cells (part of a glomoid). Stereoscopic drawing

first in the x and then in the y direction. The result is shown in the the inset in Fig. 1C. For comparison the test was carried out in other directions and gave similar results. The distances are again displayed as a histogram (Fig. 1C). Since the distances are related to the area, Fig. 1C describes the oxygen supply situation to a better approximation than does Fig. 1A. In Fig. 1C lower distances prevail.

Fig. 2B illustrates the large size of the specific tissue Sects. that must be supplied with oxygen by one vessel. The mean value is about 200 µm^2. Additionally, the radius that would have a circle of the same area (equivalent radius) is given. Fig. 2A shows the corresponding histogram of the area of the vessels.

In the very specialized tissue of the carotid body, this improved analysis still does not describe the situation correctly, since oxygen transport within the tissue depends on the actual oxygen pressure within the vessels. We have therefore started experiments in which we use our histologic Sects. to reconstruct the capillary vasculature as a three-dimensional model. This method has the advantage that the reconstruction of the vasculature together with the adjacent cells allows an individual analysis of the tissue situation. Fig. 3 shows the complicated structure of a winding capillary within the cells of a glomoid (6). The functional meaning of this morphologic structure is not known; the important rheological consequences of the bendings and their possible influence on exchange processes between blood and cells must be considered.

These morphologic data can be used to estimate the fall of PO_2 caused by the distances shown in the histogram if Krogh's tissue model were applied to a central capillary that supplies a cylindrical space with oxygen (3,4).

The capillary radius was assumed to be 6 µm, the diffusion coefficient D, 2.2×10^{-5} (cm^2/s), the solubility coefficient α, 2.2×10^{-2} (ml/ml×atm), and the oxygen consumption, 9 ml O_2/100 g×min (2). For the different vascular distances the following PO_2 differences were calculated: 34 µm, ΔPO_2 = 6.2 mmHg; 60 µm, ΔPO_2 = 19.9 mmHg; 90 µm, ΔPO_2 = 50.5 mmHg. Since in the carotid body the venous PO_2 is very close to the arterial PO_2, we would expect that only about 4% of the tissue PO_2 values would be below 40 mmHg. Acker et al. (1), however, found that the median PO_2 of the cat carotid body was about 20 mmHg and that almost 80% of PO_2 values were below 40 mmHg. We therefore concluded that large capillary distances cannot be the main reason for the low tissue PO_2 values within the carotid body. Summarizing, we would like to point out that our morphologic measurements did not reveal long diffusion distances, which could be the reason for low oxygen pressures within the tissue of the carotid body. We have therefore to look for other mechanisms that may produce low oxygen pressures such as plasma skimming and uneven flow distribution.

References

1. Acker, H., Lübbers, D.W., Purves, M.J.: Pfluegers Arch. *329*, 136-155 (1971)
2. Daly, M. de, Lambertsen, C.J., Schweitzer, A.: J. Physiol. *125*, 67-89 (1954)
3. Forster, R.E.: in: Arterial Chemoreceptors, Torrance, R.W. (ed.). Oxford and Edinburgh: Blackwell 1968, pp. 115-128
4. Krogh, A.: J. Physiol. *52*, 409-415 (1919)
5. Schur, P., Torrance, R.W.: in: Morphology and Mechanisms of Chemoreceptors, Paintal, A. (ed.). New Delhi: Navchetan Press Ltd. 1976, pp. 113-118
6. Seidl, E.L.: in: The Peripheral Arterial Chemoreceptors, Purves, M.J. (ed.). London: Cambridge U. Pr. 1975, pp. 293-299

DISCUSSION

Purves: Did you say it had been proved that the glomus caroticum is perfused with plasma and the shunt with red cells?

Lübbers: Yes, we have shown this in a number of ways. We have injected FITC, which is quenched by red cells. We could have a film that shows the carotid body lighting up like a lamp when FITC is flushed in. Second, we have measured the local decrease of PO_2 using microelectrodes after stopping perfusion of the carotid body and showed that the rate of this decrease was the same whether the carotid body was perfused with blood or equilibrated saline. We interpreted this to mean that there are no, or only a few, red cells around to give additional oxygen-carrying capacity. Third, there is the cryophotometric method I mentioned earlier. The results of all these tests show that under physiological conditions the carotid body is perfused by only a small number of red cells, i.e., that plasma skimming takes place. With low PO_2 the number of red cells available in the capillary system of the specific cells increases. We have done one measurement with 2% oxygen and the results correspond well to what I have just told you.

Whalen: We found very few low values for PO_2 in the carotid body, which corresponds rather well to your distributions of capillaries.

Lübbers: I think that the difference in the local PO_2 measurements depends upon differences in methods. For example, one must be certain that the measurements are made inside the specific tissue of the carotid body. As you know, it is not a homogeneous tissue, and if the measurement is made peripherally in the region of the shunt or surrounding vessels, the result will be a saturation of 98% and an oxygen tension of 70-90 mmHg depending upon the P_aO_2. So the measurement must be done within the specific tissue; otherwise it is very difficult to judge what the local oxygen tension really means.

Whalen: I am shocked that the carotid body is perfused only with plasma. Has no one frozen the carotid body? Are red cells seen in it?

Lübbers: Yes, we have frozen the carotid body, and no, red cells are not seen. I mentioned the cryophotometric method. We did these experiments because we had hoped to use the red cells as an indicator of the local capillary PO_2, but unfortunately we were unable to detect enough red cells in the center of the carotid body to make any sensible measurements. We were probably just measuring shunt flow.

Whalen: Could it be that the technique you used to freeze the carotid body affected the results? For example, if you did not occlude the vein first, then one might not expect to find many, if any, red cells.

Lübbers: No. The only time that red cells can be lost from the tissue is during the cutting of the carotid body. The freezing is so quick that blood is not lost from the venous end.

Wiemer: Could the distribution of mitochondria in any way affect the PO_2 profile?

Lübbers: Normally the distribution of mitochondria is rather random. In other organs we know that they may modulate but not really influence the local PO_2. There are situations, for example, in the kidney and also in some receptor organs such as the baroreceptors, in which a lot of mitochondria are present. However, within the carotid body with relatively few mitochondria, only small PO_2 gradients would be expected, and that is actually what we have measured.

O'Regan: Considering the high metabolic rate of the carotid body, do you think plasma can really supply sufficient oxygen?

Lübbers: May I in turn ask you a question? Do you know what the oxygen consumption of the carotid body is?

O'Regan: I assume the figures of Daly, Lambertsen, and Schweitzer.

Lübbers: I also have figures for this value, and from a rough calculation it seems that the carotid body oxygen requirement can be supplied with plasma.

McDonald: Do you think that the chemoreceptive nerves in the center of the carotid body, which are exposed to the lower PO_2, have a higher activity than those at the periphery?

Lübbers: I think so.

McDonald: Is there any physiological evidence from other groups to support the view that for a given P_aO_2 there is a range of chemoreceptor activity?

Lübbers: We measured the relationship between tissue PO_2 and P_aO_2. Generally, in the physiological range there is a good agreement between the two sets of values.

Torrance: From the distribution of tissue PO_2 would you expect different sensitivities of chemoreceptor fibers in relation to the arterial blood?

Lübbers: Different sensitivities are not needed because the zone of low oxygen tension changes with P_aO_2.

McDonald: But for a given P_aO_2, one portion of the carotid body is exposed to one PO_2 and another part to another PO_2. Is that correct?

Lübbers: Yes. There are quite different PO_2 values within the carotid body. It depends on the position of the capillaries. But that is also true for other tissues.

Torrance: This point has been tested by Goodman who recorded the activity simultaneously of two different chemoreceptor fibers and plotted frequency of discharge of one vs the other. The plot is a straight line that usually goes through the origin, suggesting that the fiber in the region of high PO_2 starts to fire at a high PO_2, and the fiber in the region of the lower PO_2 fiers at a lower PO_2. I think that this is the point McDonald was making.

Bingmann: Acker and I have tested the discharge of the carotid body in relation to local PO_2 and in relation to P_aO_2. First, the steepness of the curves showing the response to changes in hypoxia of individual fibers can differ. In some observations there was good agreement between tissue PO_2 and chemoreceptor discharge, while measurements of P_aO_2 did not agree with the course of the receptor discharge.

Session II

Electrophysiological Characteristics
of the Cell Elements
in the Carotid Body

Effects of Temperature and Stimulating Agents on Carotid Body Cells*

C. Eyzaguirre, M. Baron, and R. Gallego

The mechanisms of impulse generation in carotid body chemoreceptors are complex. One problem is that there is little information about the behavior of carotid body (type I or II) cells. Thus one of our aims is to see if the electrophysiology of these cells can be correlated with the generation of sensory discharges.

The membrane potential (MP) of type I cells (identified by intercellular staining with procion navy blue) ranged from 10 to 55 mV, mean 19.8 ± 0.4 mV (SEM) at 36-37°C. These values are low, which may be due in part to some injury of the membrane by the microelectrode since these cells are only 10 μm in fresh preparations. Their input resistance (R_o) varied between 10 and 170 MΩ (mean 42.2 ± 1.6 MΩ). These values are large, probably because of the small size of these cells; R_m, calculated from the mean R_o, was 132.5 Ω cm^2, which is low compared to excitable cells. The membrane capycity (C_m) gave a mean of 5.5 ± 0.3 μF/cm^2, which is similar to values obtained by others in other tissues (7). R_o was directly proportional to MP since a cell with a large MP also had a large R_o and a vice versa.

Effects of Temperature

The glomus cells were very sensitive to temperature. A temperature drop from, say, 37 to 30°C induced marked cell depolarization and decrease of R_o. These changes were accompanied by a great reduction in sensory discharge frequency (1). Cells having large MP and R_o values showed a greater change in these parameters than cells having lower initial values. Glomus cells showed a dynamic response to cooling: the faster the cooling, the larger the effects on MP and R_o; also adaptation followed this initial change.

Depolarization induced by cooling had a reversal or equilibrium potential. To study this effect the cell membrane potential was displaced (by currents injected through the microelectrode) in either direction. An analysis made in 24 cells showed a reversal potential for cooling at about -7 mV (Fig. 1). These results showed that there are, perhaps, at increase an conductance to more than one ion during a drop in temperature.

Ionic substitution experiments were then conducted. Removal of {Na}$_o$ at normal (i.e., 37°C) temperature hyperpolarized the cells by about 4 mV and R_o increased by about 25 MΩ. In Na-free solutions cooling

---

*This work was supported by grants NS 05666 and NS 07983 from the U.S. Public Health Service.

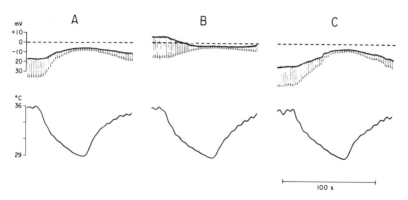

Fig. 1. Equilibrium or reversal potential of response to cooling. Preparation kept at 36°C and periodically cooled to about 29°C. (*upper traces*) Changes in MP and R_O induced by cooling applied at three membrane potential levels. (*lower traces*) Temperature changes recorded close to carotid body. (A) Cell had resting potential of -17 mV and R_O of 90 MΩ. Cooling reduced these values to -5 mV and 12 MΩ. (B) MP was displaced in a depolarizing direction by outward steady current to +6 mV and R_O increased to 100 MΩ. Cooling repolarized cell to -2.5 mV and resistance fell to 10 MΩ. (C) Membrane potential was driven to -24 mV by inward steady current (R_O = 90ΩM) and cooling reduced these values to -8 mV and 10 MΩ. During cooling equilibrium or reversal potential appears to be about -4 mV (1)

(to about 30°C) was less effective in inducing cell depolarization. These results suggest that Na ions are, probably, important in these processes. Removal of {Ca}$_o$ at normal temperature induced hyperpolarization of 3-4 nV and an unstable R_O. In the cold O{Ca}$_o$ induced hyperpolarization and an increase in R_O of 4-5 mΩ. Lack of Ca may create some membrane instability, which allows other ions to move across the cell membrane. This effects seems to disappear at low temperature. An excess (10.8mM) of either Ca or Mg depolarized the cells by about 4 mV, and Ro decreased. At low temperatures only an excess of Ca changed (in the same direction) MP and R_O, and Mg ions were ineffective.

Low external Cl (11.2mM) induced cell depolarization of about 8 mV and decreased R_O by about 23 MΩ. Cooling affected MP more markedly in low Cl than in normal solutions. Changes in R_O resulting from cold were more marked in low Cl in 65% of the cases. These experiments show that Cl ions play a role in maintaining the glomus cell membrane potential. This is not unique since chemosensitive neurons of Aplysia also operate through changes in Cl permeability during stimulation by CO_2 (2). O{K}$_o$ did not change significantly either the MP or the R_O of these cells. An excess of K (46.9mM total concentration) induced changes in MP and R_O that could be accounted for by the reduction of Na in the solution. Reducing Na was necessary to maintain constant osmolarity. This observation is in disagreement with Goodman and McCloskey (6) who obtained cell depolarization by superfusing with isotonic KCl (150 mEq/liter). The reasons for this discrepancy are not entirely clear. In addition, we have reported (4) effects of K on carotid body cells in tissue slices. Cell depolarization in high K and hyperpolarization in low K may have been due to osmotic effects (5). In fact, O{K}$_o$ was achieved by removing KCl from the solution, while high K was obtained by adding KCl. These solution were, respectively, 3.5% hypo- and hyperosmotic.

Ouabain (5 × $10^{-5}M$) induced cell depolarization at normal temperatures, and this effect was potentiated during exposure to cold. This finding is still too incomplete to postulate or negate the presence of a Na pump.

That the activity of glomus cells is dependent on the temperature is not too surprising since carotid body tissues have a high metabolism, which is reflected in its oxygen consumption. It is not known which elements are responsible for this phenomenon; but the high oxygen consumption may be located at the glomus or sustentacular cell level. The other elements in the carotid body (nerve fibers, vessels, supporting tissue) are likely to behave more conventionally.

Effects of Natural and Chemical Stimuli

Preparations bathed with saline equilibrated with 50% O_2 in N_2 (pH 7.43) were routinely used. After baselines were obtained, saline equilibrated with nitrogen was allowed to flow. An increase in sensory discharge frequency was induced by nitrogen, but there was no significant change in either MP or R_o. Likewise, MP and R_o were not significantly changed by superfusing the preparations with 5-50 mg/liter, although the nerve discharges increased markedly. The hyperpolarizing effect of NaCN (50 µg) on cells in tissue slices (4) requires further investigation. ACh (10-100 mg/liter) markedly increased sensory discharge activity without concomitant changes in MP or R_o. CO_2 (6% in 50% O_2, 44% N_2) increased the sensory discharge frequency, but there were no changes in either MP or R_o when the pH of the external saline was buffered to 7.43. However, a fall in pH of the bathing medium to 6.0-6.6 was accompanied by a more marked increase in discharge frequency, clear cell depolarization, and no significant changes in R_o. Thus the effects of CO_2 on MP appeared to be due to the fall in external pH. This was further tested by bathing the preparations in acid saline (ph 6.5-6.8), which also induced cell depolarization, although the input resistance did not change significantly. Conversely, when pH was increased to 8.2-8.5, there was an increase in MP and no significant changes in R_o.

Interruption of solution flow invariably increased the sensory discharges and the cell depolarization and decreased R_o. The mechanisms involved in the effects of lack of flow are not known. However, the phenomenon had a reversal or equilibrium potential (about 0 to -8 mV), which may indicate changes in permeability to more than one ion (Table 1).

Osmolarity Effects

When osmolarity was increased by adding sucrose, sodium glutamate, or glycerol, the discharge frequency increased, whereas MP and R_o decreased (Fig. 2A). The effects were observed with 5% increases in osmolarity. More hyperosmotic solutions (15-50%) depolarized the cells to values close to 0 mV. This change was concomitant with a marked decrease in R_o. When the control solution was again allowed to flow, MP and R_o slowly recovered to control levels (5).

To test the effects of hypoosmotic solutions, control solutions were made with a lower than normal sodium concentration (with less sodium glutamate), and osmolarity was brought to normal levels (305 mo smol/liter) by adding sucrose. Test solutions were made hypoosmotic by removing or reducing sucrose without changing the ionic composition.

Table 1. Effects of different stimuli on carotid body cells

Stimulus (pH)	$\bar{d}$ MP[a,b] (mV ± SEM)	p[c]	n	$\bar{d}$ R_O[a,b] (MΩ ± SEM)	p[c]	n
100% N_2 (7.43)	-0.4±0.8	<0.7	10	+4.3±4.7	<0.3	10
ACh (10-100 mg/liter) (7.43)	+0.3±0.3	<0.3	7	-0.4±2.2	<0.9	7
NaCN (5-50 mg/liter) (7.43)	0.2±0.2	<0.4	10	-1.6±1.8	<0.3	9
Base (8.5)	-2.9±0.3	<0.001[d]	27	-1.9±3.0	<0.5	24
Acid (6.5-6.8)	+3.4±0.3	<0.001[d]	27	-2.2±6.5	<0.7	21
6% CO_2 (6.6)	+3.6±0.4	<0.001[d]	50	-2.9±3.2	<0.3	50
6% CO_2 (7.43)	+0.1±0.3	<0.8	28	-0.5±3.2	<0.9	23
Diamox (2.5-5.0 mg/liter) (7.43)	+1.1±0.7	<0.1	11	+19.5±27.5	<0.4	11
6% CO_2, Diamox (7.43)	+0.7±0.9	<0.5	4	+3.3±6.7	<0.6	4
6% CO_2, Diamox (6.6)	+4.0±0.7	<0.01[d]	5	-8.8±8.3	<0.4	5
Zero flow (7.43)	+15±2.5	<0.001[d]	22	-49±9.3	<0.001[d]	21

[a] $\bar{d} = \dfrac{\Sigma(t_s - c_s)}{n}$, where c_s = control solution and t_s = test solution.

[b] (+), less negative MP and higher R_O; (-), more negative MP and lower R_O.

[c] Obtained from Student's t-tests, paired comparisons.

[d] Statistically significant.

Hypoosmolarity decreased the discharge frequency but increased both MP and R_O (Fig. 2B). These effects were observed with a 5% decrease in osmolarity. More hypoosmotic solutions (10-33%) induced greater increases in both MP and R_O. A return to the control medium induced a slow recovery toward baseline levels.

Osmotic changes may modify cell premeability to some ions, although at present we do not know what species are involved. Osmolarity variations may change the membrane configuration with consequent permeability changes; also this effect may be partly due to movements of water across the cell membrane, which may modify the intracellular concentration of some ions.

Conclusions

Results indicate that Na, Ca, Mg, and Cl contribute to the maintenance of MP and R_O of glomus cells. But they alone or in combination cannot be the exclusive elements involved in these processes. In fact, the cell membranes seem to be affected by H ions.

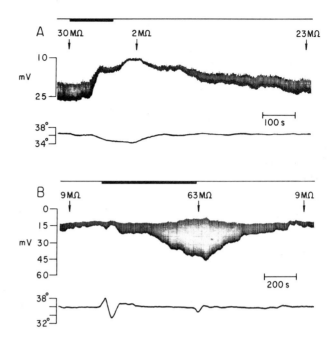

Fig. 2. Response of carotid body cells to changes in osmolarity. (*upper records*) MP and R_o recorded intracellularly from different cells. Resistance measured by passing 0.2 nA depolarizing current pulses at 0.5/s through recording micropipette (A). (B) 0.4 nA current pulses delivered at 0.2/s. (*lower records*) Temperature changes recorded near the carotid body. (A) 30% hypertonic solution (excess sucrose) is applied for 2 min (horizontal black bar above upper trace). Notice marked depolarization from 26 to 10 mV and decrease in input resistance from 30 to 2 MΩ. (B) 22% hypoosmotic solution (reduced sucrose) was passed for 9.5 min (horizontal black bar above upper trace). Notice hyperpolarization from 15 to 45 mV and increase in R_o from 9 to 63 MΩ. Changes in temperature result from a new solution in chamber and contribute somewhat to changes in MP and R_o, especially in (A) (5)

Several questions have remained unanswered. For instance, acidity, flow interruption, and hyperosmolarity induce cell depolarization and increase in the sensory discharge. An increase in temperature increases the sensory discharges but also induces cell hyperpolarization and an increase in R_o. Other stimulating agents such as CO_2 (at normal pH), NaCN, ACh, and N_2 did not change in a statistically significant way the MP or R_o of the carotid body (most likely type I) cells. Stimuli that do not ostensibly act on glomus cell membranes may have a different primary locus for their action (type II cells or nerve endings). Alternatively, if type I cells are the primary transducer elements, their action may not necessarily involve changes in MP and R_o.

References

1. Baron, M., Eyzaguirre, C.: J. Neurobiol. *6*, 521-527 (1975)
2. Brown, A.M.: Fed. Proc. *31*, 1399-1403 (1972)
3. Eyzaguirre, C., Zapata, P.: J. Physiol. *195*, 589-607 (1968)
4. Eyzaguirre, C., Fidone, S., Nishi, K.: in: The Peripheral Arterial Chemoreceptors. Purves, M.J. (ed.). London: Cambridge University Press 1975, pp. 175-194
5. Gallego, R., Eyzaguirre, C.: Fed. Proc. *35*, 404 (1976)

6. Goodman, N.W., McCloskey, D.I.: Brain Res. *39*, 501-504 (1972)
7. Hubbard, J.I., Llinás, R., Quastel, D.M.J.: Electrophysiological analysis of synaptic transmission. Baltimore: The Williams & Wilkins 1969
8. Paintal, A.S.: J. Physiol. *189*, 63-84 (1967)
9. Verna, A., Roumy, M., Leitner, L.-M.: Brain Res. *100*, 13-23 (1976)
10. Zapata, P., Stensaas, L.J., Eyzaguirre, C.: Brain Res. *113*, 235-253 (1967)

DISCUSSION

<u>Willshaw</u>: How do you know if the tissue in your experiments is alive?

<u>Eyzaguirre</u>: Dead nerves do not discharge and dead cells do not have resting potentials.

<u>McDonald</u>: Have you used antidromic nerve stimulation and recorded intracellularly from glomus cells to see if there is a change in MP or R_o?

<u>Eyzaguirre</u>: There is no change.

<u>McDonald</u>: Have you used pharmaceuticals to change the glomus cell MP?

<u>Eyzaguirre</u>: We have used NaCN and ACh and saw no changes.

<u>Speckmann</u>: How did you measure R_o?

<u>Eyzaguirre</u>: By passing current pulses through the recording electrode in conjunction with a bridge circuit. This may be inaccurate because the electrode characteristics can change while going through the tissue; but I do not know of a better way to do this.

<u>Speckmann</u>: That is the disadvantage of these cells. We have compared R_o measured with the bridge circuit and with two electrodes passing current through one and measuring MP values with the other. The R_o measured with the bridge is about 10 times higher than that measured with two electrodes. What about your very high R_o values of up to 180 MΩ?

<u>Eyzaguirre</u>: Some are that high.

<u>Speckmann</u>: But the mean R_o was between 50 and 60 MΩ?

<u>Eyzaguirre</u>: The mean was about 42 MΩ. But the calculated specific resistance is low because the cells are very small.

<u>Speckmann</u>: What is the K sensitivity of these cells?

<u>Eyzaguirre</u>: There is no response to K.

<u>Paintal</u>: Even if one has doubts about R_o, I think the change in MP is very clear. Is this open to error?

<u>Eyzaguirre</u>: We may be recording potentials lower than normal.

<u>Paintal</u>: If you are recording from cells having high metabolic activity, one can understand why the MP falls when temperature is lowered. What

happens to the cell MP if you apply sufficient amounts of NaCN for a long time to alter the metabolic properties?

Eyzaguirre: If you apply fairly large doses of NaCN for 30-60 min, the cells become depolarized, but this has nothing to do with the increase in chemoreceptor discharge.

O'Regan: Did all your cells behave in the same way? McDonald, Mitchell, and Hellström have shown two types of cells. Second, did you try this in a blood-perfused organ? If so, were these differences between that organ and the superperfused preparation?

Eyzaguirre: The answer to the first question is yes, to the second, no.

Whalen: Let me come to your defense, at least partially. As far as PO_2 in these tissues is concerned, we have duplicated your preparations and those of Fay, which were perfused. The PO_2 values were very high in both situations. I feel quite sure that in your experiments, the tissues were not hypoxic even in the core.

Eyzaguirre: The bath seems to be satisfactory for the tissues.

Zidek: You have shown that R_o decreases under low external chloride. Have you any explanations of this behavior?

Eyzaguirre: The problem is quite complicated, and I do not have any suggestions at the moment.

McDonald: With regard to O'Regan's question about two types of glomus cells in the rat carotid body, I believe that your studies all relate to the cat carotid body.

Eyzaguirre: Yes.

McDonald: Did you identify each penetrated cell using histologic procedures or do you have some other criteria that allow you to distinguish them from the other cells in the carotid body?

Eyzaguirre: We did not stain every cell. We established a criterion, which was the cooling effect. Once we found that a number of cooled cells became depolarized and were type I, we were satisfied that we could identify the cells. Every cell studied for natural stimulation was cooled to see if it was depolarized.

Acker: Is the silence of the type I cell true?

Eyzaguirre: I do not know. When people use better electrodes, they begin to see spikes. This is true for the adrenal medulla. It is a matter of resting potential. For no reason that I can think of, the adrenal medullary cells give action potential when stimulated, but the MP has to be high.

Whalen: You got membrane depolarization when you cooled or interrupted the flow, but discharges were the opposite. Is it not true then that you could not correlate depolarization with discharge?

Eyzaguirre: So far, I cannot associate depolarization or hyperpolarization with anything.

Lübbers: What is the size of your puncturing electrode? Also, I wonder if flow had a specific effect on your preparation; if you bubble nitrogen without stopping the flow, do you get something else?

Eyzaguirre: Our electrodes showed between 10 and 40 MΩ. We did not measure the tip diameter with the electron microscope, but it is probably less than 0.5 µm.

Lübbers: Acker has found that with bigger electrodes MP decreases considerably because the cells become leaky.

Eyzaguirre: That is correct. With regard to the other point, using a solution equilibrated with nitrogen, we did not see any effects on the cells; when we stopped the flow with or without nitrogen, we obtained cell depolarization.

Lübbers: Do you have any explanation for that?

Eyzaguirre: No.

Torrance: With Diamox you did not get significant R_o changes, but the change you did get was big in comparison with most of the other changes. The mean was 19.5 ± 27.5 SE. Is this a mixture of two approximately equal opposite effects of Diamox? It seems to be very active, but it averages out to zero.

Eyzaguirre: Diamox has very variable effects.

Torrance: Did you find that there were som cells that consistently went another way and they averaged out at something that was not significant?

Eyzaguirre: Of 11 cells tested Diamox did not affect R_o in 3 cells. In 3 cells the drug induced an increase in R_o. In 5 cells R_o decreased. Diamox did not change MP in 2 cases. It depolarized 7 cells and hyperpolarized 2 cells.

Speckmann: At the equilibrium potential level, there was no change in R_o?

Eyzaguirre: No, there is always a change in R_o at this E_{rev} level.

Are the Conventional Electrophysiological Criteria Sufficient to Give Evidence of an Electrogenic Sodium Transport Across Neuronal Membranes?

W. Zidek, E.-J. Speckmann, H. Caspers, and A. Lehmenkühler

In a variety of receptor systems electrogenic transport of ions, especially of sodium, is considered to be one of the elementary processes involved in the response of sensory units to stimulation (1,8, 9,10,12,13). In this context Biscoe (2) discussed an electrogenic sodium pump to be taken into account also in the function of carotid body chemoreceptors. In most of the electrophysiological investigations, however, it has not been determined directly whether the concerned ion pumps in fact excrete sodium in excess of potassium. Therefore, the question arises if the indirect methods usually employed in electrophysiology are really sufficient to give evidence of the electrogenesis of the ion transports. The aim of the present investigation was to examine this problem.

The experiments were carried out on isolated neurons, which have the advantage that they can be controlled properly in experimental conditions. Three identified neurons (B1-B3) of the buccal ganglion of *Helix pomatia* served as such model units (11). The membrane potential (MP) of the neurons was recorded by conventional techniques. For measurement of membrane resistance two separate microelectrodes were used to inject current and to record independently the MP. Sodium injections into the neurons were performed by interbarrel iontophoresis (3). The intracellular potassium activity was determined by K^+-selective microelectrodes (7,15,16).

As a substantial indicator of electrogenic transport mechanisms, the temperature sensitivity of the receptor systems in question has been considered important (5). Therefore, the temperature sensitivity of the neurons used was tested at first. Fig. 1A1 shows that lowering the temperature of the bath fluid decreased the MP when its initial value ranged at normal (40-60 mV) levels. This temperature response can be due either to changes in membrane conductance or to changes in electrogenic ion transport or to both. In a further series of experiments it was examined to what extent the temperature response was evoked by conductance changes. First, the membrane resistance of the neurons was measured during temperature changes. Fig. 1C shows the membrane resistance displayed as a current-voltage relationship. It can be seen that the resistance increased, with the delayed rectification simultaneously being reduced, when the temperature was changed from 22 to 8°C. Second, the temperature test was performed at various levels of the initial MP. As shown in Fig. 1A1 and 1A3, amplitude and polarity of the temperature response depended on the amount of the initial MP. This relationship is shown in the evaluation in Fig. 1B1 and 1B3, which demonstrates that the equilibrium potential of the temperature response was in the range of -65 mV. These findings indicate that conductance changes contributed substantially to the temperature response.

Now the question arises as to which ion currents are involved. The equilibrium potential of about -65 mV already suggests that potassium

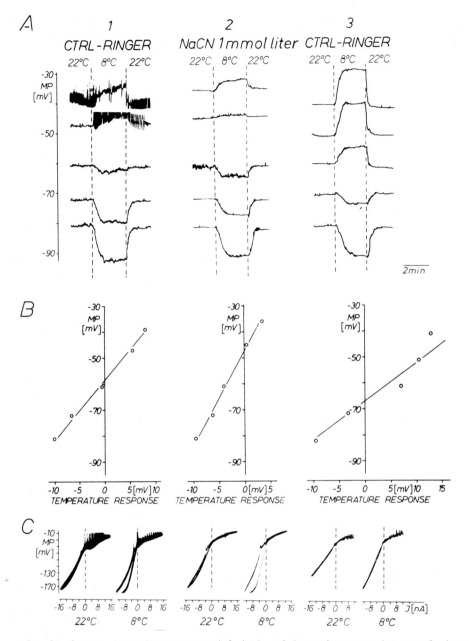

Fig. 1. Changes in membrane potential (MP) and in membrane resistance during temperature lowering of bath fluid. Neuron B2 in buccal ganglion of *Helix pomatia*. (1 and 3) Control (CTRL)-Ringer solution; (2) CTRL-Ringer solution with 1 mM/liter NaCN admixed. (A) MP changes during bath temperature lowering at different levels of initial MP. Initial MP was shifted artificially by current injection through a separate microelectrode. Inkwriter tracings. (B) Plot of temperature response as a function of initial MP. Same experiment as in (A). (C) Membrane resistance displayed as current voltage relation

and/or chloride currents may play a role. To clarify this question, the potassium gradient was reversed by a 24-h incubation in potassium-free solution. After this procedure the temperature response was reversed in polarity when normal or elevated external potassium concentrations were applied. Changing the chloride and sodium gradient across the membrane resulted in no substantial variations in the temperature response. Only the application of calcium-free solution diminished the amplitude of the temperature reaction. This effect can be explained by the simultaneous decrease in membrane resistance. From these results it can be concluded that the temperature response of the MP is evoked essentially by changes in the potassium conductance and that its equilibrium potential corresponds mainly to the potassium equilibrium potential. A potassium equilibrium potential in the same order of magnitude resulted also from measurements of the intracellular potassium activity in steady-state conditions (6).

Another criterion for the existence of an electrogenic ion transport is the rapid and distinct neuronal depolarization following the application of metabolic inhibitors (5). Fig. 3 shows that the admixture of, e.g., ouabain elicited such a depolarization. A similar decrease in MP was found after application of cyanide. However, these effects may be due to a depression of either an electrogenic or an electroneutral pump. To test these posibilities the potassium equilibrium potential after application of cyanide was determined by means of the temperature response, which had been demonstrated to be preferentially evoked by changes in the potassium conductance. Fig. 1A shows the temperature response at various levels of the initial MP in control solution and after application of cyanide. The evaluations in Fig. 1B demonstrate that cyanide caused a reversible decrease of the equilibrium potential by about 15-20 mV. Such a shift of the potassium equilibrium potential yields a sufficient explanation for the depolarization observed after application of cyanide and ouabain. Therefore, the depolarization induced by metabolic inhibitors need not automatically be attributed to an inhibition of an electrogenic ion transport but can also be due a shift of the potassium equilibrium potential.

It was also examined whether an intracellular injection of sodium ions that induces a hyperpolarization always indicates the existence of an electrogenic sodium pump (14). Fig. 2A shows that sodium injection evoked a hyperpolarization at normal MP (45 mV) also in these neurons. The membrane resistence was simultaneously reduced. Moreover, the sodium effect on MP exhibited an equilibrium potential in the range of 60-70 mV. Application of ouabain reduced this equilibrium potential by 20-30 mV (Fig. 2B). Also from these experiments no final indication of an electrogenic transport can be deduced.

Besides the criteria mentioned above, evidence of an electrogenic sodium transport was suggested by the fact that the potassium equilibrium potential was found to be lower than the actual MP if special experimental procedures were applied (4). Such a combination of findings resulted also after application of ouabain in the present experiments if the intracellular potassium activity was determined by ion-selective microelectrodes. A typical experiment is displayed in Fig. 3. It shows that the depolarization of the neuron is associated with a decrease in the intracellular potassium activity down to 20-30 mM/liter. In those experiments the time courses of changes in the MP and in the intracellular potassium activity differed to a great extent, and the MP usually exceeded the potassium equilibrium potential evaluated from these measurements when ouabain was washed out. In contrast to this finding the determination of the potassium equilibrium potential by means of the neuronal temperature response revealed a shift of the potassium equilibrium potential in parallel with the membrane potential

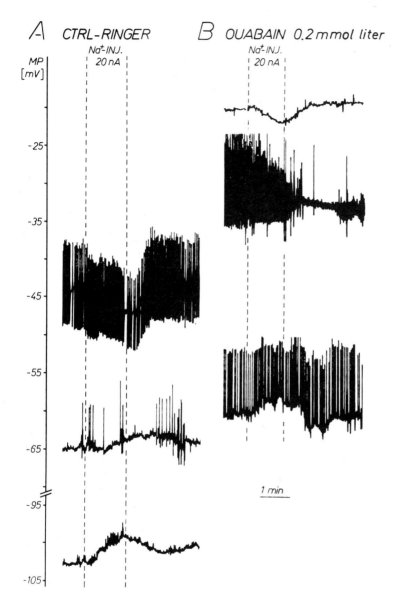

Fig. 2. Membrane potential (MP) changes induced by iontophoretic injection of Na$^+$ at different levels of initial MP. Neuron B2 in buccal ganglion of *Helix pomatia*. Inkwriter tracings. (A) Control (CTRL)-Ringer solution; (B) CTRL-Ringer solution with 0.2 mM/liter ouabain admixed

changes. The discrepancy between the results of the two methods might be caused by the fact that the method using the temperature response, on the one hand, determines the potassium activity immediately below the membrane. Ion-selective electrodes, on the other hand, probably measured the potassium activity in inner parts of the units, i.e., at sites remote from the membrane. Since the potassium activity in the space below the neuronal membrane is responsible for the MP, no reliable

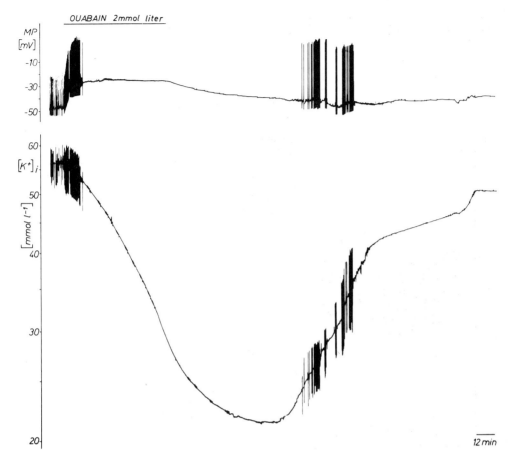

Fig. 3. Changes in membrane potential (MP) and in intracellular potassium activity $\{K^+\}_i$ induced by application of 2 mM/liter ouabain to normal bath fluid. Neuron B3 in buccal ganglion of *Helix pomatia*. Inkwriter tracings

indication of electrogenic ion pumps can be achieved by an overall determination of the intracellular potassium in non-steady-state conditions, as has been done in the literature.

In summary, our findings demonstrate that the conventional electrophysiological methods are unable to give definite evidence of the existence of an electrogenic sodium transport.

References

1. Baylor, D.A., Nicholls, J.G.: J. Physiol. *203*, 571-589 (1969)
2. Biscoe, T.J.: Physiol. Rev. *51*, 437-495 (1971)
3. Eccles, J.C., Eccles, R.M., Ito, M.: Proc. Soc. Med. *160B*, 181-196 (1964)
4. Glitsch, H.G.: J. Physiol. *220*, 565-582 (1972)
5. Kerkut, G.A., York, B.: The Electrogenic Sodium Pump. Bristol: Scientechnica 1971

6. Kostyuk, P.G., Sorokina, Z.A., Kholodova, Y.D.: in: Glass Microelectrodes. Lavallee, M.L., Schanne, O.F., Hebert, N.C. (eds.). New York: Wiley 1969, pp. 322-348
7. Lux, H.D., Neher, E.: Exp. Brain Res. *17*, 190-205 (1973)
8. Nakajima, S., Onodera, K.: J. Physiol. *200*, 161-185 (1969)
9. Nakajima, S., Takahashi, K.: J. Physiol. *187*, 105-127 (1966)
10. Pierau, F.K., Torey, P., Carpenter, D.O.: Brain Res. *73*, 156-160 (1974)
11. Schulze, H., Speckmann, E.-J., Kuhlmann, D., Caspers, H.: Neuroscience Letters *1*, 277-281 (1975)
12. Smith, T.G., Stell, W.K., Brown, J.E., Freeman, J.A., Murray, G.C.: G.C.: Science *162*, 456-458 (1968)
13. Sokolove, P.G., Cooke, I.M.: J. Gen. Physiol. *57*, 125-163 (1971)
14. Thomas, R.C.: Physiol. Rev. *52*, 563-594 (1972)
15. Vyskocil, F., Kriz, N.: Pfluegers Arch. *337*, 265-276 (1972)
16. Walker, Jr., J.L.: Analyt. Chem. *43*, 89A-93A (1971)

DISCUSSION

Eyzaguirre: Are these cells very different from those that are chemosensitive, or are these cells also chemosensitive?

Zidek: We cannot answer this question because we have not tested the influence of different PO_2 levels.

Bingmann: To a certain extent probably all neurons are chemosensitive. Chalazonitis has demonstrated that snail neurons exibit chemoreceptive properties. Speckmann and Caspers observed chemoreceptive responses in cortical and spinal neurons of mammals. Furthermore, isolated spinal sensory ganglion cells in tissue culture responded to changes of PCO_2 and PO_2.

Kessler: What would be the explanation for the lack of homogeneous distribution of potassium activity within the cytoplasm?

Zidek: We can assume that there might be diffusions of potassium in a homogeneous cytoplasm with diffusion coefficients that are much lower than in free solution, e.g., in water. Perhaps the diffusion coefficients in the cytoplasm are about 10,000 times lower than in free solution. This would explain the slow decay of the potassium activity during cooling. Another explanation might be that there are compartments in the cell that release their potassium with different velocities.

Paintal: Mammalian neurons behave quite differently to temperature; so it is a question of this particular specimen of *Helix pomatia* being peculiar as a neuron.

Speckmann: May I comment on your comment? Not all neurons in mammals behave unlike these neurons. Klee and coworkers have shown that motoneurons in the spinal cord of cats behave quite similarly to the neurons in *Helix pomatia*.

Willshaw: When you changed the outside potassium concentration, did you expect to obtain a Nernstian response in MP?

Zidek: We obtained a Nernstian response at the higher potassium concentrations from 4 or 5 mM/liter to 100 mM/liter.

Willshaw: It did not actually look like that in your data.

Zidek: The main difference is that the slope of the curve is not 58 mV/decade but is in the range of 30 mV/decade. This might result from the fact that the potassium conductance is not that of an ideal model, which would have a slope of ca. 60 mV.

Willshaw: When you calculate the change in potential, you would expect a Nernstian slope if potassium is solely responsible for creating the potential. How well did that match the changes in potential you actually got?

Zidek: We obtained a relative potassium conductance of 0.6-0.7 by using the potassium concentrations just mentioned. This is in accordance with most of the other determinations of the relative conductance of potassium. With these data we can explain the shifts of MP produced by metabolic inhibitors, etc.

Ji: I am interested in looking at your temperature change from 20 to about 10°C, whereas Eyzaguirre's work involves temperature changes from 37 to about 30°C. Have you varied this temperature gradually and tried to see at what point these membrane depolarizations come? The biological membranes have a distinct phase-transition temperature of around 17-20°C, and when there is such a transition, there may be a lot of changes in the membrane.

Zidek: We have tested the dependence on temperature in the range of 0-25°C. The MP increased with temperature. We found a curve that reached its maximum at 25°C. We could not obtain a further increase in MP by further increases in temperature.

Reaction of Cultured Carotid Body Cells to Different Concentrations of Oxygen and Carbon Dioxide

F. Pietruschka, D. Schäfer, and D. W. Lübbers

Measurements of tissue PO_2 in the cat carotid body performed by Acker, Lübbers, and Purves (1) showed that, compared to other organs, the oxygen tension is relatively low. They found that 79% of the values were below 40 mmHg. Weigelt (9) reported even lower PO_2 values with a maximum at 5 mmHg for the rabbit carotid body. We therefore decided to investigate whether and to what extent the oxygen concentration influences the differentiation of isolated carotid body cells in tissue cultures. We cultured the cells in monolayer under both hypoxic and hypercapnic conditions.

Twelve carotid bodies of 0.1-0.5 mm^3 were prepared from rabbid embryos and few-day-old rabbits. Connective tissue was removed from the organs by emzymatically dissolving the cells with trypsin, collagenase, and hyaluronidase. Because we intended to culture the cells as functional units, the dissociation was interrupted before the cells were completely isolated (6). The cell clusters were cultivated on polystyrene coverslips coated with collagen in petri dishes covered with 6 μm Teflon membranes. The preparations were placed in a CO_2 incubator supplied with commercially available trimix gas.

In one experimental series the gas mixture consisted of 5% CO_2 and various O_2 concentrations of 0, 2, 5, 7.5, 10, and 21% in N_2. In another series the cultures were gassed with 10% CO_2 and 0, 5, and 21% O_2 in N_2. The PO_2 decrease across the 2 mm-thick layer of the medium was about 4 mmHg, calculated according to the formula given by McLimans (3)*, so that the PO_2 at the monolayer incubated with 2% O_2 (15.2 mmHg) was about 11 mmHg. The cells were cultured for 3 days. They were then fixed with glutaraldehyde and prepared for electron-microscopic examination (6). After fixation the clusters of epithelioid cells were counted in the stereomicroscope.

As a measure of the cell outgrowth, the number of epithelioid clusters was determined in relation to the concentration of CO_2 and O_2 in the gas atmosphere. The values for the cell outgrowth from embryos or few-day-old rabbits were higher with 10% than with 5% CO_2. With 5% CO_2 the outgrowth of embryo cells was most pronounced at O_2 concentrations of 2-7.5% (Fig. 1a). The examination of the fine structure of the cultured cells showed that about 60% of the clusters grown under either hypoxic or hypercapnic conditions contained differentiated cells. Similar to in vivo, the type I cells had round or ovoid nuclei with peripheral aggregations of chromatin (N1 in Fig. 2). They were partly enveloped by elongate processes of type II or sustentacular cells (N2 in Fig. 2). There was a well-developed Golgi apparatus (G in Fig. 2)

---

*The authors are indebted to Mr. U. Großmann for helping with the calculation.

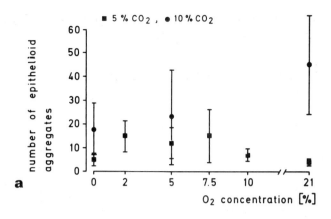

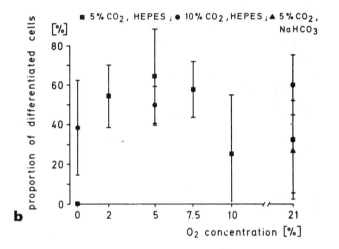

Fig. 1. (a) Number of epithelioid cell aggregates outgrown after culturing for 3 days at different O_2 concentrations. Values for outgrowth are higher with 10% CO_2 (*circles*) than with 5% CO_2 (*squares*) (mean ± SD). (b) Proportion of differentiated cells in 3-day-old cultures in relation to O_2 concentration. Examination of fine structure shows that about 60% of clusters grown under either hypoxic or hypercapnic conditions contain differentiated cells. Cells cultured in $NaHCO_3$-buffered medium (*triangle*) show that the effect of CO_2 is independent of the HCO_3-concentration in the medium

and, as a characteristic feature, numerous dense-cored vesicles (arrow in Fig. 2), whose vinicity to the dictyosomes of the Golgi apparatus indicated that the synthesis of catecholamines continued during culturing.

With respect to the maintenance of differentiation, we noticed two significant differences in the experimental series with either 5 or 10% CO_2 (Fig. 1b).

1. Under extreme hypoxic conditions induced by incubation with N_2, we did not find any differentiated cells with 5% CO_2, whereas with 10% CO_2, 40% of the clusters consisted of differentiated type I cells.

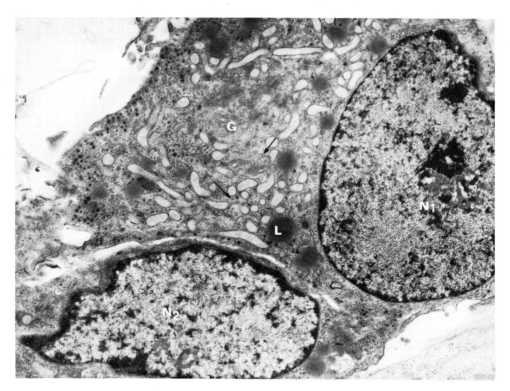

Fig. 2. Carotid body cells after culturing for 3 days. N1, nucleus of type I cell; N2, nucleus of type II cell; G, Golgi apparatus; L, lipid droplet; *arrow*, dense-cored vesicles. 900x

In the N_2-CO_2 gas mixtures used, a contamination of 0.005% O_2 is tolerable. This corresponds to 0.04 mmHg on the surface of the medium. This O_2 supply would diminish the O_2 consumption to 1% of the normal values.

2. With 21% O_2 the number of differentiated clusters was twice as high at 10 as at 5% CO_2.

In another series of experiments carried out with a $NaHCO_3$-buffered medium, the effect of CO_2 was independent of the HCO_3-concentration in the medium (Fig. 1b).

The relationship of mitochondria per surface area to cytoplasm area was determined by morphometric examination. The mitochondria were reduced by 5% CO_2 under extreme hypoxic conditions (Fig. 3a). With 2% O_2 the mitochondria population increased. This increase in the number of mitochondria is also reported from carotid bodies in vivo in chronically hypoxic rabbits (4). Further increase of the O_2 concentration decreased the mitochondria population to the values reported for the carotid body cells in vivo (7). This finding agrees with the results on rat heart cells cultivated in vivo published by Wollenberger (10), who reported a similar effect of the O_2 concentration on the activity of succinic dehydrogenase. With 10% CO_2 the O_2 concentration did not significantly influence the mitochondria population.

For the number of dense-cored vesicles calculated per cytoplasm area we found such a large difference between the single cells that a significant influence of the O_2 concentration could not be determined. It is striking that the cells grown with 5% CO_2 and 0% O_2 had no dense-

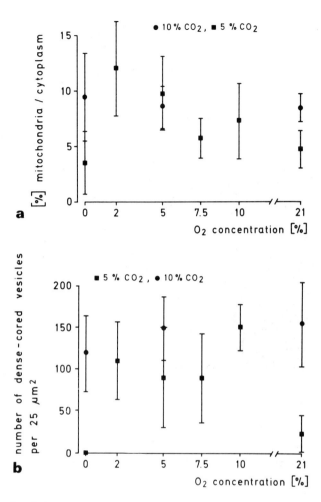

Fig. 3. (a) Relationship of mitochondria per surface area to cytoplasm area determined by morphometric examination. With 10% CO_2 (*circles*) O_2 concentration does not significantly influence mitochondria population. With 5% CO_2 (*squares*) mitochondria population augments under hypoxic condition (2% O_2). (b) Number of dense-cored vesicles calculated per cytoplasm area. A significant influence of O_2 concentration cannot be detected

cored vesicles (Fig. 3b). This lack of cell differentiation is correrlated to the reduced mitochondria population under hypoxic conditions (Fig. 3a).

The positive effects of hypoxia and hypercapnia on the differentiation of carotid body cells in culture are as follows:

1. The cells derive from perinatal animals. Embryos can tolerate lower values for tissue PO_2 and higher ones for PCO_2.

2. In vivo an increased CO_2 concentration lowers the intracellular pH, which is a stronger effect than that of $NaHCO_3$ (8). In addition, acid pH decreases the catecholamine activity and stimulates the catecholamine synthesis (5).

3. Lee (2) found with histochemical methods that the activity of carbonic anhydrase in carotid body type I cells in vivo was comparable to that of renal tubule cells. He suggested that carbonic anhydrase might work as a H^+ pump or be responsible for the penetration of CO_2 into the cells.

References

1. Acker, H., Lübbers, D.W., Purves, M.J.: Pflügers Arch. *329*, 136-155 (1971)
2. Lee, K.D.: in: Arterial Chemoreceptors. Torrance, R.W. (ed.). Oxford and Edinburgh: Blackwell 1966, pp. 133-141
3. McLimans, W.F., Blumenson, L.E., Tunnah, K.V.: Biotechnol. Bioeng. *10*, 741-763 (1968)
4. Moeller, M., Moellgard, K., Soerensen, S.C.: J. Physiol. *238*, 447 (1974)
5. Nahas, G.G.: in: Carbon Dioxide and Metabolic Regulations. Nahas, G., Schaefer, K.E. (eds.). New York-Heidelberg-Berlin: Springer-Verlag 1974, pp. 107-134
6. Pietruschka, F., Schäfer, D.: Cell Tissue Res. *168*, 55-63 (1976)
7. Seidl, E., Schäfer, D., Zierold, K., Acker, H., Lübbers, D.W.: Light microscopic and electron microscopic studies on the morphology of the cat's carotid body. (This symposium)
8. Waddell, W.J., Bates, R.G.: Physiol. Rev. *49*, 285-329 (1969)
9. Weigelt, H.: Der lokale Sauerstoffdruck im Glomus caroticum des Kaninchens und seine Bedeutung für die Chemorezeption. Dissertation, Bochum 1975
10. Wollenberger, A., Karsten, U., Kössler, A.: Cardiology *56*, 224-230 (1972)

DISCUSSION

McDonald: What is the relationship between the dense-cored vesicle concentration and catecholamine content as determined by fluorescence microscopy?

Pietruschka: We have studied the number of dense-cored vesicles in relation to O_2 concentrations, but because of the great variation discovered, we found no relation.

McDonald: Maybe I did not make my question clear. Was the number of dense-cored vesicles in the glomus cells in culture correlated with the amount of catecholamines the cells contained?

Pietruschka: We cannot measure catecholamines in the glomus cells. We have only fluorescence-microscopic investigation of the cells but not quantitative, only qualitative.

McDonald: Yes, but when the dense-cored vesicles were absent, was the fluorescence of catecholamine absent as well? And when?

Pietruschka: No, we did find fluorescence when we could not find dense-cored vesicles. But we did not measure this on the same cell because it is not possible.

McDonald: Did you find the reverse - the absence of catecholamines when the dense-cored vesicles were present?

Pietruschka: We did not try.

Hanbauer: Have you studied enzymes in order to use another parameter besides the growth of cells to estimate the viability of the cells in culture?

Pietruschka: We cannot do biochemical analysis with these cells for there are too few. What we can do is use histochemical methods, and we have done catecholamine fluorescence of the glomus cells and also of the cells and nuclei.

Kobayashi: I would like to mention three points. First, I got the impression that the dense-cored granules in vitro are smaller in size than those in vivo. Second, the Golgi complex in vitro is not well developed - I thought it was smaller than in vivo. Third, from your electron micrographs it seems that in vitro the dense-cored vesicles tend to accumulate around the surface of the cell. Am I correct in these three points?

Pietruschka: If you compare our cells with cells of embryonal carotid bodies, you will find that there are many dense-cored vesicles as in the embryo. If you compare our cells with those of adult carotid bodies, there are more dense-cored vesicles in the adult. I do not know if it is the same with the Golgi apparatus.

Pallot: This question about the number of electron dense-cored vesicles - is it not a fact that there is not a terribly good correlation between catecholamine content and number of dense-cored vesicles?

Pietruschka: Yes, but the number of dense-cored vesicles is the only thing that we can see and measure with electron microscopy.

Meaning of the Type I Cell for the Chemoreceptive Process — An Electrophysiological Study on Cultured Type I Cells of the Carotid Body

H. Acker and F. Pietruschka*

The carotid body as a chemoreceptor is able to transduce changes of PO_2, PCO_2, and pH into nerve impulses. Several elements in the carotid body can be responsible for this process: type I cells, type II cells, small nerve fibers, and mitochondrial bags. We have started to investigate, by intracellular measurements, the electrophysiological characteristics of type I cells and their dependence on changes of PO_2 to determine their role in the chemoreceptive mechanism. The measurements were done on cultured type I cells, as described by Pietruschka (2,3). These measurements on cultured cells have the following advantages: {1} it is always possible to identify the cell to be punctured; and {2} because the cells are cultured in a monolayer, PO_2 gradients can be neglected, and a direct correlation between changes in PO_2 and electrophysiological parameters can be established. The disadvantage of this method is that the growing cells are very flat, and it is extremely difficult to puncture them.

For intracellular recordings double-barrel microelectrodes were used. We use this type of electrode because it can be filled very easily by injecting electrolyte solutions. The microelectrode tip diameter was about 0.2 μm, controlled by scanning electron microscopy**. The electrodes were filled with $3M$ KCl, had a resistance between 7 and 20 MΩ, and had a tip potential that varied between 0 and 5 mV. Membrane resistance was measured by injecting negative current pulses of 0.4 nA at a frequency of 1 Hz through the recording microelectrode by means of a bridge circuit. From the evoked changes in membrane potential (MP) the membrane resistance was calculated. The cells were superfused with a buffered solution equilibrated with different O_2, CO_2, and N_2 mixtures. During recordings PO_2, pH, and temperature of the superfusion medium were controlled continuously. In addition the vibrations of the perfusion chamber were registrated.

Fig. 1. shows the MP and input resistance of a type I cell. The MP is about 60 mV and the input resistance about 30 MΩ. The type I cell is not able to generate spikes either spontaneously or by electrical intracellular stimulation, i.e., the type I cell is a silent cell. As shown in Fig. 2, the membrane potentials of the type I cells vary between 10 and 180 mV with a mean of 37 mV (n = 43). The input resistance varied between 20 and 150 MΩ. These findings are in general accordance with the measurements of Baron and Eyzaguirre (1). Even though the method for measuring input resistance is open to criticism, we had to

---

*We are indebted to Prof. E.-J. Speckmann and Dr. D. Bingmann for their helpful discussions about electrophysiological problems.

**We thank Dr. K. Zierold for controlling the tip diameter of the microelectrodes by scanning electron microscopy.

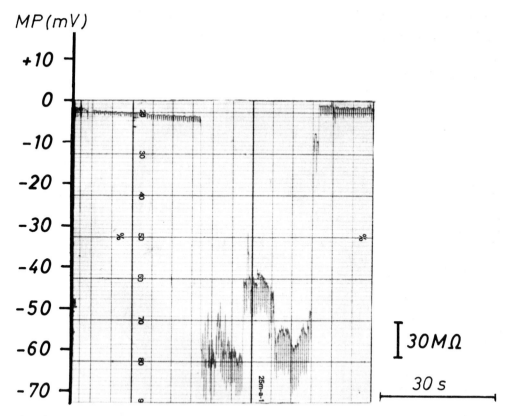

Fig. 1. Membrane potential (MP) of 60 mV of cultured type I cell. By injecting current of 0.8 nA at a frequency of 1 Hz through the recording microelectrode, an input resistance of about 30 MΩ can be calculated. Time, 30 s

use it because these cells are too small to be penetrated with a second electrode for passing current. However, to prove that these high input resistances are possible, we investigated cultured fibroblasts and skeletal muscle fibers. We found input resistance values of 10-20 MΩ for the fibroblasts, which have nearly the same size as the type I cell. The muscle fibers have resistance values between 2 and 7 MΩ. With this method different cells show different resistance values, so the high input resistance of the type I cells must have a particular meaning.

By varying PO_2 of the superfusion medium the MP and membrane resistance were influenced significantly. Fig. 3 shows that by decreasing the PO_2, the MP and the input resistance increased by about 10%. This was observed in all cells in which it was possible to maintain the recording electrode for 4-5 min (6 cells). Also when PO_2 decreased from lower values, the type I cell reacted similarly. For comparison, we investigated the dependence of the MP of fibroblasts and skeletal muscle fibers on PO_2 and did not find any distinct influence on the MP of a change in PO_2.

In conclusion, we can say that {1} cultured type I cells of the carotid body are not able to generate spikes; {2} the MP occurs in a range of 10-80 mV; {3} the input resistance is between 20 and 150 MΩ; {4} during a PO_2 decrease the MP and input resistance increase. The experiments

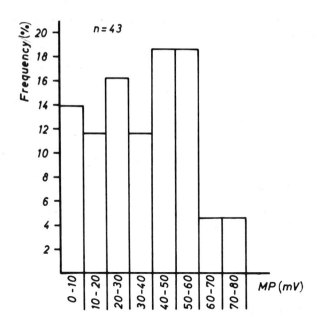

Fig. 2. Frequency distribution of MP values of cultured type I cells. MP values range from 10-80 mV with mean value of 37 mV. n = number of measurements

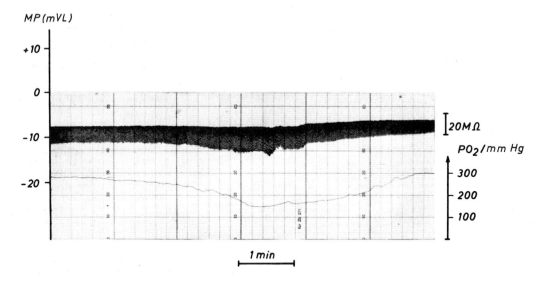

Fig. 3. Relationship between input resistance, MP of cultured type I cell, and PO_2. By decreasing PO_2 of superfusion medium, input resistance and MP of the cultured type I cell increase. Time, 1 min

may demonstrate that the type I cells are PO_2-dependent secretory cells and that spike generation may occur in another part of the carotid body tissue, i.e., type II cells or small nerve fibers. Since it is known that type I cells contain dopamine and that dopamine diminishes chemoreceptive nervous activity in the carotid body (4), we may infer from our experiments the following steps in the chemoreceptive process: decreasing PO_2 → higher input resistance and MP of the type I cell → decreased dopamine release → increased spike activity and vice versa.

References

1. Baron, M., Eyzaguirre, C.: J. Neurobiol. *6*, 521-527 (1975)
2. Pietruschka, F.: Cell Tissue Res. *151*, 317-321 (1974)
3. Pietruschka, F., Acker, H., Gattermann, S., Seidl, E., Lübbers, D.W.: Arzneim. Forsch. *23* (No. 11), 1610 (1973)
4. Sampson, S.R.: in: The Peripheral Arterial Chemoreceptors. Purves, M.J. (ed.). London: Cambridge U. Pr. 1975, pp. 207-220

DISCUSSION

Speckmann: Are your cells sensitive to potassium?

Acker: I do not know. We had difficulties in puncturing these cells. The PO_2-dependent changes of MP were found in only six cells because it was only in these six cells that we were able to maintain the microelectrode for 5 or 6 min. Because of this difficulty we have not been able to prove the potassium sensitivity.

Speckmann: In one Fig. you show an increase in MP and an increase in input resistance. What do you think about these two phenomena?

Acker: Our conclusion is that the cells are secretory. By an increased MP and resistance, it becomes more difficult for vesicles to come out of the cells, e.g., for dopamine-rich vesicles that may inhibit the spontaneous activity of small nerve fibers. Thus, by decreasing PO_2 more and more nerve fibers are able to fire.

Eyzaguirre: I am very interested in Acker's results. The fact that we did not find significant changes in MP or resistance during very low oxygen is not necessarily in contradiction to what Acker has presented. I am sure that the level of oxygenation of his tissue was a lot higher than ours because of being so exposed to oxygen in the medium. So I would not be surprised if in our tissue slices we eventually find the same result. Also I sympathize with him about the difficulty in impaling these flat cells against a piece of glass. We are having similar problems with tissue slices.

Wiemer: The recordings you showed were in the range of normoxia or even slightly hyperoxic. What happens in the hyperoxic state and did you try variations of PCO_2?

Acker: The lowest value of PO_2 was in the range of 30 mmHg. It was surprising to us to see an effect in this range and not under real hypoxic conditions. Until now we have not tried variations of PCO_2.

McDonald: Were all the cells in which you measured a change in membrane potential during exposure to hypoxia in culture for the same period of time? Was there any change in the sensitivity of the glomus cells to hypoxia from the time the carotid body was removed and therefore in effect denervated?

Acker: We compared 3-day-old cells and nearly 14-day-old cells, and they had the same behavior toward PO_2 changes.

Caspers: Just a short question. In response to the title of your paper, my question is what is the meaning?

Acker: One meaning is that they are not able to generate spikes, and another is that they perhaps influence some other nerve structures by secreting transmitters, for instance, dopamine. That means that these cells may control the level of activity of spike-generating nerve structures. When the PO_2 is decreased, the membrane may close, the secretory function is diminished, and the spike activity can increase. But this model is open to discussion.

Nishi: The high resistance of these cells, which both you and Eyzaguirre found, seems to indicate there is no electrical connection beween type I cells. The electron-microscopic picture shows many, dense tight junctions. I thought there would be a lot of interconnection between these cells. What do you think about this point?

Acker: That is a very difficult question. Eyzaguirre has a lot of experience and does not know; but by comparing type I cells with fibroblasts that are of the same size, we obtained different resistance values, so the high resistance value of the type I cells must have a special meaning. In the tissue culture we have the special condition that the type I cell is isolated and not surrounded by the type II cell. But this cannot explain the high resistance values of the type I cell because Eyzaguirre could show the same values in tissue measurements.

Eyzaguirre: I would not be surprised if there are gap junctions between cells. It is a little difficult to envision a tissue of this type to be totally isolated in terms of units and cells. I would not be a bit surprised if there is some communication. The trick is to prove it. It is very difficult to prove because you would have to inject current into one cell and record from the other - and that is not easy.

Bingmann: Are there any data available concerning the relation between membrane resistance of cells and the secretory rate of these cells? You concluded there was a decrease of the membrane resistance - some pharmacologic substances like dopamine might be secreted more easily.

Acker: I know only the values from Sampson about this problem. If I remember correctly, he measured the MP of the adrenal medulla with and without ACh. He could show that the MP differed markedly; and if ACh has something to do with the secretory function of these cells, ACh by increasing the secretory rate would decrease the MP.

Session III

Histochemical and Biochemical Investigation of the Transmitters in the Carotid Body

Dopamine Beta-Hydroxylase Activity in the Cat Carotid Body

C. Belmonte, C. González, and A. G. Gracia

Dopamine beta-hydroxylase (DBH), the enzyme that catalyzes the conversion of dopamine to norepinephrine, is present within sympathetic storage vesicles (6,12). Similar dense-cored vesicles are found in the glomus cells of the carotid body. Glomus (7) cells store dopamine and norepinephrine (13), and thus DBH should be one of the biosynthetic enzymes present in the glomus cells. However, there also exist in the carotid body sympathetic fibers from the superior cervical ganglion, which contain norepinephrine in their dense-cored vesicles. The purpose of this work was to measure DBH activity in normal and sympathectomized carotid bodies to determine what proportion of the DBH activity corresponds to that present in the carotid body cells versus the sympathetic nerve endings. Furthermore, the possible influence of the sinus nerve endings on the DBH enzymatic activity in the glomus cells was studied by cutting the sinus nerve in sympathectomized animals.

Twenty-seven cats of both sexes weighing 1.5-3 kg were anesthetized with sodium pentobarbital (30 mg/kg); in one group of animals the carotid artery and its accompanying tissue were removed and placed in a plastic chamber filled with ice-cold Locke's solution. There the carotid body was cleaned under a dissecting microscope. In the remaining animals the dissection was made in situ and the carotid body directly removed from the animal. In an all-glass homogenizer each carotid body was homogenized in 1 ml of 5 mM Tris buffer, pH 6.8, containing 0.1% Triton X-100 and 0.25% bovine serum albumin. The homogenate was centrifuged at 27,000×g for 10 min at 0°C, and aliquots (0.1 ml) of the supernatant were used to assay DBH activity according Goldstein et al. (3).

The enzyme assay involves a two-step coupled reaction. In the first reaction, tyramine is converted to octopamine by DBH at pH 5.5. In the second reaction, the enzymatically formed octopamine is further converted by an excess of purified bovine adrenal phenylethanolamine N-methyltransferase (PNMT) to N-methyloctopamine (synephrine) using {^{14}C}S-adenosylmethionine as a methyl donor. The reaction mixture for the first step was identical to that used by Goldstein et al. (3) except that a 100 times higher concentration of N-ethylmaleimide (NEM, 10 μM) was used to inactivate endogenous inhibitors of the enzyme. NEM was chosen instead of copper because it gave maximal enzyme activity over a wide range of concentrations.

The first step of the reaction was run for 2 h and was linear for 2.5 h. At the end of the incubation period, 10 μM dithiothreitol (Cleland's reagent) was added in order to neutralize the excess of NEM, and the pH of the reaction was abruptly changed to 8.6. The second step of the reaction was started by adding an excess of partially purified bovine adrenal PNMT (10) to convert the octopamine formed to N-methyloctopamine using {^{14}C}S-adenosylmethionine ({^{14}C}-SAM) as a methyl donor. The second step of the reaction was run for 45 min, after which organic extraction of the end product was carried out as described by Goldstein et al. (3). A boiled sample (95°C for 5 min) served as blank (400-500 cpm). An octopamine standard (0.2 nM) was included in each assay in order to convert counts per minute to units of DBH activity. The PNMT step of the reaction was linear for a wide range of octopamine concentrations (12 pM to 2 nM). DBH activity is expressed as picomoles of octopamine formed per hour per carotid body. All but one of 31 normal carotid bodies directly removed from the animals presented DBH activity; values showed a wide distribution from 71 to 2400 pM/h per carotid body. The mean value was 526±104 (SE) pM/h per carotid body. This value is not very high but proved the presence of enzymatic activity in carotid body tissue.

To confirm the adequate inactivation of endogenous DBH inhibitors a known amount of partially purified bovine adrenal DBH was added to a duplicate of each of the samples used for DBH determination. As shown in Fig. 1, the recoveries were around 80-100%, thus indicating that inactivation of endogenous inhibitors was effective; data were not corrected for recovery. Differences among carotid bodies of the same animals were not significant.

The proportion of the measured DBH activity associated with the sympathetic innervation of the carotid body was determined in cats whose right superior cervical ganglion was removed under Nembutal anesthesia. Seven to fifteen days later, when nervous degeneration was presumably complete, the activity of DBH was measured separately in both carotid bodies and in the nictitating membrane of both eyes; the reduction to significant values of DBH activity in the nictitating membrane of the operated side was taken as proof of the completeness of the denervation. The sympathectomized carotid bodies had a mean DBH activity

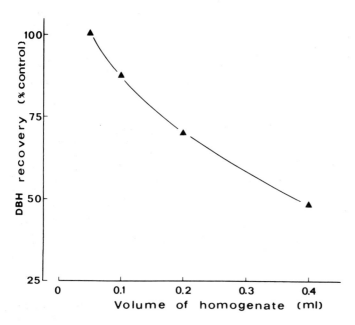

Fig. 1. Inhibition of partially purified DBH from bovine adrenal medulla with increasing amounts of carotid body homogenate. Aliquots of tissue homogenates were added to a fixed amount of DBH and enzymatic activity estimated as described in text

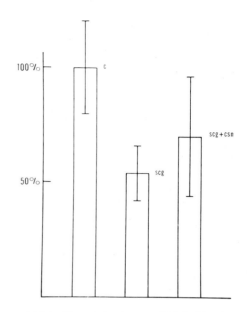

Fig. 2. Percent of DBH activity in control carotid bodies (c), in carotid bodies after removal of superior cervical ganglion (scg), and after removal of superior cervical ganglion and transection of carotid sinus nerve (scg+csn)

of 280±65 (SE) pM/h per carotid body, which is significantly lower ($p<0.05$) than in normal carotid bodies, as is shown in Fig. 2. In a group of 6 animals both superior cervical ganglia were removed, and the carotid sinus nerve on the right side was cut; 2 weeks later DBH activity was measured in the nictitating membranes and the carotid bodies of both sides. The carotid bodies of the sympathectomized and deafferented side had a mean DBH activity of 369±137 (SE) pM/h per carotid body, a value not significantly different from that found in the contralateral carotid bodies, which were subjected only to sympathectomy. Lower values were obtained in the carotid bodies dissected in vitro, both in the normal and in the sympathectomized animals*.

One of the criteria used in the identification of a substance as a neurotransmitter has been the presence in the tissue of the synthesizing enzymes (9). Our study describes the existence in the carotid body of DBH activity, the enzyme that converts dopamine into norepinephrine. The persistence of enzymatic activity in the carotid body after sympathetic denervation indicates that DBH is associated not only with the sympathetic fibers entering the carotid body but also with its parenchymal cells. Chiocchio, King, and Angelakos (2) suggested on the basis of pharmacologic evidence that many of the glomus cells contained only dopamine, while others contained norepinephrine or a combination of norepinephrine and dopamine. More recently, two types

---

*These values were 293.2±27.7 pM/h per carotid body (SE) in 12 normal carotid bodies and 192±20 pM/h per carotid body in 8 sympathectomized carotid bodies; the smaller difference between DBH activity in the normal and the sympathectomized carotid bodies when these are dissected in vitro is probably due to the more complete elimination of synthetic fibers that pass on the surface of the carotid body to other areas (13).

of glomus cells have been described based on the size of their dense-cored vesicles (4,8). Hellström and Koslow (5) suggested that small vesicle cells could contain norepinephrine, whereas large vesicle cells could store dopamine; if this is the case, DBH activity found in our experiments after sympathectomy would correspond to the enzyme occurring in the glomus cells that contain norepinephrine. Transsection of the sinus nerve did not modify DBH activity, which suggests that the enzymatic activity is not related to the sensory innervation of the carotid body, at least under conditions of normoxia.

References

1. Brimijoin, S.: J. Neurochem. *19*, 2183-2193 (1972)
2. Chiocchio, S., King, M.P., Angelakos, E.T.: Histochemie *25*, 52-59 (1971)
3. Goldstein, M., Freedman, L.S., Bonnay, M.: Experientia *27*, 632-633 (1971)
4. Hellström, S.: J. Neurocytol. *4*, 77-86 (1975)
5. Hellström, S., Koslow, S.H.: Brain Res. *101*, 245-254 (1976)
6. Kaufman, S., Friedman, S.: Pharmacol. Rev. *17*, 71-99 (1965)
7. Lever, J.D., Boyd, J.D.: Nature *179*, 1082 (1957)
8. McDonald, D.M., Mitchell, R.A.: in: The Peripheral Arterial Chemo-receptors. Purves, M.J. (ed.). London: Cambridge 1975, pp. 101-131
9. McLennan, H.: Synaptic Transmission. Philadelphia: W.B. Saunders 1963
10. Molinoff, P.B., Weinshilboum, R., Axelrod, J.: J. Pharmacol. Exp. Ther. *178*, 425-431 (1971)
11. Morgado, E., Llados, F., Zapata, P.: Neurosci. Lett. (1977) (in press)
12. Potter, L.T., Axelrod, J.: J. Pharmacol. Exp. Ther. *142*, 299-305 (1963)
13. Zapata, P., Hess, A., Bliss, E.L., Eyzaguirre, C.: Brain Res. *14*, 473-496 (1969)

DISCUSSION

Starlinger: Did you use cupric ion?

Belmonte: No. According to Udenfriend the concentration of copper ions must be adjusted very carefully, and for that reason we used NEM.

Starlinger: We were not successful in estimating the enzyme. We used cupric ions.

Belmonte: Yes, I think that the problem is the endogenous inhibitors in the tissue.

Starlinger: How do you make the blanks?

Belmonte: We boil a sample at 95°C for 5 min.

O'Regan: You suggest here that about half the DBH present in the carotid body is contained in the sympathetic nervers. It is not possible that if you sympathectomize the carotid body you can interfere with the amount of enzyme in the glomus cells themselves?

Belmonte: Well, that could be a possibility, but perhaps the difference could also result from the sympathetic fibers both inside and on the surface of the carotid body.

Lübbers: Did you try any physiological stimulus to see if you could influence the amount of activity of the enzyme, for example, by nitrogen or carbon dioxide?

Belmonte: No, we tried with reserpine which produces enzymatic activation in the adrenal medulla, but still the results are not clear. The problem is that the variability among animals is quite large, and we prefer to have a number of carotid bodies before making a statistical compilation.

Lübbers: How many carotid bodies did you use?

Belmonte: 29.

Hess: How about ganglion cells? Even sympathectomy would not affect their number in the carotid body.

Belmonte: Of course, that could explain it. But perhaps the morphologists could tell us better what the quantitative importance of the ganglion cells in the carotid body is. Perhaps they can account for the activity, but I doubt it.

Hess: Yes, probably it is too much activity to be accounted for by the number of ganglion cells.

Zapata: We have done determinations of DBH activity in normal and in sympathectomized carotid bodies of the cat also. Our values are a little lower than those shown by Belmonte, but we used a different technique - the one-stage method of Kirshner. There is some accordance with the view presented by Belmonte in the sense that we found a very high value of DBH activity in the ganglioglomerular nerves, whereas the carotid nerve had a very low value for DBH, which was not significant at all.

Acker: The contents of the vesicles in the type I cell are still a mystery to me. Do you know from direct evidence that the vesicles contain DBH?

Belmonte: No, but the fact that there is still DBH in the carotid body after sympathectomy means that it probably is in the dense-cored vesicles.

Acker: Is it absolutely necessary that DBH be in the vesicles? Could it not be located in the cytoplasm?

Belmonte: No. There is a part that is soluble and another that is bound to the membrane of the vesicles, but it is not outside the vesicles.

Torrance: If the amount of this enzyme present in the sympathetic nerve fibers of the carotid body is equal to or greater than the amount present in the type I cells, which have presumably a very much greater volume than the sympathetic nerve endings, does this mean that the substance, whatever it is, with which this enzyme is concerned turns over very slowly in the type I cell?

Belmonte: Well, I do not think it is a problem of volume. The first point is that this 50% that corresponds to the sympathetic fibers is open to discussion. The problem in the nonsympathectomized animals is that there is very high activity in the ganglioglomerular nerves. If you do not clean the carotid body very well, then you get more activity in the control animal. So this difference of 50% could be less than that, it depends on how well you clean the carotid body.

Torrance: What I am suggesting is that there is not a very rapid dopamine metabolism in the type I cells, if this is the necessary enzyme for dopamine metabolism. The total amount in the cells, which is many times the volume of the sympathetic nerve endings, is equal to that in the sympathetic nerve endings.

Belmonte: But I do not think you can take the amount that is in the sympathetic nerve endings as a control. Some fibers run over the surface of the carotid body.

McDonald: Did you examine the effect of sympathectomy only at a single interval following the operation, or did you follow the sequence of events following sympathectomy? How did you ascertain that all of the nerves were gone and that there was no regeneration?

Belmonte: First, we made the measurements a minimum of 7 days after sympathectomy. Second, we used the nictitating membranes as control; these are very rich in DBH, so we were sure that the sympathectomy was complete when these had very low levels of the enzyme.

McDonald: Did sympathectomy mean removal of the ganglion or cutting the nerves coming out of the ganglion?

Belmonte: Removal of the ganglion.

Zapata: We got very high values of DBH activity in the ganglioglomerular nerves, and the only explanation for the high activity in these nerves and the smaller amount in the glomus itself is that perhaps most of the fibers in this nerve are not directed toward glomus tissue but past and go to the sinus.

Paintal: In another paper in this volume Zierold showed that the percentage of type I and type II cells in glomoid tissue amounts to about 75%. The nerves are about 7%. Now if this 7% includes both the carotid sinus nerve and the sympathetics, and you assume that the sympathetics are 20% of that value, then 1% of the tissue has more than 60% of the dopamine activity, which means that the concentration of DBH is very low in the type I cell.

Belmonte: The amount of DBH activity that disappears after sympathectomy in our experiments does correspond to only 1% of the glomus tissue formed by its sympathetic supply. We surely keep some of the ganglioglomerular nerve fibers running on the surface of the carotid body that go to the sinus area, and these contribute to the DBH activity measured in the control experiments. Thus, the activity that disappears after sympathectomy, and that you are attributing exclusively to the symapthetic fibers in the carotid body, probably corresponds in part to DBH present in the sympathetic nerves running over the carotid body surface. But what you say could be basically correct.

Fidone: It is possible that the norepinephrine that the DBH is manufacturing is actually in only a very small percent of the glomus cells? If that were the case, the concentration would not be so disturbingly low. For example, the percentage of large and small vesicle-containing cells could be important here.

McDonald: The large and small vesicle-containing cells have been demonstrated for the rat carotid body. In the cat there are several different cell types, but we do not have the quantitative analysis as in the rat carotid body, and I think the situation may be a little different in the cat.

Endogenous Acetylcholine Levels in Cat Carotid Body and the Autoradiographic Localization of a High Affinity Component of Choline Uptake*

S. Fidone, S. Weintraub, W. Stavinoha, C. Stirling, and L. Jones

It was first noted by Schweitzer and Wright that acetylcholine (ACh) might be involved in carotid body chemoreception, but it was Eyzaguirre and co-workers (4) who later studied this problem systematically and concluded that ACh might be a sensory transmitter released from the glomus (type I) cells to excite neighboring afferent nerve terminals. However, ACh has never been chemically identified in carotid body tissue, although bioassays have suggested the presence of an ACh-like substance in carotid body extracts (5,8). Furthermore, Osborne and Butler (9), proceeding from Biscoe's (1) suggestion that the synapses on the glomus cells are efferent, not afferent, have recently proposed that ACh is the transmitter at this junction and that consequently this substance is contained in the nerve terminals, not in the glomus cells.

In light of this controversy, we have undertaken to determine whether chemically identifiable ACh is present in the carotid body and in what cellular constituents of the organ it is contained. Specifically we have {1} measured the endogenous ACh levels in the tissue using pyrolysis gas chromatography and mass spectometry (GS/MS), {2} determined whether the ACh in the tissue is lost following chronic denervation by Sect. of the carotid nerve, as might be expected if the ACh were contained in the carotid nerve terminals, and {3} studied the kinetics of choline uptake by the carotid body and the autoradiographic localization of the high affinity component of this uptake process.

GS/MS Measurements of Carotid Body ACh

Carotid bodies were removed from the animals and quickly cleaned of their connective tissue in a chamber filled with ice-cold Locke's solution equilibrated with 100% O_2 and containing 30μM eserine (Sigma). In some animals the carotid body on one side had previously been denervated by excision of the carotid nerve 14 days prior to its removal for ACh analysis. The tissue was weighed (mean carotid body weight = 719 μg) and then frozen in liquid nitrogen until analysis. Significant degradation of tissue ACh did not occur during sample preparation because carotid bodies cooled in situ with ice-cold Locke's solution prior to removal from the animal, or cooled and then quickly frozen in liquid nitrogen, gave ACh values that were not significantly different from those obtained without these procedures ($P > 0.2$, non-paired, double-tailed Student's t-test). A detailed description of the sample preparation and the analytical conditions employed for pyrolysis GS/MS measurements of ACh have been published elsewhere (7).

---

*This work was supported by U.S. Public Health Service grants NS-12636, NS-05666, NS-07938, and MH-25168.

Table 1. ACh Content of Normal and Denervated Cat Carotid Bodies (7)

Condition	ACh content[a]	
	nM/g tissue	pM/organ
Normal	17.0 ± 1.1 (7)	11.7 ± 1.1 (7)
Normal[b]	18.1 ± 2.8 (4)	11.3 ± 0.5 (4)
Denervated[c]	22.9 ± 5.4 (4)	12.0 ± 2.0 (4)

[a] Results are expressed as the mean value ± SEM with the numbers of carotid bodies in parentheses.
[b] Normally innervated carotid bodies from the side contralateral to the carotid bodies that were denervated in the operated animals.
[c] Carotid bodies were denervated by removal of the carotid nerve 14 days prior to excision of the tissue for analysis.

The results of the ACh determinations in normal and denervated carotid bodies are shown in Table 1. The values in each group are not significantly different from one another ($P > 0.2$, using nonpaired, double-tailed Student's t-test for all samples, or paired Student's t-test for normal vs denervated carotid bodies from operated animals). Earlier reports (5,8) of the ACh content of cat carotid bodies obtained using bioassays have claimed values as much as 10 times higher than these obtained using pyrolysis GS/MS. We cannot explain this large discrepancy: however, pyrolysis GS/MS is a sensitive and highly specific method for measuring tissue ACh levels and is free of many of the difficulties and uncertainties that accompany bioassays. It is interesting to note here also that Christie (2) reported many years ago that extracts from human carotid body tumors induced hypotension in decapitated cats, and this effect was not blocked by atropine in doses that completely abolished the effects of administered ACh. These extracts also produced contraction of the virgin guinea pig uterus. By other tests the effects of histaminelike reactions were ruled out. Christie labeled this unidentified substance carotidin and noted that it was also present in the carotid body of the elasmobranch. Whether carotidin or some other substance may have interfered with the bioassay determinations of ACh is unknown.

At 14 days following denervation nearly all the nerve terminals and preterminal fibers of the carotid nerve were either completely degenerated or in a very advanced state of degeneration. Our data therefore suggest that carotid body ACh is not contained in the carotid nerve innervation to this organ. The remaining cellular constituents of the carotid body that are possible sources of ACh include the glomus (type (type I) cells, sustentacular (type II) cells, ganglion cells, and the sympathetic innervation to the blood vessels. The ganglion cells can probably be discounted because they are too few in number. Also, since the ganglion cells are usually found around the perimeter of the organ, it is our experience that after carotid bodies have been thoroughly cleaned of their connective tissue capsules, they rarely show histologic evidence of ganglion cells. Regarding the sympathetic innervation to the blood vessels, preliminary data from our laboratories show that the ACh levels in the carotid body are also unchanged after chronic superior cervical ganglionectomy.

Localization of ACh in the Carotid Body

In recent years numerous reports have appeared describing two kinetically different processes for choline (Ch) uptake: a low affinity, saturable uptake common to all neurons (and probably most other tissue as well), and a high affinity, saturable uptake, which is reported to be specific for cholinergic neurons (12). To test for high and/or low affinity uptake of Ch, carotid bodies were incubated at 37°C in {^{3}H}Ch (1-70μM), and the time course of accumulation of total radioactivity was determined. Uptake was linear for approximately 10 min, then gradually plateaued over the next 15-20 min. Thereafter, carotid bodies were incubated for 10 min in 1-70μM{^{3}H}Ch. The uptake showed saturation with increasing {^{3}H}Ch concentrations. The data were corrected for a small, passive, nonsaturable component of Ch entry, which was determined in separate experiments by incubating carotid bodies in 1-70μM{^{3}H}Ch together with a much higher concentration (10mM) of unlabeled Ch. In this situation the uptake of radioactivity by the specific high and low affinity uptake processes is nearly completely abolished by the presence of the large excess of unlabeled Ch, whereas nonspecific leakage of the labeled Ch into the cells is unaffected (12).

Kinetic analysis of the corrected data using a Lineweaver-Burk double reciprocal plot resulted in a curvilinear distribution of the data points, which could be resolved into two distinct components. The K_m and V_m values for the high affinity (K_{mH}) and low affinity (K_{mL}) components of Ch uptake in the carotid body were estimated from the double reciprocal plot by assuming that {1} at low concentrations only the high affinity component contributes to Ch uptake, which then allows K_{mH} and V_{mH} to be obtained directly from the graph, and {2} at high concentrations both the high and low affinity components contribute to Ch uptake, and that the graphically determined K_{mc} and V_{mc} are combined values representing both components, whereby $V_{mL} = V_{mc} - V_{mH}$ and $K_{mL} = K_{mc} \times V_{mc}/V_{mL}$ (3). The K_m values obtained for Ch uptake in the carotid body are $K_{mH} = 3.6 \mu M$ and $K_{mL} = 49.5 \mu M$, which falls within the ranges of 1-8μM for K_{mH} and 25-100μM for K_{mL} reported for other putative cholinergic systems.

It has been suggested that the high affinity uptake of Ch might furnish a heuristic approach to the labeling of cholinergic cells, since at low Ch concentrations negligible accumulation of Ch occurs via the low affinity component (3,10,12). To determine which neuronal or cellular constituent (s) of the carotid body are responsible for the high affinity uptake of Ch, two experimental approaches have yielded the following results: {1} Chronic total denervation of the carotid body by Sect. of both the carotid nerve and the ganglioglomerular sympathetic nerve for periods of 1 week to 3 months did not result in the loss of the high affinity component of Ch uptake (nor the low affinity component). This suggests, in agreement with our GS/MS data described above, that these nerve fibers may not be cholinergic, but that one or more of the remaining cellular elements in the carotid body might have this property. {2} To determine which cells of the carotid body have the capacity for the high affinity uptake of Ch, carotid bodies incubated in 1-2μM{^{3}H}Ch were prepared for autoradiography using rapid freezing and freeze-drying techniques in order to maintain histologic localization of the labeled Ch (6,11).

Following a short in vitro incubation with {^{3}H}Ch, the carotid bodies were rinsed briefly in ice-cold Locke's solution and quickly immersed in a Dewar flask of liquid propane cooled to its melting point (-190°C) with liquid nitrogen. The tissue was then transferred at -196°C to the sample holder of a high vacuum (5 × 10^{-8}mmHg) freeze dryer. The sample holder, fitted with a temperature controller, was brought slowly to -100°C over several hours. The temperature was further increased at 10°C/day to -70°C, where it remained for 2 days. It was then raised to -40°C within 8 h and to 0°C at 20°C/day. Finally, the temperature was raised to +40°C for 2 h, prior to breaking the vacuum. (The zone between -70°C and -40°C was passed through quickly because it has been reported that ice crystal formation is accelerated in this temperature zone after freezing.) The tissue was then transferred to a Thunberg tube, fixed in osmium vapor, and vacuum-embedded in Araldite. Thick Sects. (1 μm) were cut and coated with Kodak NTB-2 emulsion using a constant-rate withdrawal apparatus (48 mm/min). Exposure time was 1-2 weeks.

Autoradiographs of carotid bodies incubated in low (1μM) and high (30μM) concentrations of {^{3}H}Ch are shown in Fig. 1A and 1B, respectively. At low {^{3}H}Ch concentrations, when only the high affinity component should contribute significantly to the uptake (3,10,12) the autoradiographic label appears to be concentrated principally over the glomus cells. Sustentacular cells are difficult to distinguish with the light microscope, and hence their degree of labeling is uncertain.

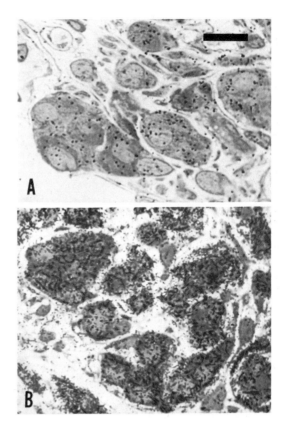

Fig. 1. Light-microscopic autoradiographs of carotid bodies incubated for 10 min in Locke's solution containing 1μM (A) and 30μM (B) {^{3}H}Ch. Tissue was quick-frozen and freeze-dried prior to application of nuclear track emulsion, as described in text (*Marker*) 10 μm

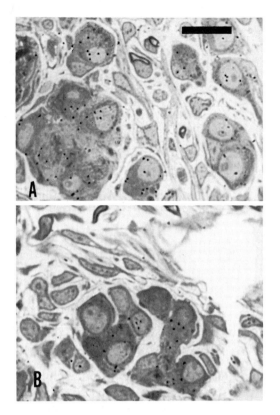

Fig. 2. Light-microscopic autoradiographs of carotid bodies incubated with 1μM {^{3}H}Ch for 10 min in (A) 15mM Na$^+$-Locke's solution and (B) normal Locke's solution with 10μM HC-3. Tissue was quick-frozen and freeze-dried prior to application of nuclear track emulsion, as described in text. (*Marker*) 10 μm

In contrast to the low affinity component of Ch uptake, the high affinity component is reduced or abolished in low Na$^+$ (Fig. 2A) and by low concentrations of hemicholinium-3 (HC-3) (Fig. 2B). These same conditions also greatly reduce or abolish the synthesis of {^{3}H}ACh from {^{3}H}Ch in the carotid body. This was determined in separate experiments in which {^{3}H}ACh was separated and measured using high-voltage paper electrophoresis, radiochromatogram scanning, sample oxidation, and liquid scintillation spectrometry. Also, as the {^{3}H}Ch concentration is reduced, the percent of total radioactivity found as {^{3}H}ACh is increased. Similar findings have been reported in other cholinergic systems (3,10,12). These observations suggest that in the carotid body, as elsewhere, the high affinity component of Ch uptake is coupled to the synthesis of ACh. Since the autoradiography of carotid bodies incubated in low concentrations of {^{3}H}Ch indicates that the glomus cells may be the principal site of the high affinity component of Ch uptake, it is reasonable to conclude that the glomus cells are capable of ACh synthesis. Of course, our data cannot rule out the possibility that other cellular constituents of the carotid body may also have this property.

References

1. Biscoe, T.J.: Physiol. Rev. *51*, 437-495 (1971)
2. Christie, R.V.: Endocrinology *17*, 421-432 (1933)
3. Dowdall, M.J., Simon, D.J.: J. Neurochem. *21*, 969-982 (1973)
4. Eyzaguirre, C., Zapata, P.: in: Arterial Chemoreceptors. Torrance, R.W. (ed.). Oxford: Blackwell 1968, pp. 213-247
5. Eyzaguirre, C., Koyano, H., Taylor, J.R.: J. Physiol. *178*, 463-476 (1965)
6. Eyzaguirre, C., Nishi, K., Fidone, S.: Fed. Proc. (Symposium) *31*, 1385-1393 (1972)
7. Fidone, S., Weintraub, S., Stavinoha, W.: J. Neurochem. *26*, 1047-1049 (1976)
8. Jones, J.V.: in: The Peripheral Arterial Chemoreceptors. Purves, M.J. (ed.). New York: Cambridge University Press 1975, pp. 143-162
9. Osborne, M.P., Butler, P.J.: Nature *254*, 701-703 (1975)
10. Simon, J.R., Atweh, S., Kuhar, M.L.: J. Neurochem. *26*, 909-922 (1976)
11. Stirling, C.E., Kinter, W.B.: J. Cell Biol. *35*, 585-604 (1967)
12. Yamamura, H., Snyder, S.: J. Neurochem. *21*, 1355-1374 (1973)

DISCUSSION

McDonald: In your autoradiographic studies analyzing the distribution of labeled choline in the carotid body, there were grains over some cells, the identify of which was in question. Did you use some quantitative means of determining if there are more grains over glomus cells than over other types of cells?

Fidone: The low density of grain overlying structures other than glomus cells suggested to us that these cells are principally responsible for the high affinity uptake of choline in the carotid body. But with the light microscope it is difficult to distinguish with certainty all of the structures in which you find any uptake of choline. To do so, we would have to work at the electron-microscopic level, which is difficult with water-diffusible substances. Presently we are attempting to use Van Harreveld's technique, in which the tissue is brought quickly against a polished copper surface at liquid nitrogen temperature. The result is a thin 10µm layer of tissue that is histologically preserved. In the meantime, however, we are approaching the present data from the standpoint that the high affinity choline uptake is a marker for acetylcholine synthesis, a relationship that has been demonstrated in numerous other preparations. Of course, if this relationship is not valid for this tissue, it would not be possible to localize acetylcholine in the carotid body with the technique I have described. So, to answer your question, we have not done a quantitative autoradiographic analysis of how much grain is located over cells other than glomus cells, and we do not know whether, in addition to glomus cells, we also have labeled fibroblast cells, mast cells, or sustentacular cells. I think you must admit, though, that the glomus cells here are clearly labeled, and if there is a relationship between the high affinity uptake of choline and acetylcholine synthesis, we are faced with the obvious possibility that acetylcholine is contained in the glomus cells.

Lübbers: Could you give us some information about the physiological function of this acetylcholine you have found?

Fidone: I think we must be careful not to go beyond the data, especially in regard to any possible role for acetylcholine in sensory transmission. A very important question remains to be answered and that is whether acetylcholine release is causally related to chemoreceptor discharge. If we can, for example, see a change in chemoreceptor discharge that closely parallels changes in acetylcholine release, it would build a stronger case for the role of acetylcholine as an afferent transmitter in chemoreception. We are presently studying this relationship. It is interesting to note here that the carotid body contains receptor sites that bind the cholinergic ligand α-bungarotoxin. We have observed this using a relatively low concentration of this substance, namely, 20nM. Low concentrations of curare reduce the level of binding, and this specific binding is not lost following total denervation of the carotid body.

Pietruschka: What proof do you have that the traces you found in your autoradiography are really bound to choline and not built up into other molecules?

Fidone: It is undoubtedly not only choline. Much of the choline is converted to acetylcholine, phosphorylcholine, and phospholipid. In these experiments we are using the high affinity choline uptake as a marker for cholinergic systems.

Torrance: When Eyzaguirre spoke of membranes behind which acetylcholine "hid" so that blockers could not act upon it, it seemed to be acceptable that there was a low turnover of acetylcholine because it might be hiding there also and be taken back in stores. However, Woods has shown that peroxidase with penetrate into the clefts of the carotid body from the blood and one can, in fact, measure a sort of half-time for things in and out of the space. For bicarbonate it is 20 s or so.

Fidone: We have also measured the washout half-time of radiolabeled inulin and sucrose and found it to be very, very short.

Torrance: If you could measure the volume of the space and the sensitivity of the carotid body to the particular molecule you are considering, acetylcholine or dopamine, then you could estimate the rate at which the substance must be made if it is going to have a significant effect. I suspect that for dopamine this estimated rate of formation would be much larger than what would be expected from the small turnover that you and others have discussed.

Fidone: First, the turnover rate of acetylcholine in the carotid body is not slow; it compares favorably with that of the superior cervical ganglion, i.e., about 85% of the stores can be labeled in 1 h. The endogenous *level* of acetylcholine is somewhat low, but its turnover rate is reasonably high. Second, although it is true that the amount of transmitter released and the space it occupies is important, another factor that is also very important is the proximity and density of the receptors that are able to bind the transmitter. You might be able to effect a very powerful cholinergic response with a minimal amount of acetylcholine but with a high density of receptors.

O'Regan: What a role do you propose for your nicotinic receptors on the glomus cell?

Fidone: They are not necessarily on the glomus cell. They could also be on the type II cell. Jones has demonstrated myosin ATPase activity and has pointed to the presence of actinlike filaments in the type II cell, and I think Paintal has proposed that the carotid body is a modified mechanoreceptor. We do not know where the nicotinic receptors are located, and that is why we are doing the autoradiography of the labeled bungarotoxin. But if they are on the type I cell and this cell contains both acetylcholine and dopamine, perhaps the acetylcholine released from the type I cell acts back onto the cell to release dopamine, Burn and Rand style. However, such speculation is clearly going beyond the data.

Molecular Biology of Chemoreceptor Function: Induction of Tyrosine Hydroxylase in the Rat Carotid Body Elicited by Hypoxia

I. Hanbauer

Dopamine is the most abundant catecholamine stored in glomus cells of the rat carotid body (9,13,19). Recent evidence (10) suggests that dopamine participates in the early response to hypoxia; however, the molecular mechanism whereby dopamine is released and participates in the chemoreceptor function has not yet been elucidated. In order to improve our present understanding of these mechanisms, the changes in the activity and properties of tyrosine hydroxylase - the rate-limiting enzyme in the synthesis of dopamine - have been monitored at various times after exposure to hypoxic conditions. It is generally accepted that in nervous tissue the activity of tyrosine hydroxylase increases when the rate of firing of catecholaminergic neurons is increased. The duration of the enhanced neuronal activity and the cellular localization of tyrosine hydroxylase determine whether the affinity of tyrosine hydroxylase for its cofactor (20) or the synthesis rate of tyrosine hydroxylase is increased (2). When the increase in activity is long-lasting, tyrosine hydroxylase synthesis is increased. Since the $T_{1/2}$ of tyrosine hydroxylase is greater than 20 h, a long-term increase of tyrosine hydroxylase activity ensues.

Changes in Tyrosine Hydroxylase Activity Elicited by Hypoxia

McDonald and Mitchell (12) proposed that the carotid sinus nerve establishes reciprocal synapses with the glomus cells, which suggests that this nerve not only conveys information from the carotid body to the brain but also may bring information to the carotid body concerning a central control of the trophism and excitability of the glomus cells. Moreover, Sampson et al. (16) have shown that an increased centrifugal impulse flow in the carotid sinus nerve enhances the relaese of catecholamines from the storage sites in the carotid body of cats. These authors have also observed that transection in the carotid sinus nerve diminished the excitability of the glomus cells by hypoxia or electrical stimuli applied to the nerve. Our own experiments showed that hypoxia depleted the dopamine content of rat carotid body and that the duration of this depletion was shortened following transection of the carotid sinus nerve (10). These findings prompted us to study whether hypoxia ecilited a change in tyrosine hydroxylase activity and whether these changes required an intact sinus nerve.

When rats were exposed for 30 min to 5% O_2, the P_aO_2 dropped to 30-40 mmHg (8; see also Table 1). At this time the dopamine content of the carotid body was reduced by about 70% (10; see Table 1); it remained below normal levels for longer than 2 h (10). Although the dopamine content of the carotid body was lowered, the V_{max} and affinity of tyrosine hydroxylase for its cofactor remained unchanged immediately after hypoxia (8). However, two or more exposures to hypoxic conditions for 30 min caused a delayed increase in the V_{max} of the enzyme (Table 1) without a change in the K_m of tyrosine hydroxylase for its cofactor

Table 1. Biochemical changes elicited in carotid body by exposure to hypoxia[a]

Parameter	Hypoxia (min)	Interval following hypoxia (h)	Room air[b]	5% O_2 + 95% N_2[b]
PO_2 (mmHg)	20	0	117 (108-125)	37 (29-45)[c]
Dopamine (pM/pair carotid bodies)	30	0	30 ± 1.9 (6)	9.3 ± 1.9 (6)[c]
cAMP (pM/pair carotid bodies)	2 × 30	1	45 ± 3 (8)	57 ± 3 (3)[c]
RNA synthesis (cpm[^{3}H]uridine/pair carotid bodies)	2 × 30	4	1832 ± 201 (4)	3155 ± 93 (4)[c]
α-Amanitin-sensitive RNA synthesis (cpm[^{3}H]uridine/pair carotid bodies)	2 × 30	4	278 ± 19 (6)	629 ± 55 (6)[c]
Tyrosine hydroxylase (pM/pair carotid bodies/h)	2 × 30	23	4.3 ± 0.2 (6)	6.1 ± 0.4 (6)[c]

[a] Male rats were exposed to 5% O_2 (two 25-min periods interrupted by 15 min at room air). cAMP concentration was assayed by radioimmunoassay (Collaborative Research Inc., Waltham, Mass.). Incorporation of [^{3}H]uridine (15 µCi/ml culture medium) into RNA was determined by precipitation with 10% trichloracid containing 100 µg/ml heparine. The α-amanitin, (20 µg/ml) sensitive RNA synthesis was calculated by substracting α-amanitin-resistant from total [^{3}H]uridine incorporation. Tyrosine hydroxylase activity was assayed as described by Hanbauer et al (7).

[b] Values expressed as mean ± SEM; number is in parentheses.

[c] Different from room air ($P < 0.02$).

(8). An increase in V_{max} of tyrosine hydroxylase may occur when the number of enzyme molecules is increased because of either an acceleration in the rate of synthesis or a decrease in the rate of degradation of tyrosine hydroxylase. Therefore, it was important to resolve whether the increase in tyrosine hydroylase caused by hypoxia was related to a change in protein synthesis.

When rats were treated with cycloheximide immediately after two 30-min exposures to hypoxia, the long-term increase in the V_{max} of tyrosine hydroxylase was curtailed (8), suggesting that the increase in V_{max} of tyrosine hydroxylase elicited by hypoxia required new protein synthesis. This notion was supported by studies on RNA synthesis. Table 1 shows that the rate of incorporation of {^{3}H}uridine into total RNA was markedly increased between 2 and 5 h after the termination of exposure to hypoxia and reached a maximum at 4 h. In the carotid body of rats exposed to hypoxic conditions, the stimulation of RNA synthesis was blocked by α-amanitin, suggesting that, in part, the stimulation of RNA synthesis was dependent on RNA polymerase II activity (Table 1). These findings were taken as an indication that the enhancement of RNA synthesis caused by hypoxia included an increase in mRNA synthesis.

The carotid sinus nerve appeared to play a role in the regulation of tyrosine hydroxylase synthesis because the increase in V_{max} of tyrosine hydroxylase and the stimulation of RNA polymerase II activity required the presence of an intact carotid sinus nerve (7). Since the basal tyrosine hydroxylase activity in intact and denervated carotid bodies was identical, it was inferred that the carotid sinus nerve regulates the synthesis of tyrosine hydroxylase formation only in response to specific hypoxic conditions.

Effect on Hypoxia on the Molecular Form of Cyclic-Nucleotide Phosphodiesterase

Phosphorylation of nuclear acidic proteins mediates the regulation of gene expression by hormones and neurotransmitters (3,15,17). This phosphorylation follows the activation and dissociation of the catalytic subunit of cAMP-dependent protein kinase and the uptake of the activated catalytic subunit by the nucleus. The activation of cAMP-dependent protein kinase can be used as an indicator for an increase in cAMP content (14). A highly specialized tissues - such as the carotid body - the increase in cAMP and activation of protein kinases occur only in a limited cell population; therefore, the second messenger response may not be easily detected by direct measurements of the cAMP content in the whole carotid body. However, it is possible to detect a second messenger response by measuring either the changes in the molecular forms of cyclic-nucleotide phosphodiesterase or the dissociation of the catalytic subunit of cAMP-dependent protein kinase.

Because of the large proportion of cAMP-independent protein kinase (unpublished observations) present in the carotid body, it was difficult to study the kinetics of this enzyme. However, we have obtained evidence that after hypoxia there is a change in the equilibrium of the various molecular forms of cyclic-nucleotide phosphodiesterase present in the carotid body. Table 1 shows that a small but significant increase in the cAMP content of carotid body occurs within 1 h after the termination of the exposure to hypoxic conditions. Studies on the kinetic properties of the molecular forms of cyclic-nucleotide phosphodiesterase present in the carotid body are shown in Table 2. In the carotid body of rats kept in room air, two molecular forms of cyclic-nucleotide phosphodiesterase can be measured in the presence of an optimal Ca^{2+} concentration. Exposure to hypoxia shifts this equi-

Table 2. Changes in the K_m forms of Cyclic-nucleotide phosphodiesterase elicited by exposure to Hypoxia[a]: Role of Ca^{2+}

Experimental condition	Addition	Number of molecular forms	Low K_m (μM)	High K_m (μM)
Room air	$10^{-5} M\ Ca^{2+}$	2	4.0	69
5% O_2 + 95% N_2	$10^{-5} M\ Ca^{2+}$	1	6.5	–
Room air	$10^{-4} M$ EGTA	2	3.3	44
5% O_2 + 95% N_2	$10^{-4} M$ EGTA	2	4.3	45

[a] Male rats were exposed to 5% O_2 (two 25-min periods interrupted by 15 min at room air). Kinetics of phosphodiesterase were studied using cAMP as a substrate, as described by Filburn and Karn (5).

librium and increases the abundance of the low K_m form of cyclic-nucleotide phosphodiesterase. This change results from a lowering in the K_m of the high K_m form of cyclic-nucleotide phosphodiesterase. Actually, following hypoxia only one molecular form of cyclic-nucleotide phosphodiesterase was detected in the rat carotid body (Table 2). The data given in Table 2 also show that the increase in the affinity of the enzyme for cAMP depended entirely on the presence of Ca^{2+}. These results indicate very strongly that an endogenous Ca^{2+}-dependent activator was released from its binding sites during hypoxia.

Preliminary experiments have shown that the activator content in cytosol of the carotid body was increased after exposure to hypoxia. Moreover, when the carotid body homogenate was incubated with $1\mu M$ cAMP, $0.7\mu M$ ATP, and type I protein kinase isolated from rat brain, the apparent K_m of cyclic-nucleotide phosphodiesterase was lowered from 133 to $29\mu M$ (see Table 3). This observation was in line with reports in the literature showing that the release of a Ca^{2+}-binding activator from synaptic membranes can be achieved by a cAMP-dependent phosphorylation process (6). When the Ca^{2+} activator is released into the cytosol, it associates with the high K_m form of cyclic-nuceotide phosphodiesterase, provided optimal Ca^{2+} concentrations are present (11).

Table 3. Decrease in K_m of phosphodiesterase (high K_m form) caused by in vitro phosphorylation of carotid body homogenate[a]

Additions to carotid body homogenate	Phosphodiesterase K_m (μM)
cAMP ($1\mu M$) + ATP ± boiled protein kinase	133 ± 1.0 (2)
cAMP ($1\mu M$) + ATP + protein kinase	29 ± 5.3 (5)[b]

[a] Carotid body homogenate was incubated for 3 min in presence of $1\mu M$ cAMP, $0.7\mu M$ ATP, and cAMP-dependent protein kinase isolated from rat brain. After centrifugation at 7000 rpm for 20 min, the kinetics of phosphodiesterase were studied in the supernatant using cAMP as a substrate, as described by Filburn and Karn (5). Number is in parentheses.

[b] $P < 0.01$.

At present, an increase of the Ca^{2+}-binding activator in the cytosol of carotid body following hypoxia can be interpreted in a number of ways. One possibility is that the activator was released by membrane phosphorylation involving a catalytic subunit of cAMP-dependent protein kinase. This hypothesis emphasizes the role of the modest increase in the cAMP content of the carotid body caused by hypoxia; the required link would be the direct evidence for activation of a cAMP-dependent protein kinase. Another possibility is an interaction of the Ca^{2+}-dependent modulator protein with mitochondria. It has been shown that these organelles can transport large amounts of Ca^{2+} (1) and may play an important role in the regulation of many physiological and biochemical processes that depend on Ca^{2+}. The possibility deserves consideration that during hypoxia a change in the intracellular Ca^{2+} concentration in glomus cells brings the Ca^{2+}-binding protein into an activated state. It has been demonstrated that several Ca^{2+}-binding proteins are structurally and functionally similar (18); one of them, troponin-C, which participates in the activation of actomyosin ATPase (4), could be involved in the receptor mechanisms for hypoxia. Experiments to elucidate whether the increase in the function of the Ca^{2+}-binding protein is due to a membrane phosphorylation or to a release of Ca^{2+} are currently being conducted in this laboratory.

References

1. Chance, B.: J. Biol. Chem. *240*, 2729-2748 (1965)
2. Chuang, D.M., Costa, E.: Proc. Nat. Acad. Sci. *71*, 4570-4579 (1974)
3. Chuang, D.M., Hollenbeck, R., Costa, E.: Science *193*, 60-62 (1976)
4. Ebashi, S., Endo, M., Ohtsuki, I.: Rev. Biophys. *2*, 351-384 (1969)
5. Filburn, C.R., Karn, J.: Analyt. Biochem. *52*, 505-516 (1973)
6. Gnegy, M.E., Nathason, J.A., Uzunov, P.: Mol. Pharmacol. (1976) (in press)
7. Hanbauer, I.: in: Adv. Biochem. Psychopharmacol. Costa, E., Gessa, G.L. (eds.). New York: Raven Press 1977 (in press)
8. Hanbauer, I., Lovenberg, W., Costa, E.: Neuropharmacology (1976) (in press)
9. Hellström, S., Koslow, S.H.: Acta Physiol. Scand. *93*, 540-547 (1975)
10. Hellström, S., Hanbauer, I., Costa, E.: Brain Res. *118* (1976) (in press)
11. Lin, Y.M., Liu, Y.P., Cheung, W.Y.: FEBS Lett. *49*, 356-360 (1975)
12. McDonald, D.M., Mitchell, R.A.: J. Neurocytol. *4*, 177-230 (1975)
13. Mollmann, H., Niemeyer, D.H., Alfes, H., Knoche, H.: Z. Zellforsch. *126*, 104-115 (1975)
14. Russel, D., Byus, C.V.: in: Adv. Biochem Psychopharmacol. Costa, E., Giacobini, E., Paoletti, R. (eds.). New York: Raven Press 1976, Vol. XV, pp. 445-454
15. Sahib, M.K., Jost, Y.C., Jost, J.P.: J. Biol. Chem. *246*, 4539-4545 (1971)
16. Sampson, S.R., Nicolaysen, G., Jaffe, R.A.: Brain Res. *85*, 437-446 (1975)
17. Stein, G.S., Spelsberg, T.C., Kleinsmith, L.J.: Science *183*, 817-823 (1974)
18. Stevens, F.C., Welsh, M., Ho, H.C., Teo, T.S., Wang, J.H.: J Biol. Chem. *251*, 4495-4500 (1976)
19. Zapata, P., Hess, A., Bliss, E.L., Eyzaguirre, C.: Brain Res. *14*, 473-496 (1969)
20. Zivkovic, B., Guidotti, A., Costa, E.: Mol. Pharmacol. *10*, 727-735 (1974)

DISCUSSION

Willshaw: You have said that transection of the sinus nerve blocks tyrosine hydroxylase induction. Did you subject your animals to hypoxia after you transected the sinus nerve?

Hanbauer: Yes.

Belmonte: Perhaps you should cut the sympathetic nerves to be sure that you are eliminating any unspecific effects through changes in vascularization of the carotid body or through efferent sympathetic fibers that go to the glomus cells, as has been described in the rat.

Hanbauer: I have preliminary studies where I have removed the superior cervical ganglion, and the activation of tyrosine hydroxylase in carotid body does not occur.

McDonald: Did you cut both carotid sinus nerves?

Hanbauer: I cut the nerve only on one side and used the other side as control.

O'Reagn: Your results are very much in keeping with those of Sampson and his co-workers, who show that the synthesis of catecholamines was increased by hypoxia, depending on whether the nerve was intact.

Acker: You show this influence from calcium. Do you know how much calcium the carotid body usually consumes during hypoxia?

Hanbauer: No. However, to obtain the shift in cyclic-nucleotide phosphodiesterase K_m, I have to work in a medium that contains $1 \mu M$ calcium.

Acker: Do you also need oxygen in this reaction?

Hanbauer: Yes. In these experiments the oxygen content from saturation with room air is sufficient. However, RNA synthesis was done under constant oxygenation with 95% O_2.

Lübbers: We have found that under our experimental conditions the oxygen comsumption of the carotid body depends very much on the calcium content of the medium. If we withdraw calcium, the oxygen consumption stops. Do you have any explanation why the calcium plays a key role in carotid body function?

Hanbauer: Calcium plays a role in the maintenance of the membrane permeability and also is a cofactor for specific enzyme reactions.

Trzebski: I wonder if sympathectomy really prevents nonspecific effects of the catecholamine in such severe hypoxic conditions, if we take into account that the catecholamine level in the blood is increased in this condition just from the adrenal.

Thorn: Hypoxia acts on all body cells, and you have described an unspecific effect that can occur all other tissues.

Hanbauer: The length of hypoxic exposure was not so great, and except for hyperventilation the rats showed no abnormal behavior. If the exposure lasted longer, it might also affect other cells.

Nishi: Is the content of AMP increased in the carotid body that had been exposed to hypoxia?

Hanbauer: Yes, I found an increase at 1 h *after* hypoxia

Nishi: Have you ever tried to block phosphodiesterase activity, as with aminophylline? I have observed that aminophylline increases chemoreceptor discharge.

Hanbauer: No.

Ji: I am very interested in the observation that Hellström reported about the increase in the mitochondrial volume in hypoxia. This observation, coupled with your observations that in hypoxia the two K_m forms of enzyme go to one form and that calcium is essential for this shift, makes me wonder whether in hypoxia mitochondrial volume swells because it takes up calcium. This happens in vitro, and under hypoxic conditions the mitochondria remove calcium from cytoplasm, with triggers this kind of enzymatic intraconversion.

Pallot: Mills and Jöbsis show there is an unusual cytochrome within the carotid body, but, unlike the idea of Ji that the mitochondria take up calcium when they are hypoxic, I think the normal biochemical belief is that if a mitochondrion becomes reduced, than it actually loses calcium rather than the other way around. The question then becomes, what is the actual PO_2 at the mitochondrion? The normal cytochromes maintain their oxidized state down to extremely low tissue PO_2 values, and the release of calcium depends upon its coming from this mitochondria that contained the usual affinity cytochrome.

Torrance: If you expose the carotid body to a falling P_aO_2, the discharge first goes up, but below a PO_2 of 20 or 30 mmHg, it may start to fall. We tried perfusing the carotid body with blood of very low PO_2, and the response went quickly to a high level and then it dropped to a lower level.

Hanbauer: The P_aO_2 in the carotid artery before exposure to hypoxia was 108; after 10 or 20 min of hypoxia, it dropped to 35.

Torrance: But surely the gas mixture you are putting them in was 5% O_2, which is an inspired PO_2 of 37.5.

Hanbauer: Of course, it is possible that the PO_2 of inspired air was greater than 5%, considering that gases can penetrate rubber tubing, which we used for flushing the desiccator. However, measurements with a blood gas analyzer showed a mean value for P_aO_2 of about 35.

Torrance: Is it not quite extraordinary that the inspired PO_2 is 37.5? This, I assume, is the PO_2 of 5% O_2. In addition, because of the wetting of the O_2 when it gets into the lungs, 47 mmHg must be subtracted from the P_aO_2 so that a P_aO_2 of 35 mmHg would be astonishing. You would surely expect to have a P_aO_2 of 25 or 20 mmHg in an animal breathing 5% O_2, which is the sort of level at which we found that the discharge of chemoreceptors seemed to fall off, as if there might be some damage.

McDonald: If the animal ceased to be hyperventilating in response to the hypoxic stimulus, the PCO_2 would then rise. Did you measure the PCO_2 in these animals as well as the PO_2?

Hanbauer: The PCO_2 dropped slightly, but it was not significant.

McDonald: Did the animals cease hyperventilating in response to hypoxia?

Hanbauer: No

Torrance: Was it necessary to use as intense a hypoxic stimulus as 5% inspired O_2? Could you get this to happen on 10% O_2, which is after all the sort of inspired PO_2 at which you get a sustained activity of chemoreceptors?

Hanbauer: No, I have not yet done this determination with 10%. But from the literature I found that the rat can stand this low oxygen concentration without losing consciousness.

Effects of Hypoxia on Carotid Body Type I Cells and Their Catecholamines. A Biochemical and Morphologic Study

S. Hellström

The carotid body of different species contains a high concentration of catecholamines, mainly dopamine and norepinephrine (5,6,8,10,12,16). Electrophysiological studies have shown that these catecholamines may play a role as modulators of the chemoreceptor discharge (2,13,15). The dominating cells, the type I cells, are believed to store the catecholamines in their granulated vesicles. Morphologic studies of these cells and their granulated vesicles during stimulation, e.g., by hypoxia for shorter or longer periods of time, have shown rather conflicting results (1,3,4). In our attempts to further understand the chemosensory mechanisms of carotid body we have focused our interest on the involvement of the carotid body catecholamines, the type I cells, and their granulated vesicles in the receptor machinery during hypoxia. The catecholamine analysis was performed by aid of mass fragmentography. As has been recently shown (8,9), the mass fragmentography technique is extremely suitable to analysis of the catecholamines of the carotid body, where in most species only small amounts of tissue are available. Quantitative electron microscopy or morphometry was used to study the type I cells and their granulated vesicles.

Methodology

We exposed rats to a hypoxic gas mixture consisting of 5% O_2 and 95% N_2 for time periods of different lengths. The rats were then rapidly removed from the hypoxic chamber and instantly killed by cervical dislocation. Then they were either assayed by mass fragmentography or used for the electron-microscopic studies.

For the catecholamine determinations, the carotid bodies were dissected out, homogenized in formic acid (0.1N formic acid plus 10 mg ascorbic acid/ml) containing the appropriate internal standards (α-methylnorepinephrine and deuterated dopamine), and further prepared for the mass fragmentography according to conditions described earlier by us (8,9).

For the morphologic studies the chests of the killed animals were immediately opened, a cannula inserted through the left ventricle into the ascending aorta, and the rats fixed by perfusion. The fixative was composed of a mixture of glutaraldehyde (4%), paraformaldehyde (3%), and picric acid (0.5 mg/liter) in a sodium cacodylate buffer (0.1N). After 15 min of perfusion, the carotid body was dissected out, rinsed, postfixed in osmium tetroxide (2% in sodium cacodylate buffer), and further prepared for electron microscopy. Sects. from 4 different levels of the carotid body were examined at a magnification of 8000×. The volume density of the granulated vesicles mitochondria and nuclei was measured from prints magnified 24,000× with the aid of a point-counting method (14). The size of the granulated vesicle profiles was measured with a Zeiss Particle Analyzer TG3Z on prints magnified 48,000×. The morphometric procedure has been described in detail earlier (7).

Levels of Carotid Body Catecholamines Under Normal and Hypoxic Conditions

The control carotid bodies contained large amounts of dopamine and to a minor extent norepinephrine, 30.0 pM and 7.2 pM per pair of carotid bodies, respectively (Table 1). During hypoxia the level of dopamine increased to about two-thirds of the controls after 15 min and to one-third of the basic level after 30 min of treatment. When rats were exposed to room air after a 30-min period of hypoxia, the control level of dopamine was reached after about 2 h. In addition, the catecholamine determinations were carried out on carotid bodies where the carotid sinus nerve had been removed 14 days before the analysis. The decrease in dopamine level during hypoxia occurred in the denervated carotid body in a ratio similar to that recorded for an innervated carotid body.

Morphometry Studies

As the decrease of dopamine was most pronounced after 30 min of hypoxia, this point of time was selected for the morphometric analysis. The morphometry revealed, for controls, two kinds of type I cells with respect to the size and the amount of their granulated vesicles. These kinds of cells have been designated small vesicle cells and large vesicle cells (7) (Fig. 1). Compared to the controls, the size of the granulated vesicles from the hypoxic tissue was identical; thus the vesicles of the large vesicle cells were about 25% larger than those of the small vesicle cells (Table 2). After hypoxia for 30 min the volume density of the granulated vesicles of the large vesicle cells was reduced about 20%, whereas that of the small vesicle cells was unchanged (Table 2). The mitochondrial volume density was increased for the large vesicle cells as well as for the small vesicle cells, and the volume occupied by the nuclei was slightly reduced for the large vesicle cells (Table 2). When the pictures were examined quantitatively, the mitochondria of the small vesicle cells and the large vesicle cells seemed enlarged and their interior less electron dense than that of the controls. The granulated vesicles were mostly scattered throughout the cytoplasm, but in some cells the granulated vesicles seemed to be preferentially distributed close to the plasma membrane. The number of these latter cells differed between the controls and the hypoxic tissue. In controls, one cell profile of eight exhibited peripherally distributed vesicles, while the ratio in the hypoxic rats was one of six.

Discussion

Our mass fragmentography study showed that hypoxia selectively depletes the rat carotid body content of dopamine. The depletion may be due to either a change in synthesis or degradation of dopamine or to a release of dopamine from the believed dopamine-containing type I cells. An approximate turnover time for carotid body dopamine is more than 90 min (8), so even if the synthesis were completely blocked, it would be unlikely that the dopamine content could decrease by two-thirds in 30 min. Furthermore, with a blocked synthesis the amount of norepinephrine should have decreased. The second interpretation, a depletion due to a release of dopamine from the carotid body cells, seems more reasonable. Dopamine has been shown to have an inhibitory action on the

Table 1. Catecholamine concentrations[a] in carotid body of rats after exposure to hypoxic conditions for various times

Conditions	Length of exposure (min)	Intact		Sinus nerve cut	
		Dopamine	Norepinephrine	Dopamine	Norepinephrine
Controls		30 ± 1.9	7.2 ± 0.25	30 ± 3.3	10.7 ± 0.56
5% O_2	15	20 ± 1.5[b]	7.8 ± 0.82	18 ± 2.2[b]	8.4 ± 0.55
5% O_2	30	9.3 ± 1.9[b]	8.8 ± 1.5	12 ± 1.9[b]	8.1 ± 1.12
5% O_2 + room air	30 + 120	24 ± 1.8[b]	6.8 ± 0.82	30 ± 5.9	9.4 ± 0.88

[a] Catecholamine concentrations are given in pM/pair carotid bodies ± SEM.
[b] $P < 0.05$ when compared with control rats kept in room air.

Table 2. Morphometric data on the type I cells of carotid body during hypoxic and control conditions

Parameter	Small vesicle cells		Large vesicle cells	
	Control	Hypoxia (30 min)	Control	Hypoxia (30 min)
Number of cells	107	87	84	104
Mean diameter of vesicle profiles (nm)	55.9 ± 0.90 (61.5)[a]	55.7 ± 0.84 (61.3)[a]	68.7 ± 1.12 (77.3)[a]	68.7 ± 0.93 (77.3)[a]
Volume densities of mitochondria	8.8 ± 0.38	11.0 ± 0.91[b]	11.5 ± 0.47	14.2 ± 0.67[b]
Vesicles	1.02 ± 0.071 (0.66)[b]	0.89 ± 0.063 (0.58)[b]	1.95 ± 0.129 (1.35)[b]	1.55 ± 0.104[b] (1.08)[b]
Nuclei	45.9 ± 1.60 (42)[d]	42.9 ± 2.40 (39)[d]	39.9 ± 1.40 (35)[d]	34.4 ± 1.34[b] (28)[d]
Mean cell profile areas (μm^2)	30.1 ± 0.97	30.0 ± 1.64	34.0 ± 2.71	31.6 ± 1.14

[a] True mean vesicle diameters calculated according to Froesch (1973).
[b] Significantly different from the control ($P < 0.05$).
[c] Volume densities of the dense-cored vesicles corrected for error due to Holmes effect (Weibel, 1969).
[d] Volume densities of the nuclei compensated for overestimated according to Konwinski and Kozlowski (1972).

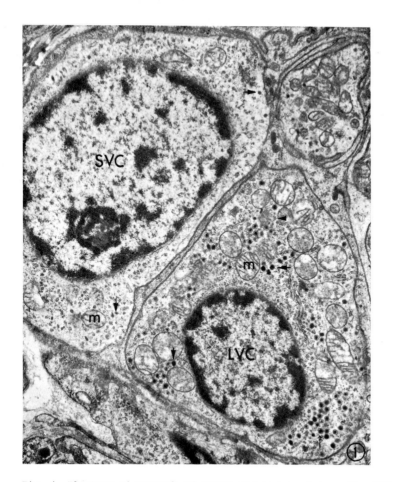

Fig. 1. Electron micrograph of parts of large vesicle cells (LVC) and of small vesicle cells (SVC) in carotid body from rat exposed to hypoxic gas mixture for 30 min. In addition to several granulated vesicles (↓), cells exhibit nuclei and mitochondria (m). Note rounded mitochondria and ion electron density of their interior. 24,000×

chemosensory discharge (13,15). Thus a selective depletion of dopamine during hypoxia may reflect events similar to those that occur in the small intensity fluorescence cells of sympathetic ganglia where dopamine plays a role as a modulator of the neuronal activity (11). Suggesting that dopamine is released from the type I cells during hypoxia leads to the question, what cellular components are involved in the dopamine release? Earlier morphologic studies on the carotid body type I cells during acute hypoxia are rather conflicting. Authors report changes in numbers of granulated vesicles, some an increase in number (1) and others a decrease in number (3). With the morphometric methods used in this study the cytoplasmic volume occupied by granulated vesicles in the large vesicle cells was reduced 20%. Because the size of the vesicle profiles was unchanged, the decrease in volume density reflects a decrease in number of granulated vesicles, which points towards an exocytotic release of the vesicle contents. This could explain in part the dopamine depletion. The mitochondria seemed to be changed during hypoxia. A poor tissue preservation can be excluded, this findings seems to be specific for the hypoxic conditions.

Conclusions

The carotid body content of dopamine is decreased during hypoxia without any change in the norepinephrine level. The change in dopamine content occurs regardless of whether the carotid sinus nerve is intact. There is a decrease in volume density of granulated vesicles of the large vesicle cells during hypoxia. Because the size of the vesicles is unchanged, the decrease reflects a decrease in number of granulated vesicles. We suggest that dopamine release during hypoxia at least partly involves an exocytotic procedure. The released dopamine may act as a modulator of the chemoreceptor discharge similar to the supposed role of dopamine in the small intensity fluorescence cells of the sympathetic ganglia.

References

1. Al-Lami, F., Murray, R.G.: Anat. Rec. *160*, 697-718 (1968)
2. Black, A.M.S., Comroe, J.H., Jacobs, L.: Am. J. Physiol. *223*, 1097-1102 (1972)
3. Blümcke, S., Rode, J., Niedorf, H.R.: Z. Zellforsch. Mikrosk. *80*, 52-77 (1967)
4. Chen, I-Li, Yates, R.D., Duncan, D.: J. Cell Biol. *42*, 804-816 (1969)
5. Chiocchio, S.R., King, M.P., Carballo, L., Angelakos, E.T.: J.: Histochem. Cytochem. *19*, 621-626 (1971)
6. Dearnaley, D.P., Fillenz, M., Woods, R.I.: Proc. R. Soc. Lond. (Biol.) *170*, 195-203 (1968)
7. Hellström, S.: Neurocytol. *4*, 77-86 (1975)
8. Hellström, S., Koslow, S.H.: Acta Physiol. Scand. *93*, 540-547 (1975)
9. Hellström, S., Koslow, S.H.: Brain Res. *102*, 245 (1976)
10. Kobayashi, S.: Arch. Histol. Jpn. *33*, 319-339 (1971)
11. Libet, B., Tosaka, T.: Proc. Natl. Acad. Sci. U.S.A. *67*, 667-673 (1970)
12. Möllman, H., Niemeyer, D.H., Alfes, H., Knoche, H.: Z. Zellforsch. Microsk. Anat. *126*, 104-115 (1972)
13. Sampson, S.R.: Brain Res. *45*, 266-270 (1972)
14. Weibel, E.R.: Int. Rev. Cytol. *26*, 235-302 (1969)
15. Zapata, P.: J. Physiol. *244*, 235-251 (1975)
16. Zapata, P., Hess, A., Bliss, E.L., Eyzaguirre, C.: Brain Res. *14*, 473-496 (1969)

DISCUSSION

Belmonte: Did you control your experiments with sympathectomized animals?

Hellström: No, not during hypoxia, but such studies are in preparation.

Lübbers: What was the P_aO_2?

Hellström: About 40 mmHg.

Lübbers: You measured mitochondrial volume. What is the reference cell volume?

Hellström: The cytoplasmic volume of the type I cell.

Lübbers: For the type I cell you found 10-11% - is this correct?

Hellström: Yes, between 8 and 12%.

Lübbers: Approximately the same value as we reported earlier?

Hellström: Yes.

O'Regan: Both you and McDonald and Mitchell divide the cells into large vesicle cells and small vesicle cells. The vesicle diameter of McDonald and Mitchell is about twice yours. Can you use vesicle diameter as a criterion for dividing the cells at all? Also, your results with norepinephrine are completely different from Mills and Slotkin, who found that with hypoxia they got a decrease in norepinephrine that depended upon whether the sinus nerve was intact.

Hellström: The difference in vesicle size between our results and those of McDonald might result from a different fixative, but still we have the same difference between the smaller and larger types of vesicles when comparing their means - about 25%. Comparing the definite vesicle size, not the mean of a Sect. profiles, gives about the same difference. Regarding the second point, I have no explanation. They are not using the same assay as we are.

Acker: If dopamine is released from the type I cell during hypoxia, what then do you do with the finding of Sampson and others that dopamine inhibits chemoreceptor activity?

Hellström: It might be that dopamine is released to modulate the activity.

Ji: It is not necessary to suppose that, since you observed reduced dopamine vesicles, dopamine is released in hypoxia. Instead, the rate of synthesis of dopamine could be reduced in hypoxia, which, in turn, might reduce the release of dopamine. In normoxia it might be released rapidly.

Hess: Both Hellström and McDonald have two sizes of vesicles in the rat carotid body. Hellström has suggested in a previous study that the vesicles of one kind contained the norepinephrine and the vesicles of the other kind have the dopamine. McDonald's figures indicate that 50% of the cells are the small vesicle kind and 50% are the large vesicle kind. This would mean that half the cells have norepinephrine as a transmitter and the other half have dopamine, which does not work out in terms of the content of norepinephrine compared to dopmaine in the carotid body.

Hellström: I do not want to stress these two types of cells because we do not know if there is any functional significance between them. But there is more than double the amount of vesicles in the large vesicle cell type compared to the small vesicle type, and we do not know anything about the concentration of the amine in each type of granule.

Kobayashi: We have some evidence that in the adrenal medulla the epinephrine-secreting cell also contains norepinephrine, dopamine, and dopa. An analogy might be possible with the carotid body.

Hess: The norandrenergic granules in the adrenal medulla of the rat do not look at all the noradrenergic or dopaminergic, or whatever the granules are in the carotid body, so it seems very difficult to compare sizes of granules using the adrenal medulla.

Belmonte: Since you have not cut the sympathetic nerves in your preparation, I do not think you can say there is a specific modulation of dopamine. Perhaps there is just a general sympathetic response that affects those cells and they liberate dopamine.

Torrance: Why do you say dopamine is discharged and norepinephrine is conserved, rather than norepinephrine is discharged and dopamine is used up to resynthesize or replace the discharged norepinephrine?

Hellström: We know that the turnover for this amine is at least 90 min, so I do not think that such a large discrepancy between the amount of dopamine and norepinephrine present could explain the large decrease in dopamine.

Torrance: Are these resting turnover times or hypoxic turnover times? They might be very different.

Hellström: Yes, they might be, but it is known from studies of brain neurons that the activity of the enzyme in synthesizing the catecholamines is only very slightly changed during hypoxia.

Hanbauer: I have measured dopamine beta-hydroxylase at various times after hypoxia and could not find any increase in the activity by either activation or, at a later time point, increased synthesis of the enzyme. An increased turnover of the neurotransmitter should result in an activation of the enzyme.

Torrance: I would have expected a steady-state with some activity of the enzyme. If there is hypoxia and the norepinephrine is depleted, the activity of the enzyme would, by simple mass action, cause an increased rate of turnover of substances without any new enzyme formation necessary.

Zapata: Perhaps Hellström's interpretation is more plausible since several investigators have shown that dopamine easily produces chemosensory inhibition. But on the other hand, Sampson's report that norepinephrine also reduces chemosensory discharges has not been repeated by others. I tried to demonstrate chemosensory inhibition induced by norepinephrine, but was unable in even a single experiment to show this.

Fidone: I think perhaps the data are at hand to resolve Torrance's question. Is the dopamine beta-hydroxylase activity, which is not all that markedly high, sufficient to convert that amount of dopamine to convert that amount of norepinephrine in that period of time? I think the data already presented here could answer this question.

Loss of Histochemically Demonstrable Catecholamines in the Glomus Cells of the Carotid Body After α-Methylparatyrosine Treatment

M. Grönblad and O. Korkala

Summary

A statistically significant decrease in the intensity of catecholamine fluorescence of some carotid body glomus cells was observed after inhibition of the enzyme tyrosine hydroxylase by injection of 80 mg/kg α-methylparatyrosine. The intensity of the formaldehyde-induced fluorescene was measured in individual glomus cells. The maximum decrease in the intensity was observed 4-6 h after the α-methylparatyrosine injection. This suggests a rapid turnover in the catecholamines of carotid body.

Introduction

The inhibition of tyrosine hydroxylase is the rate-limiting step of the catecholamine biosynthesis (8). The use of the tyrosine hydroxylase inhibitor α-methylparatyrosine (α-MPT) as a tool for studies on catecholamine depletion rates and turnover in the peripheral and central nervous systems have been widely accepted (10,13). The catecholamine-storing (glomus) cells of the carotid body probably act as modulating, inhibitory interneurons in the chemosensory reaction (7,9,12). The possibility of humoral release of catecholamines by the glomus cells is, however, in no way excluded.

The aim of this study is to examine the effect of α-MTP on the catecholamine concentration of individual glomus cells of the rat carotid body. The intensity of the formaldehyde-induced fluorescence (FIF) reaction for catecholamines (3) was measured photometrically.

Material and Methods

Thirteen adult male rats of the Sprague-Dawley strain, each weighing 150-200 g, were used for the experiments. Four rats were kept as uninjected controls. The rest were given a single intraperitoneal injection of 80 mg/kg α-MPT (F. Hoffmann-La Roche & Co. AG Basel). The rats were killed 1, 2, 4, 6, 8, and 10 h, respectively, after the injection, together with the untreated control rats. The entire carotid bifurcation area was removed bilaterally. The tissue blocks were immediately frozen in propane cooled with liquid nitrogen and dried for at least 7 days at -40°C in vacuum with a phosphorus trap close to the tissue holder.

For the histochemical demonstration of catecholamines, the FIF method was used (3). The tissue specimens were then embedded in Epon and cut at 5 μm with a Pyramitome (LKB). For fluorescence microscopy a Leitz Ortholux microscope equipped with the Ploem (11) epi-illuminator was used. The filter combination employed was BG 38, BG 3, TAL 408, and K 470. As a light source an Osram HBO 200 mercury lamp was used.

From Sects. cut out of the middle of the carotid bodies of treated and untreated rats, the fluorescence intensity of individual glomus cells was estimated visually and photometrically. The photometric measurement was done in an area of 8.3 µm diameter using the linear cell sampling method, as described by Costa and Eränkö (2). In order to avoid subjective selection the measurement was done from the cytoplasm of those glomus cells that came into view when the stage of the microscope was moved along the longitudinal axis of the organ. The cytoplasmic intensity of 100 glomus cells was measured in each carotid body. Fluorescence intensity was expressed in intensity units (IU) obtained by dividing 100 by the exposure time.

The fluorescent cells were grouped into four subclasses according to their photometrically estimated FIF intensity: intensely fluorescent (30-50 IU), moderately fluorescent (20-30 IU), weakly fluorescent (10-20 IU), and nonfluorescent (less than 10 IU) (Table 1). The nonspecific background fluorescence photometrically estimated in 20 Sects. was 5-8 IU and photometric values below 10 IU were therefore judged to indicate lack of fluorescence.

Table 1. Intensity of formaldehyde-induced fluorescence in the glomus cells of the carotid body of adult rats

Time after injection of α-MTP	Fluorescence intensity[a]			
	10 IU	10-20 IU	20-30 IU	30-50 IU
Controls	8	61	24	7
1 h	3	44	36	17
2 h	41	51	8	0
4 h	75	23	2	0
6 h	75	23	2	0
8 h	33	55	9	3
10 h	2	39	46	13

[a] Results are expressed as a percent of the total number of glomus cells.

Results

One hour after the α-MPT treatment, no change was observed in the glomus cell fluorescence, and the specimens were comparable to the control ones (Table 1). After 2 h, a clear tendency toward a decrease in the FIF of the glomus cells was observed, although the change was not statistically significant (Table 1) Four hours after the treatment, the decrease in FIF intensity was clear, and the number of intensely or moderately fluorescent cells decreased and that of the weakly fluorescent and nonfluorescent cells increased. No further change was observed between 4 and 6 h after the α-MTP treatment (Table 1 and Fig. 1). Eight hours after the treatment, a shift toward normal had started (Table 1 and Fig. 1), and after 10 h the ditribution of the FIF intensity was similar to that of the controls (Table 1 and Fig. 1).

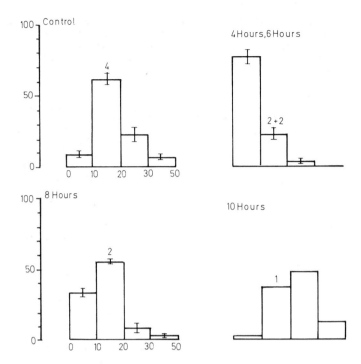

Fig. 1. Distribution of fluorescence intensity (expressed in IU) of 100 glomus cells in adult rat carotid body in controls and in rats 4-6, 8, and 10 h after α-MPT treatment. Number above columns indicates number of animals in each group; vertical bars indicate ± SEM

The parenchymal *adrenergic nerves* were seen to lose their FIF almost totally, 4-6 h after the treatment. After 8 h, however, some clear fluorescent nerve profiles were seen. The *mast cells*, often observed as yellow granular cells between the glomus cell collections, remained unchanged after the α-MPT treatment.

Statistical Analysis

The difference between the photometric values of the controls and those of the experiments (carotid bodies of rats 4 or 6 h after the drug treatment) was statistically analyzed. Since the specimens 4 or 6 h after the treatment were identical, they were analyzed here simultaneously. The test of Kolmogorov and Smirnov was used in the comparison of two empirical distributions, which were orderly (4). The calculated D is 0.67. Theoretical distribution of Qn gives the random values of the variable D_n n ($n = 4$). The probability corresponding to D_n $n = 1.34$ is $0.9449 > 1-α$. Thus the change between the controls and the experiments (4-6 h after the treatment) is highly significant.

Discussion

A pair of adult carotid bodies contains considerable amounts of three biogenic amines: dopamine (an average of 28.9 pM), norepinephrine (an average of 16.9 pM), and serotonin (an average of 7.5 pM), while epinephrine is missing (6). The glomus cells are the sites for dopamine and norepinephrine storage (5), while the interlobular mast cells probably serve as serotonin storage sites (1). The glomus cells can be divided by electron microscopy into large and small granule-containing cells, the former probably storing dopamine and the latter, norepinephrine (5,9).

The tendency of the glomus cells to react differently to the catecholamine-depleting action of α-MPT, some cells remaining even moderately fluorescent, while many of them lost the FIF totally, may depend on the amine contained in the individual cells. For example, the dopamine-containing glomus cells might be more readily depleted of their dopamine content, while the norepinephrine-containing cells would be more resistant to the catecholamine-depleting action of α-MPT. The unchanged fluorescence of the mast cells after the α-MPT treatment is not surprising, since α-MPT does not affect serotonin synthesis (14). As a whole the carotid body catecholamines were depleted relatively quickly after the α-MPT treatment. α-MPT was shown in an earlier study to reduce the FIF intensity in small intensely fluorescent (SIF) cells of the rat superior cervical ganglion 18 h after a 500 mg/kg injection (15). In the perikarya of adrenergic neurons in the superior cervical ganglion, no fluorescence could be detected 12 h after the treatment, and a reduction could already be seen after 4 h (15). Thus, the turnover of catecholamines in the carotid body glomus cells could be more rapid than that of SIF cells in the superior cervical ganglion or the catecholamine stored in them more accessible to synthesis inhibition. Electron-microscopic studies of the effect of α-MPT on the catecholamine-storing cells of the carotid body have shown no significant differences in the number of granular vesicles after α-MPT but did show increased variation in their electron density (Grönblad and Korkala, unpublished observations).

References

1. Böck, P.: Z. Mikrosk. Anat. Forsch. *82*, 461-476 (1970)
2. Costa, M., Eränkö, O.: Histochem. J. *6*, 35-52 (1974)
3. Eränkö, O.: J. R. Microsc. Soc. *87*, 259-276 (1967)
4. Fisz, M.: in: Wiley Publications in Statistics. New York: John Wiley 1963, p. 394
5. Hellström, S.: J. Neurocytol. *4*, 77-86 (1975)
6. Hellström, S., Koslow, S.: Brain Res. *102*, 245-255 (1976)
7. Korkala, O., Waris, T.: Cell Tissue Res. *158*, 355-362 (1975)
8. Levitt, M., Spector, S., Sjoerdsma, A., Udenfriend, S.: J. Pharmacol. Exp. Ther. *148*, 1-8 (1965)
9. McDonald, D., Mitchell, R.A.: J. Neurocytol. *4*, 177-230 (1975)
10. Moore, K., Dominic, J.: Fed. Proc. *30*, 859-870 (1971)
11. Ploem, J.S.: Prog. Brain Res. *34*, 27-38 (1971)
12. Sampson, S.R.: Fed. Proc. *31*, 1383-1384 (1972)
13. Sharman, D.F.: Med. Bull. *29*, 110-119 (1973)
14. Spector, S., Sjoerdsma, A., Udenfriend, S.: J. Pharmacol. Exp. Ther. *147*, 86 (1965)
15. Van Orden, L.S.III, Burke, J.P., Geyer, M., Lodoen, F.V.: J. Pharmacol. Exp. Ther. *174*, 56-71 (1970)

DISCUSSION

McDonald: When you were reviewing some of your earlier work, you commented on the fact that hemorrhage led to an increased number of large dense-cored vesicles next to the limiting membrane of glomus cells. Was this due to hypotension? Why did you use hemorrhage and what did you conclude from the morphologic changes you found?

Korkala: I referred to the work of Landgren and Neil, who did hemorrhage experiments and received a very much increased output of the chemosensory nervous drive. What we found was a statistically increased number of contacts between the dense-cored vesicles and the nerve ending membrane, so that the dense-cored vesicles seemed to be going toward the synaptic specializations. But I do not know how specific this is, whether it is hypovolemia or hypoxia.

McDonald: Is it correct that in the control state in the rat carotid body some glomus cells are not fluorescent or are only weakly fluorescent?

Korkala: There are cells that are surely type I cells that do not fluoresce at all. They are a minor group, but such cells do exist.

McDonald: Is this true in other species as well as the rat? Did you find the same variability in fluorescence intensity in glomus cells in the human fetus?

Korkala: As you could see in the postnatal development pictures, in the new born rat, as in the human fetus, the fluorescent intensity is rather marked - rather homogenous too. But after the postnatal development the intensity of the type I cell fluorescence goes down in many cells; some cells exist that do not fluoresce at all.

Eyzaguirre: Does the fact that some cells do not fluoresce mean that those cells are different in terms of their content of catecholamines or is it that the method is not sensitive enough to detect only a small amount?

Korkala: The method is sensitive enough to detect the varicose nerves. Perhaps it would be possible to show catecholamines in these nonfluorescent cells with some sensitive methods. But in electron-microscopic studies I sometimes found type I cells with no dense-cored vesicles at all, but this was rather rare in the adult rats.

Zapata: What is your criterion to identify those cells without dense-cored vesicles as type I cells?

Korkala: The structure of the cell in general - it is round and in most cases ovoid, although occasionally, it has long cytoplasmic processes, and the nucleus looks quite different from that of the type II cell. Mitochondria are typicall, and ribosomal rosettes are also quite typical and help to differentiate between the two cell types. I did not find any difficulty in that.

Lübbers: May I ask Böck if he has made similar observations in the carotid body?

Böck: Yes, we have studied carotid bodies of guinea pigs, rats, mice, and cats, and we found differences in the fluorescene intensity within various type I cells, some being, one could say, nonfluorescent. But I believe that if the precursor of the catecholamine were injected into the animal, we would see that all cells of the carotid body take up the precursor and decarboxylate it. Then under these conditions, all cells would be fluorescent.

Lübbers: Maybe it depends on the level of the Sect.. I am really a little afraid to categorize cells in the carotid body without any catecholamine vesicles.

Böck: With the electron microscope I have never found type I cells without vesicles, and I agree with Zapata that it would be very hard to identify the cells, for the dense-cored vesicles are the marker for us. But I agree with you concerning the fluorescence. Often it is seen that the specific fluorescence is strong at the cell periphery or in small cell processes and relatively low in the perinuclear regions. Therefore, the level of the Sect. is an important consideration.

McDonald: What is the effect of the dose of α-methylparatyrosine on blood pressure in these rats?

Korkala: I did not examine that.

McDonald: Did it deplete the catecholamines from the vasomotor noradrenergic nerve endings?

Korkala: I have only examined the carotid body, but it seems that in this tissue parenchymal noradrenergic nerves also lost their fluorescence intensity after treatment.

McDonald: You suggested that hemorrhage and perhaps hypotension leads to exocytosis from the glomus cells. If there is hypotension associated with the administration of this drug, could part of the effect be secondary rather than primary, which directly influences the catecholamine metabolism of the cell?

Korkala: Yes, I agree that is possible.

Enzymes and Inhibitors of the Catecholamine Metabolism in the Cat Carotid Body

H. Starlinger

Enzyme activities of catecholamine metabolism in the carotid body of the cat were determined. First, experiments on monoamine oxidase (MAO) and second, experiments on dopamine beta-hydroxylase (DBH) are described.

The starting point of these experiments was the postulation by Acker and Lübbers (1) of a very strong oxidase activity in the cat carotid body, which, by its characteristics, had to be different from the enzyme system of mitochondrial respiration. The substrate for this oxidase could not be determined by measurements on oxygen consumption, and therefore biochemical experiments were carried out, which did not show a strong respiratory activity of carotid tissue homogenates (13). Thus, a search for specific oxidases was started. Of the low molecular weight compounds in the carotid body of the cat, the catecholamines are important fraction. Moreover, these molecules have been implicated in the chemoreceptor function of this organ (7,8,10,16). Therefore, it seemed of additional interest to look into the properties of enzymes that oxidize these compounds or their intermediates. MAO is an enzyme that oxidizes dopamine to aldehydes. DBH is the enzyme that oxidizes dopamine to norepinephrine. The main finding reported here is the absence of both MAO and DBH activity in the carotid body tissue of the cat. This lack of activity probably does not result from the absence of the enzyme molecule itself, but from the presence of potent inhibitors of enzyme activity.

Monoamine Oxidase

Experiments were performed to show that the failure to detect MAO activity was not a result of insufficient assay conditions. Since the amount of carotid body tissue of one cat is approximately 1 mg, a micromethod had to be used. The method of Roth and Stjärne (11) was adapted to the microliter range and was calibrated with rat liver mitochondria. Fig. 1 shows that in 5 µg of mitochondrial protein, the enzyme is easily detected. This corresponds to a fresh weight of 70 µg of liver tissue. In some experiments with carotid tissue homogenate, 1.8 mg of carotid body tissue were used. The enzyme should have been detected, even if its activity has been lower than in liver tissue by a factor of 25, but no measurable activity was observed.

Thus, lack of sensitivity was no explanation for the failure to detect MAO. Complete absence of the enzyme is also not very probable, because MAO is a marker enzyme for the outer mitochondrial membrane, and mitochondria are present in the cells of the carotid body. Therefore, the presence of an enzyme inhibitor was considered. An inhibitor is indeed detected in mixing experiments in which carotid body homogenate is incubated together with rat liver mitochondria (Table 1). When analogous experiments were performed with tissue of the superior cervical ganglion, no inhibitory effect was observed. Attempts to characterize the inhibitor are still preliminary, because the amounts of

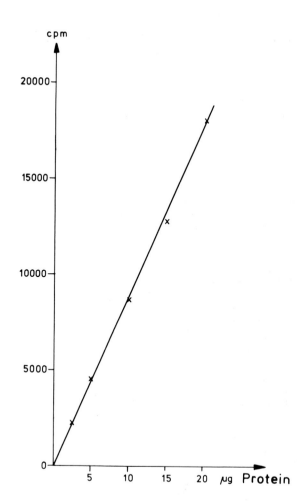

Fig. 1. Monoamine oxidase activity in rat liver mitochondria (37°C for 1 h); 1000 cpm corresponds to 120 pM of oxidized dopamine

tissue available are very small. The inhibitor is destroyed in 5 min at 100°C. Fractionation experiments have not yet yielded clear-cut results but, apparently the inhibitor is not firmly bound to coarse particles (Table 1).

We now have indications that a similar inhibitor may be present in the adrenal medulla of the cat, and since we can get larger amounts of this tissue, we will concentrate our future efforts on this inhibitor. We hope that its characterization will be helpful in the understanding of the physiology of the carotid body. If catecholamines play a role in the chemoreceptor function of this organ, their level must be regulated, and it is not unlikely that molecules influencing the activity of enzymes of catecholamine metabolism have a function in the regulatory processes.

Dopamine Beta-Hydroxylase

DBH was assayed by incubation with {^{14}C}dopamine as a substrate and separation of the norepinephrine formed from dopamine on silica gel thin-layer plates by a modification of the method of Seiler and Wiechmann (12). The details of this method are described in the legend

Table 1. Mixing experiments with rat liver mitochondria and carotid body tissue (37°C for 1 h)

Experimental conditions	n^a	Mean MAO activity (cpm)
Rat liver mitochondria (7 µg protein) without addition	21	5361
Rat liver mitochondria + cat carotid body homogenate (1-2 carotid bodies = 0.6 mg)	21	3302
Rat liver mitochondria (11 µg protein) without addition	5	7581
Rat liver mitochondria + cat carotid body sediment after centrifugation at 39,000 × g (min) (1 carotid body = 0.7 mg)	5	3230
Rat liver mitochondria + cat carotid body supernatant after centrifugation at 39,000 × g (min) (1 carotid body = 0.7 mg)	5	5500
Rat liver mitochondria (9 µg protein) without addition	5	5192
Rat liver mitochondria + cat carotid body sediment after centrifugation at 500 × g (min) (2 carotid bodies = 0.9 mg)	5	4867
Rat liver mitochondria + cat carotid body sediment after centrifugation at at 27,000 × g (min)	5	3818
Rat liver mitochondria + cat carotid body supernatant after centrifugation at 27,000 × g (min) (2 carotid bodies = 0.9 mg)	5	2924

[a] n = number of experiments

to Fig. 2, which shows that the amount of norepinephrine formed in a given time increases linearly with the amount of protein added. We detected 200 pM of norepinephrine. This sensitivity was sufficient to detect DBH activity in the particulate fraction from bovine adrenal medulla corresponding to 45 µg of fresh tissue. The enzyme could also be detected in cat adrenal medulla and in bovine carotid body*

From Table 2 it can be seen that under the same conditions, in general, cat tissues have lower activities than bovine tissues, and that the bovine carotid body has a lower activity than the bovine adrenal medulla. From these observations we would expect an activity in the cat carotid body lower than 0.05 nM. An activity of this magnitude is below the level of detection by our method, and thus the lack of DBH activity in cat carotid tissue observed by us is not surprising.

---

*The latter experiment could not be performed in Dortmund, since it is difficult to get the fresh material from the slaughterhouse. It was performed once thanks to the cooperation of Prof. Thorn and his laboratory in Hamburg, who provided the material. I want to thank Mr. Bartels, especially.

Table 2. Activity of dopamine beta-hydroxylase (37°C for 1 h)

Tissue	Beef Norepinephrine formed (nM/4 mg-h)	Cat Norepinephrine formed (nM/4 mg-h)
Adrenal medulla	32.0	4.0
Carotid body	0.4	<0.2

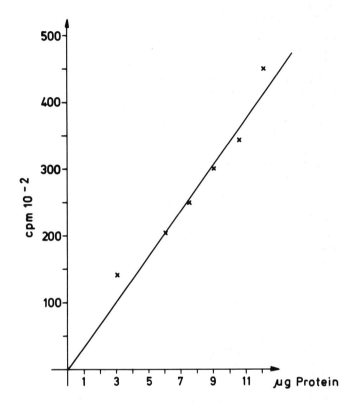

Fig. 2. *Dopamine beta-hydroxylase activity*. 5-20 µl of beef adrenal medulla particulate fraction (prepared by the method of Levin et al. (6)), adjusted with buffer to 20 µl, was added to 50 µl reaction mixture, whose composition was as follows: ascorbate 4mM, fumarate 4mM, ATP 5mM, pargyline 0.5mM, $CuSO_4$ 10µM, $MgCl_2$ 1.2mM, catalase 670 mg/liter, phosphate buffer (pH 6.0) 100mM, ^{14}C-labeled dopamine 1.4mM (10 µCi/µM). After incubation (25 min, 37°C under 95% O_2 + 5% CO_2), catecholamines were reacted with dansyl chloride and separated on thin-layer silica gel plates with mixture of ethylacetate and cyclohexane (75 : 50). Spots were visualized under UV light, and scraped off the plate, and radioactivity of norepinephrine spot was measured in scintillation counter. 1000 cpm corresponds to 260 pM norepinephrine

The reason for this lack of activity is probably the presence of inhibitors. Such inhibitors have been described for DBH in other organs and species by several authors (3,4,9,14,15). The presence of inhibitors in the cat adrenal medulla is also indicated by our observation of a total lack of DBH activity, when total tissue homogenates were used instead of the particulate fraction. This is interpreted to mean that an inhibitor is (partially) removed during the preparation of the particulate fraction. That this removal is not complete is indicated by the stimulation of DBH activity by Cu^{2+} in the particulate fraction, because Cu^{2+} ions have been found to inactivate endogenous inhibitors of DBH activity (3,5,9,14,15).

The mixing experiments, with the carotid body and the particles prepared from the adrenal medulla and performed to detect the presence of the putative inhibitor, showed a clear inhibition in only eight out of twelve experiments. This will need further investigation. Since several authors have shown that the conditions for the inactivation of endogenous DBH inhibitors vary from tissue to tissue (9,10,14,15), it is still possible that under appropriate conditions the inhibition can be relieved. We hope to continue these studies, to characterize the inhibition, if possible, and to see whether this inhibitor plays a role in the regulation of norepinephrine formation.

References

1. Acker, H., Lübbers, D.W.: in: The Peripheral Arterial Chemoreceptors. Purves, M.J. (ed.). New York: Cambridge U. Pr. 1975, pp. 325-343
2. Belpaire, F., Laduron, P.: Biochem Pharmac. *19*, 1323-1331 (1970)
3. Duch, D.S., Viveros, O.H., Kirshner, N.: Biochem. Pharmacol. *17*, 255-264 (1968)
4. Goldstein, M., Freedman, L.S., Bonnay, M.: Experientia *27*, 632-633 (1971)
5. Hamprecht, B., Traber, J., Lamprecht, F.: FEBS Lett. *42*, 221-226 (1974)
6. Levin, E.Y., Levenberg, B., Kaufman, S.: J. Biol. Chem. *235*, 2080-2086 (1960)
7. Mills, E., Slotkin, Th.A.: Life Sciences *16*, 1555-1562 (1975)
8. Mills, E., Slotkin, Th.A.: Nature *258*, 268 (1975)
9. Molinoff, P.B., Brimijoin, St. Weinshilboum, R., Axelrod, J.: Proc. Nat. Acad. Sci. USA *66*, 453-458 (1970)
10. Osborne, M.P., Butler, P.J.: Nature *254*, 701-703 (1975)
11. Roth, R.H., Stjärne, L.: Acta Physiol. Scand. *68*, 342-346 (1966)
12. Seiler, N., Wiechmann, M.: Experientia *21*, 203-204 (1964)
13. Starlinger, H., Lübbers, D.W.: Pflügers Arch. *366*, 61-66 (1976)
14. Viveros, O.H., Arqueros, L., Connett, R.J., Kirshner, N.: Mol. Pharmacol. *5*, 60-68 (1969)
15. Weinshilboum, R., Axelrod, J.: Circul. Res. *28*, 307-315 (1971)
16. Zapata, P.: J. Physiol. *244*, 235-251 (1975)

DISCUSSION

Belmonte: I think we agree that probably the problem in your case is the method. I think your method, even if you have the inhibitors well controlled, has a sensitivity that is too low for the amounts of DBH that Hanbauer, Zapata, and I found. We were working in the range of picomoles.

Starlinger: No, we can estimate 200 pM.

Belmonte: Then you think the difference is probably due to inhibitors only?

Starlinger: Yes, the method is sensitive enough to estimate your amount. I shall try with your conditions and no doubt will get the same result. Your preparations also were different from ours. We estimated the enzyme in the particles, and you estimated it after treatment with Tritin X-100 and saponate.

Belmonte: Yes, that is true.

Hanbauer: Dis you assay separately the enzyme preparation on Sephadex or cellulose columns to see whether you could separate the inhibitor in case it is a smaller protein, or have you tried dialysis of the extract?

Starlinger: No.

Lübbers: It may be that the inhibitors play an important role in the regulation of the catecholamine metabolism. But on the other hand, is there a special reason to use cupric ion to get the information about the amount of this enzyme? If I understood the other papers correctly, it is rather easy to find whether they inhibit a certain amount of this enzyme. This is still not necessarily the correct amount of enzyme, because as we have seen, it depends on the use of the inhibitor. So my question is, was there a special reason to use the cupric ion and not the other assays, because cupric ion seems much more difficult to work with?

Starlinger: We used cupric ion because it is in the enzyme itself and plays a role in the reaction, and Kirshner et al. and others use cupric ion. In fact, they all found that cupric ion, even in the same tissue, is different with different fractions.

Lübbers: Yes, but if it is known that the other inhibitor gives a higher concentration of DBH, it would be useful to try the other inhibitor.

Starlinger: Yes. I think it is of interest whether there is an inhibitor in vivo and what the role of this inhibitor is.

Zapata: We should note that exactly the amount of DBH that you expected, according to your calculation, was the amount we found in our determination in the cat carotid body, namely, 50 pM/mg-h. Now, in terms of the inhibitors our experience is different. We used this modified Kirshner and Viveros method (using {^{3}H}tyramine and measuring {^{3}H}octopamine) in which copper is replaced by tramylcyclamine and by parahydroxy mercuric benzoate. We found that by using these inhibitors, there were big changes when determining the plasma DBH, but no changes for the amount of DBH in the carotid body. So regardless of whether tramylcyclamine was in the medium, the values were the same for the carotid body. We were disturbed that in this tissue the presence of inhibitors was apparently not important.

Starlinger: I shall try the two methods.

Ji: Have you tried to determine the effect of ascorbic acid on DBH activity?

Starlinger: Yes, but there was no measurable effect on DBH activity in the carotid body tissue.

Session IV
The Afferent and Efferent Chemoreceptive Pathway of the Carotid Body

A Pharmacologic Study on a Possible Inhibitory Role of Dopamine in the Cat Carotid Body Chemoreceptor

K. Nishi*

Catecholamines, mostly dopmaine, are normal constituents of the carotid body of different mammalian species (4,5,7,9,10,11,17). Recent studies on the carotid body chemoreceptor have shown that exogenously applied dopamine induces a marked depression on chemoreceptor discharge frequency (2,15). Consequently, it has been proposed by Zapata (16) that dopamine contained in the carotid body would act as a modulator substance of chemoreceptor activity and would be a part of local regulatory mechanisms. If this is the case in the cat carotid body, drugs affecting dopaminergic receptors in the central nervous system or other organs would also modify the chemoreceptor activity. Therefore, effects of dopaminergic receptor blockking agents (haloperidol and trifluperidol) and a dopaminergic receptor stimulating agent (apomorphine) (8) on cat receptor responses to various stimuli were examined in vivo. Results obtained in the present experiments have provided further evidence for the presence of dopaminergic receptors in the carotid body chemoreceptor, and suggested a possible role of an inhibitory transmitter substance of dopamine in the chemoreceptor function.

The methods employed were essentially similar to those described in previous papers (12,13). Chemoreceptor discharges were recorded in vivo from fine nerve filaments containing a single or a few active units in the carotid sinus nerve of the cat, and their frequency was used as an index of chemoreceptor activity. The animal anesthetized with sodium pentobarbitone was artificially ventilated by a positive respirator. Both the vagosympathetic nerve trunks and sympathetic nerve going to the carotid sinus region were severed to avoid possible efferent effects. Chemical substances dissolved in 0.25 or 0.5 ml of Tyrode's solution at pH 7.4 were injected into the carotid artery through a fine muscular branch of the artery.

Effects of Dopamine on Chemoreceptor Discharges

Chemoreceptor discharges were identified according to the criteria described in previous papers (12,13). Effects of intra-arterial injections of dopamine showed considerable variation, depending upon the dose injected, the rate of the injection, or the individual animals. However, the most frequently observed effect of the agent was simple inhibition of spontaneously occurring chemoreceptor discharges; the minimum doses required to induce the inhibitory effect was 0.2-0.5 μg. In some instances, when small doses of dopamine (0.2-0.5 μg i.a.) were

---

*The author expressed his sincere gratitude to Dr. F. Takenaka, Department of Pharmacology, Kumamoto University Medical School, Japan for his valuable advise, and to Mr. K. Iwasaki for his technical assistance.

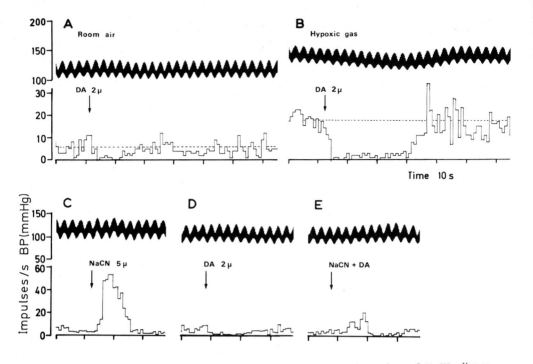

Fig. 1. Effects of dopamine on chemoreceptor responses to hypoxia and NaCN. *Upper traces*: arterial blood pressure changes. *Lower traces*: discharge frequency in a few active chemoreceptor units. (A and B) from the same animal; (C-E) from a different animal. (A) Animal was ventilated with room air. Dopamine (DA) (2 μg i.a. in 0.25 ml Tyrode's solution) injected (*arrow*). An interrupted line indicates level of average discharge frequency. (B) Animal was ventilated with hypoxic gas mixture (90% N_2 + 10% O_2). Dopamine (2 μg) injected (*arrow*). (C) Chemoreceptor response to NaCN (5 μg i.a. in 0.25 ml Tyrode's solution) (*arrow*). (D) Dopamine (2 μg i.a. in 0.25 ml Tyrode's solution) injected (*arrow*). (E) NaCN (5 μg) and dopamine (2 μg) in 0.5 ml Tyrode's solution injected simultaneously (*arrow*)

injected, an initial brief increase in chemoreceptor discharges, followed by inhibition of the discharges, occurred. Upon increasing the dose applied, the excitatory effect of the agent was not observed, chemoreceptor discharges being markedly depressed. The effect of dopamine on chemoreceptor discharges, whether there was initial excitation or inhibition, occurred within 1 s after the start of the intra-arterial injection and lasted for a short period (10-30 s). Fig. 1 illustrates effects of dopamine on chemoreceptor activity under various conditions. During the control period an intra-arterial injection of dopamine (2 μg) induced a decrease in chemoreceptor discharges (Fig. 1A). When the animal was ventilated with a hypoxic gas mixture, the resting rate of discharge frequency increased, and marked chemoreceptor inhibition was induced by the same dose of dopamine (2 μg) as that used in the control period. The chemoreceptor response to chemical substances such as NaCN or nicotine was also markedly depressed by dopamine. An intra-arterial injection of NaCN (5 μg) induced an increase in chemoreceptor discharges (Fig. 1C). This excitatory action of NaCN on chemoreceptor activity was depressed by dopamine added to the NaCN solution (Fig. 1E). Thus, exogenously applied dopamine exerts the main inhibitory effect on chemoreceptor activity. The results were consistent with those obtained in vitro by Zapata (16).

Effects of Dopaminergic Receptor Blocking Agents

Because part of the vascular effects of dopamine is mediated through the alpha-adrenergic receptor sites (11), effects of various alpha-adrenergic receptor blocking agents on chemoreceptor responses to dopamine were examined. In most instances pretreatment with phenoxybenzamine (1-5 mg/kg i.v.), phentolamine (0.5-1 mg/kg i.v.), or hydroxyergotamine (1 mg/kg i.v.) did not affect the dopamine-induced chemoreceptor depression. The results indicate that the effects of dopamine on chemoreceptor activity are not mediated through alpha-adrenergic receptor sites in the carotid body receptor.

Specific dopaminergic receptor blocking agents in the central nervous system, haloperidol and trifluperidol, were employed to see whether chemoreceptor responses to dopamine and various stimuli could be modified by these agents. Intravenous injections of haloperidol (1-5 mg/kg) or trifluperidol (1-5 mg/kg) induced a marked increase in chemoreceptor discharges and a slight fall in arterial blood pressure. Chemoreceptor discharge frequency started to increase after a latent period of 10-20 s after the start of the injection, and the excitatory effects of the agent on chemoreceptor activity lasted for 30-40 min, the discharge frequency gradually returning to the control level. A fall in arterial blood pressure following the intraveous injections of these agents lasted for 20-30 min.

Effects of haloperidol and trifluperidol on chemoreceptor responses to dopamine were examined. In most cases haloperidol (1-5 mg/kg i.v.) completely abolished or depressed the inhibitory effect of dopamine on chemoreceptor discharges and converted the response to an excitatory one; after the agent, dopamine induced an increase in chemoreceptor discharges and the number of chemoreceptor discharges increased proportionally to the dose of dopamine injected. Trifluperidol (1-5 mg/kg i.v.) also blocked the dopamine-induced chemoreceptor depression. Fig. 2 shows chemoreceptor responses to dopamine before and after haloperidol. The response pattern to dopamine after haloperidol was different from that to acetylcholine (ACh) or 5-hydroxytryptamine (HT); chemoreceptor discharges started to increase with a latent period of 1-2 s after the start of the intra-arterial injection, attained their peak frequency within 3-5 s, and gradually returned to the control level.

After the administration of dopaminergic receptor blocking agents, chemoreceptor responses to various stimuli were studied. Chemoreceptor responses to NaCN, ACh, and hypoxia were augmented in some degree. However, the augmentation of the responses was not so marked and consistent. The dose-responses curves to NaCN obtained from a few active chemoreceptor fibers shifted toward the left after haloperidol in all cases examined, whereas in some cases, the response to ACh was not affected and the responses to hypoxia also showed considerable variation in individual chemoreceptor fibers. In some fiber responses to hypoxia were augmented after haloperidol, but in others the magnitude of the chemoreceptor response after haloperidol was similar to that in the control period.

Effects of Dopaminergic Receptor Stimulating Agent

Apomorphine is an effective dopaminergic receptor stimulating agent in the central nervous system (1,3,6). Because of its ability to stimulate dopaminergic receptors, the effects of this agent on chemoreceptor discharge were examined. Either intra-arterial injections of apomorphine (50-100 µg) or intravenous injections (0.2-0.4 mg/kg)

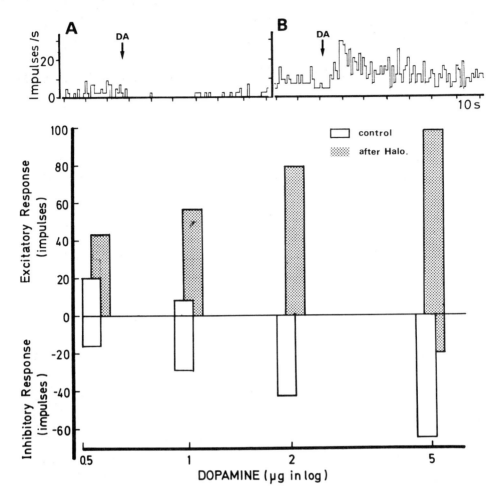

Fig. 2. Effects of haloperidol on chemoreceptor responses to dopamine. Effects of dopamine (2 μg i.a. in 0.25 ml Tyrode's solution) injected during the control period (A) and after haloperidol (2 mg/kg i.v.) (B). *Lower graph*: dose-response curves for different doses of dopamine before and after haloperidol (2 mg/kg i.v.) obtained from chemoreceptor discharges in a few active units of a different animal. Total number of chemoreceptor discharges occurring after injections of dopamine was counted and expressed as an excitatory response, while the total number of discharges depressed from the level of the resting frequency was counted and expressed as an inhibitory response

induced a marked decrease in chemoreceptor discharge frequency, and this inhibitory effect lasted for about 30 min. Fig. 3 shows a typical example of effects of apomorphine. Chemoreceptor responses to NaCN and hypoxia were markedly depressed, whereas the response to dimethylphenylpiperazinium (DMPP), which is a potent nicotinic receptor stimulating agent, was not affected by pretreatment with apomorphine. The inhibitory effect of apomorphine on chemoreceptor responses to NaCN and hypoxia lasted for about 30 min. Thus, the depressant action of apomorphine is possibly mediated through dopaminergic receptor sites in the carotid body chemoreceptor.

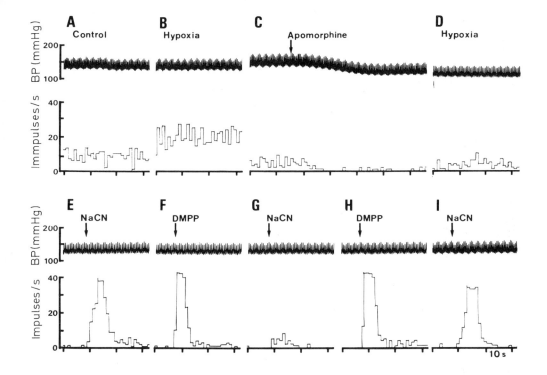

Fig. 3. Chemoreceptor responses to hypoxia and chemical substances before and after apomorphine. *Upper traces*: arterial blood pressure changes; *Lower traces*: discharge frequency in a few active chemoreceptor units. (A-D and E-I) obtained from different animals. (A) Animal was ventilated with room air during control period. (B) Animal was ventialted with hypoxic gas mixture (90% N_2 + 10% O_2) during control period. (C) Apomorphine (0.4 mg/kg i.v.) was injected (*arrow*) under normoxic conditions. (D) Animal was again ventilated with hypoxic gas mixture after apomorphine. Note that the chemoreceptor response to hypoxia is markedly depressed after apomorphine. (E) Chemoreceptor response to NaCN (2 µg i.a. in 0.25 ml Tyrode's solution) during control period. (F) Dimethylphenylpiperazinium (DMPP) (1 µg i.a. in 0.25 ml Tyrode's solution) injected (*arrow*) during control period. (G) NaCN (2 µg in 0.25 ml Tyrode's solution) injected (*arrow*) 3 min after apomorphine (100 µg i.a.). (H) DMPP (1 µg in 0.25 ml Tyrode's solution) 5 min after apomorphine. (I) NaCN (2 µg in 0.25 ml Tyrode's solution) injected 20 min after apomorphine

These experimental observations as well as earlier ones (16) clearly indicate the presence of dopaminergic inhibitory receptors in the carotid body of the cat {1} exogenously applied dopamine depresses spontaneously occurring chemoreceptor discharges and also discharges induced by NaCN or hypoxia; {2} dopaminergic receptor blocking agents, haloperidol and trifluperidol, induce an increase in the resting rate of chemoreceptor discharge frequency; {3} after administration of the dopaminergic receptor blocking agent, the inhibitory effect of dopamine is abolished, and chemoreceptor responses to NaCN and hypoxia are augmented; {4} a dopaminergic receptor stimulating agent, apomorphine, induces a marked depression of spontaneously occurring chemoreceptor discharges and also inhibits chemoreceptor responses to NaCN and hypoxia.

With regard to the site of action of dopamine and location of dopaminergic receptors in the carotid body, the present experiments have not been conclusive. However, it can be postulated that exogenously applied dopamine acts directly on afferent nerve terminals making contact with glomus type I cells (14), hyperpolarizes the nerve terminals, and hence, inhibits initiation of impulse discharges at the nonmyelinated portion of the sensory nerve fibers. After the administration of dopaminergic receptor blocking agents, the resting rate of chemoreceptor discharges even under normoxic conditions markedly increased. This suggests that a continuous release of dopamine stored in the glomus cells to the sensory nerve endings may control chemoreceptor activity by exerting a tonic inhibition through dopaminergic receptor sites on the nerve endings. Thus, endogenously stored dopamine may participate in chemoreceptor function as an inhibitory transmitter substance upon stimulation of the carotid body chemoreceptor.

References

1. Aden, N.-E., Bubenson, A., Fuxe, K., Hokfelt, T.: J. Pharm. Pharmacol. *19*, 627-629 (1967)
2. Black, A.M.S., Comroe, Jr., J.H., Jacobs, L.: J. Physiol. *233*, 1097-1102 (1972)
3. Bunny, B.S., Aghajanian, K., Roth, R.H.: Nature *245*, 123-125 (1973)
4. Chiocchio, S.R., Biscardi, A.M., Tramezzani, J.H.: Nature *212*, 834-835 (1966)
5. Chiocchio, S.R., King, M.P., Carballo, L., Angelacos, E.T.: J. Histochem. Cytochem. *19*, 621-629 (1971)
6. Ernst, A.M.: Acta Physiol. Pharmacol. Neerl. *15*, 141-154 (1969)
7. Fillenz, M., Woods, R.I.: J. Physiol. *186*, 39-40 (1966)
8. Goldberg, L.I.: Pharmacol. Rev. *24*, 1-29 (1972)
9. Kobayashi, S.: Arch. Histol. Jpn. *33*, 319-339 (1971)
10. Lever, J.D., Lewis, P.R., Boyd, J.D.: J. Anat. *93*, 478-490 (1952)
11. Niemi, M., Ojala, K.: Nature *203*, 539-540 (1964)
12. Nishi, K.: J. Pharmacol. *154*, 303-320 (1975)
13. Nishi, K., Eyzaguirre, C.: Brain Res. *33*, 37-56 (1971)
14. Nishi, K., Stensaas, L.J.: Cell Tissue Res. *154*, 303-320 (1974)
15. Sampson, S.R.: Brain Res. *45*, 266-270 (1972)
16. Zapata, P.: J. Physiol. *244*, 235-251 (1975)
17. Zapata, P., Hess, A., Bliss, E.L., Eyzaguirre, C.: Brain Res. *14*, 473-495 (1969)

DISCUSSION

Weigelt: How do you prove your specific effect of haloperidol on dopamine release and how can you exclude the possibility that haloperidol acts partly on the respiratory chain and therefore produces a hypoxic effect?

Nishi: When injecting haloperidol, you see a fall in blood pressure, which causes an increase in the chemoreceptive discharge. In order to avoid this type of chemoreceptor response, I inject barbiturate to produce the fall in blood pressure. Then I inject haloperidol, which induces a further increase in chemoreceptor discharge. It is well known that haloperidol has a specific effect in the central nervous system on the dopaminergic receptor, so I applied haloperidol to just the peripheral nervous system. This was the first demonstration of the effect of haloperidol in the peripheral nervous system on the dopaminergic response.

O'Regan: I would agree that haloperidol increases the excitability of the carotid body. Why did you use 4 mg/kg, since the usual dose is 1 mg/kg? I found that some chemoreceptor preparations were not excited after haloperidol and that norepinephrine can suppress chemoreceptor discharge in vivo, in contrast to Zapata's findings.

Nishi: By injecting catecholamine into the carotid body you can get an increase in chemoreceptor discharge and sometimes an inhibition. But the most potent drug for producing inhibition is dopamine. I also used 1 mg/kg haloperidol, but to see a marked inhibition I used 4 mg/kg.

Zapata: Peripheral dopamine receptors have also been shown by Goldberg (Adv. Neurol. 9, 53-56, 1975), who used the renal artery, where dopamine induced vasodilatation was abloished by haloperidol. I agree with you that dopamine is the most potent of the inhibitors. But as you showed also, apomorphine seems to have a less potent but more prolonged effect. Do you think that dopamine acts by inducing changes in AMP?

Nishi: That might be possible; however, I do not know exactly.

Torrance: Have you or anyone else studied the effect of dopamine on the vascular resistance to the carotid body?

Nishi: No, I have not.

Torrance: It is essential that this be done. McCloskey injected dopamine at the carotid body, and as soon as he thought it had got there, he stopped blood flow through it and followed the development of discharge during stagnant asphyxia. He found that there was no difference between the development of discharge when dopamine was included in the injection and when there was no dopamine at all. He concluded that dopamine has a purely vasomotor effect. I do hope that somebody will study the effect of dopamine on blood flow.

Paintal: It should be said that the results shown by Nishi have very short latency, less than half a second.

Torrance: So there must be a complicated effect extending over several seconds within which vasomotor effects could occur.

O'Regan: Can you have suppression on a superfused preparation where there is no vascular change by dopamine?

Torrance: There is the possibility that dopamine has a vasomotor effect on the carotid body as it has in other tissues.

McDonald: Libet found in the sympathetic ganglion that haloperidol had effects that were not always reproducible and that some of the effects were similar to those of a local anesthetic. It was found unreliable in its ability suppress the slow inhibitory postsynaptic potential produced by dopamine.

Nishi: Drugs that depress the central nervous system possess the action of local anesthetics. In the doses that I used, there is no local anesthetic action blocking the nerve fiber activity.

Blockade of Carotid Body Chemosensory Inhibition*

P. Zapata and F. Llados

There is a high content of dopamine in the carotid body (21), located in glomus type I cells by fluorescence microscopy (9). Intrastream injections of dopamine produced a transient depression of the frequency of chemosensory discharges in cat carotid bodies superfused in vitro (19). This effect was attributed to direct inhibition, since no vasomotor changes could be induced in this preparation. Furthermore, the action seems to be specific for this chemoreceptor organ, since dopamine applied to carotid sinus preparations produced only a weak and long-lasting increase in the frequency of barosensory discharges (19).

The inhibitory effect of dopamine on chemosensory discharges from carotid bodies superfused in vitro was not shared by norepinephrine (19), thus suggesting that dopamine could be acting on a different receptor site than the typical alpha-adrenoceptor. However, the carotid body superfused in vitro becomes desensitized to the inhibitory effect of dopamine after repeated injections of the drug (19), thus preventing its pharmacologic characterization through changes in sensitivity after administration of blocking agents. This prompted us to perform such a study on cat carotid bodies in situ (Zapata and Llados, unpublished).

Intracarotid injections of dopamine (1 µg) induced a transient arrest of chemosensory discharges followed by a brief rebound, while the same dose of apomorphine - a well-known dopamine agonist - elicited a less marked but more prolonged depression (20). Chemosensory inhibition was also induced by i.v. dopamine, with an ED50 of ca. 3.7×10^{-10} M/kg. Large doses of this drug were required to modify systemic blood pressure in the cat. Thus, the possibility was provided to study dose-response curves for chemosensory effects of dopamine (applied by intracarotid or i.v. route) before and after the administration of catecholaminergic blocking agents.

Sampson (13) reported that phenoxybenzamine (Dibenzyline) (20 mg/kg i.v.) was an effective blocker of dopamine-elicited chemosensory inhibition, suggesting that dopamine acted on alpha-adrenoceptor sites of chemosensory nerve endings. We tested the changes in sensitivity of carotid body chemoreceptors to dopamine injections after different doses of this blocker. As illustrated in the lower part of Fig. 1, after an i.v. dose of 3 mg/kg of phenoxybenzamine hydrocholide no significant changes were observed in the chemosensory inhibition elicited by dopamine. In the same animal this dose of phenoxy-

---

*This work was supported by grant 211/75 from the Catholic University Research Fund. Ergot derivatives were kindly provided by Sandoz Pharmaceuticals. Thanks are due to Mrs. Carolina Zapata and Ms. Luz-Gloria Oportus for technical assistance.

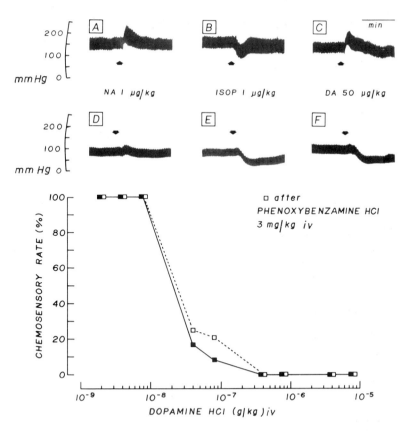

Fig. 1. Effects of phenoxybenzamine hydrochloride 3 mg/kg i.v. Upper part, blood pressure recordings from femoral artery. I.v. injections of norepinephrine bitartrate 1 µg/kg (A and D), isoproterenol hydrochloride 1 µg/kg (B and E) and dopamine hydrochloride 50 µg/kg (C and F), before (A, B and C) and after (D, E and F) administration of the blocker. Lower part, dose-response curves for dopamine effects on carotid nerve chemosensory discharges, before (solid squares) and after (open squares) administration of the blocker. Ordinate, maximal changes of chemosensory frequency expressed as percent of control frequency before each dopamine injection. Abscissa, in logarithmic scale, doses of dopamine hydrochloride in g/kg injected i.v.

benzamine produced a considerable reduction of the systemic hypertensive effect of norepinephrine, prolonged the hypotensive effect of isoproterenol (an expression of the blockade of sympathetic reflexes restoring blood pressure), and reversed the hypertensive effect of dopamine (50 µg/kg) into a hypotensive reaction, as illustrated in the upper part of Fig. 1. The reversal of systemic pressure reactions to dopamine can be attributed to blockade by phenoxybenzamine of alpha-adrenoceptors mediating its vasoconstrictor effects and the unmasking of its vasodilating actions in renal and mesenteric arteries (18). However, in experiments in which larger doses of the blocker were used (15 mg/kg), a displacement to the right of the dose-response curve of chemosensory inhibition elicited by dopamine and apomorphine was observed, but this was coincident with production by the blocker of profound hypotension. A similar apparent blockade of dopamine elicited chemosensory inhibition was established after treatment with dibenamine hydrochloride (10-20 mg/kg i.v.), but this also occurred in association with a dramatic fall in arterial pressure. When blood pressure was transiently restored to near normal values by slow i.v. injections of angiotensin or dextran, the blockade of dopamine-elicited chemosensory inhibition by dibenamine was no longer evident. Thus, our observations indicate that alpha-adrenergic blocking agents are not directly effective in abolishing the inhibitory actions of dopamine on chemosensory discharges.

In order to examine the possibility that dopamine effects on the carotid body could be mediated by beta-adrenoceptors, two blocking agents, propranolol and dichloroisoproterenol, were studied. Administration of these substances, in doses that blocked the hypotensive reactions to isoproterenol injections, did not modify the dose-response curves for chemosensory inhibition induced by dopamine or apomorphine.

It has been reported that dihydroergotamine blocks dopamine-induced chemosensory inhibition (12,14). We confirmed such an effect, which is shared by methylergometrine (20). However, the interference caused by ergot alkaloids can be ascribed to blockade of either alpha-adrenoceptors or specific dopamine-receptors (3).

Since butyrophenones have been described as powerful dopaminergic blockers (8), we studied the effects of two agents in this group. Haloperidol (20-30 µg/kg i.v.) produced a pronounced displacement to the right of dose-response curves to dopamine- and apomorphine-induced chemosensory inhibition. A marked blockade was also observed with smaller doses of spiroperidol (19). As illustrated in Fig. 2, the i.v. administration of 6 µg/kg of this drug produced complete blockade of chemosensory inhibition induced by dopamine injected into the carotid artery, but also there was chemosensory excitation in response to larger doses of dopamine. These excitatory responses elicited by dopamine when its inhibitory actions are blocked could be secondary to alpha-mediated vasoconstriction provoked by this range of doses. These responses could also result from the direct excitation of chemoreceptor mechanisms, since long-lasting excitation in response to dopamine can be observed in carotid bodies superfused in vitro after desensitization to the initial inhibitory actions (19). In that preparation, dopamine-elicited inhibition was also blocked by spiroperisol (19).

Another group of neuroleptics, the phenothiazines, present dopaminergic blocking properties (5) and, therefore, were tested for their efficacy in blocking chemosensory inhibition. Chlorpromazine hydrochloride (1 mg/kg i.v.) was able to provoke a significant decrease in the sensitivity of carotid body chemoreceptors in situ to the inhibitory actions of dopamine and apomorphine. Fig. 3 illustrates the changes in the dose-response curve of dopamine-elicited chemosensory inhibition after administration of cumulative doses of 10-100 µg/kg of perphenazine; doses of 400 µg/kg or more produced the complete blockade of inhibition and the appearance of chemosensory excitation in response to the larger doses of dopamine.

It should be pointed out that when spiroperidol and methylergometrine were injected into the carotid artery, a short-term decrease followed by a prolonged increase in the frequency of chemosensory discharges was observed. Otherwise, the first of a series of small doses of perphenazine injected i.v. also induced a transient decrease of chemosensory activity. These observations suggest that some of these substances can initially act as dopamine analogs on the receptor sites, but afterwards they behave as competitive antagonists for dopamine or apomorphine.

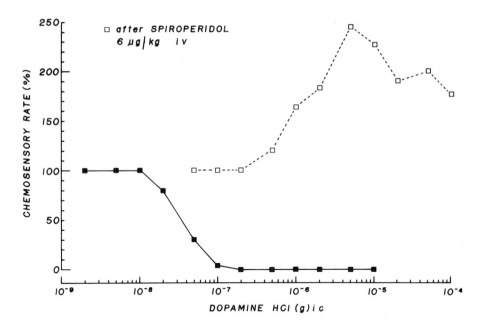

Fig. 2. Dose-response curves for dopamine effects on carotid nerve chemosensory discharges, before (*solid squares*) and after (*empty squares*) administration of spiroperidol (6 μg/kg i.v.). *Ordinate*: maximal changes of chemosensory frequency expressed as percent of control frequency before each dopamine injection. *Abscissa*: in logarithmic scale, doses of dopamine hydrochloride in g injected into the common carotid artery

To provide a comparison between different blockers regarding their efficacies as antagonists of dopamine-elicited chemosensory inhibition, their inhibition constants (K_i) were calculated from the relationship $K_m' / K_m = 1 + l/K_i$, where K_m and K_m' are the concentrations of dopamine required to give half-maximal inhibition of chemosensory discharges before and after administration of the blocker, respectively, and l is the concentration of the blocking agent. This procedure was devised by Clement-Cormier et al. (5) to measure the antagonistic efficiencies of neuroleptics for dopamine-sensitive adenylate cyclase of caudate nucleus. Application of this procedure to our observations gave the following approximate values of K_i for blockers whose concentrations are given in parentheses:

Spiroperidol	(1×10^{-8} M/kg)	$K_i = 1.0 \times 10^{-11}$ M
Haloperidol	(8×10^{-8} M/kg)	$K_i = 8.9 \times 10^{-10}$ M
Fluphenazine	(2×10^{-9} M/kg)	$K_i = 1.2 \times 10^{-9}$ M
Perphenazine	(7.4×10^{-8} M/kg)	$K_i = 3.6 \times 10^{-8}$ M
Chlorpromazine	(1.7×10^{-5} M/kg)	$K_i = 1.7 \times 10^{-7}$ M

Inhibition constants were also calculated for dibenamine and phenoxybenzamine, from experiments in which large doses of these alpha-adrenoceptor blocking agents displaced to the right the dose-response curves for dopamine-elicited chemosensory inhibition. Values obtained were at least two orders of magnitude larger than that for chlorpromazine, the least active of the dopaminergic blocking agents. However, one must take into consideration that doses used of the neuroleptics listed above produced minimal or no hypotension for the period during which dose-response curves to dopamine injections were assayed, while

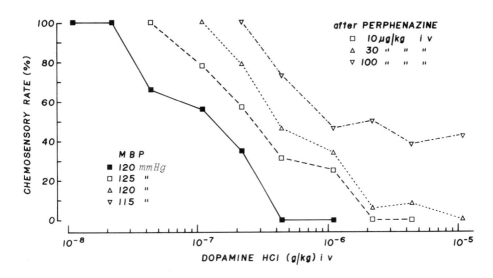

Fig. 3. Dose-response curves for dopamine effects on carotid nerve chemosensory discharges, before (*solid squares*), and after perphenazine 10 (*empty squares*), 30 (triangles with lower base) and 100 (triangles with upper base) µg/kg i.v. *Ordinates*: maximal changes in chemosensory frequency expressed as percent of control frequency before each dopamine injection. *Abscissa*: in logarithmic scale, doses of dopamine hydrochloride in g/kg injected i.v. *Inset*: mean blood pressure in mmHg, before and after the cumulative doses of the blocker

haloalkylamines produced a dramatic hypotension, near to the critical closing pressure for carotid body flow (which is otherwise elevated by catecholamines), a criticism already raised by Acker (1).

Results reported here indicate that dopamine-induced chemosensory inhibition in carotid bodies in situ is blocked by phenothiazines and butyrophenones in doses that do not modify the alpha-adrenoceptor-mediated hypertensive responses to norepinephrine injections, or the beta-adrenoceptor-mediated hypotensive effects of isoproterenol injections. Furthermore, our observations indicate that alpha- or beta-adrenoceptor blocking agents do not directly antagonize dopamine- or apomorphine-elicited chemosensory inhibition. Thus, a specific and distinct dopamine-receptor would be responsible for the inhibition of carotid body chemosensory discharges.

It is interesting to note that specific dopamine-receptors have been described elsewhere: mammalian renal artery (18), caudate nucleus (5, 11), olfactory tubercle (5), hypothalamus (7), retina (4,15), and autonomic ganglia (10), as in the ganglia of *Aplysia* (2,17) and *Helix* (16). In most of these sites dopamine effects are inhibitory (2,10,11, 15,16,18). Furthermore, butyrophenones and phenothiazines are potent and selective blockers of dopamine receptors in all these loci, as in the carotid body.

As to the functional role of dopamine in the generation of chemosensory discharges, the problem has been discussed in some detail elsewhere (20), but the most attractive hypothesis is that dopamine can be released from glomus cells to modulate the excitability of adjacent sensory nerve terminals. As in mammalian sympathetic ganglia (10) and in the abdominal ganglion of *Aplysia* (17), dopamine can evoke not only transient inhibition (probably due to hyperpolarization) but also late potentiation of a presumably cholinergic junction between glomus cells and sensory nerve endings, as previously postulated (6)

References

1. Acker, H.: Comment in: The Peripheral Arterial Chemoreceptors. Purves, M.J. (ed.). London: Cambridge U. Pr. 1975, p. 220
2. Ascher, P.: J. Physiol. *225*, 173-209 (1972)
3. Bell, C., Conway, E.L., Lang, W.J., Padanyi, R.: J. Pharmacol. *55*, 167-172 (1975)
4. Brown, J.H., Makman, M.H.: J. Neurochem. *21*, 477-479 (1973)
5. Clement-Cormier, Y.C., Kebabian, J.W., Petzold, G.L., Greengard, P.: Proc. Natl. Acad. Sci. U.S.A. *71*, 1113-1117 (1974)
6. Eyzaguirre, C., Zapata, P.: in: Arterial Chemoreceptors. Torrance, R.W. (ed.). Oxford: Blackwell 1968, pp. 213-251
7. Fuxe, K., Agnati, L.F., Corrodi, H., Everitt, B.J., Hökfelt, T., Löfström, A., Ungerstedt, U.: in: Dopaminergic Mechanisms. Calne, D.B., Chase, T.N., Barbeau, A. (eds.). New York: Raven Press 1975, pp. 223-242
8. Janssen, P.A.J.: in: Psychopharmacological Agents. Gordon, M. (ed.). New York: Academic Press 1967, Vol. II, pp. 199-248
9. Kobayashi, Sh.: Arch. Histol. Jpn. *33*, 319-339 (1971)
10. Libet, B., Tosaka, T.: Proc. Natl. Acad. Sci. U.S.A. *67*, 667-673 (1970)
11. McLennan, H., York, D.H.: J. Physiol. *189*, 393-402 (1967)
12. Mitchell, R.A., McDonald, D.M.: in: The Peripheral Arterial Chemoreceptors. Purves, M.J. (ed.). London: Cambridge Univ. Press 1975, pp. 269-291
13. Sampson, S.R.: Brain Res. *45*, 266-270 (1972)
14. Sampson, S.R., Aminoff, M.J., Jaffe, R.A., Vidruk, E.H.: J. Pharmacol. Exp. Ther. *197*, 119-125 (1976)
15. Straschill, M., Perwein, J.: Pflügers Arch. *312*, 45-54 (1969)
16. Struyker-Boudier, H.A.J., Gielen, W., Cools, A.R., Van Rossum, J.M.: Arch. Int. Pharmacodyn. Ther. *209*, 324-331 (1974)
17. Tremblay, J.P., Woodson, P.B.J., Schlapfer, W.T., Barondes, S.H.: Brain Res. *109*, 61-81 (1976)
18. Yeh, B.K., McNay, J.L., Goldberg, L.I.: J. Pharmacol. Exp. Ther. *168*, 303-309 (1969)
19. Zapata, P.: J. Physiol. *244*, 235-251 (1975)
20. Zapata, P.: in: Non-Striatal Dopaminergic Neurons. Costa, E. (ed.). New York: Raven Press 1977 (in press)
21. Zapata, P., Hess, A., Bliss, E.L., Eyzaguirre, C.: Brain Res. *14*, 473-496 (1969)

DISCUSSION

Willshaw: I was interest in trying to block the effects of physiologically induced depression of chemosensory activity. At that time phenoxybenzamine was put forward as one of the major blocking agents to use. I found that it is not necessary to have severe hypotension even when you inject quite large quantities of phenoxybenzamine. I did find a phenoxybenzamine block of the naturally induced inhibition of the chemoreceptor discharge.

O'Regan: I have tried some of these relatively high doses of phenoxybenzamine, and I found very great variation in blood pressure responses.

Some cats with small doses showed dramatic falls in blood pressure. In other cats the fall in blood pressure was transient. Did you find this? With the high doses did you always get these marked hypotensive effects?

Zapata: Yes, it is possible to find variable changes in blood pressure. However, I went up only to a dose of 20 mg/kg of phenoxybenzamine and to doses of 30 mg/kg of dibenamine. But I found in my experiments a consistent block to dopamine-induced chemosensory inhibition when blood pressure had gone down. In those circumstances when I restored blood pressure by injecting dextrose or angiotensin, the dose-response curve of dopamine returned to near control values.

Eyzaguirre: I wonder whether different anesthetics could explain the differences in the effects of dopamine or the other blockers on blood pressure.

Purves: I find it very difficult in these pharmacologic studies to know how you really distinguish vascular changes within the carotid body from a direct action upon the receptors. Since we do not know whether or how these drugs actually act upon blood flow in quantitative terms, I find it rather misleading that the responses are all interpreted in terms of a direct action on the receptor.

Zapata: I agree completely that the effect of dopamine or several of these blockers is partly modified by vasomotor changes. I am sure of that. But I feel confident about the chemosensory inhibition induced by dopamine because it is also found in the in vitro superfused preparation (Zapata: J. Physiol., 1975). There is also a recent report by Sampson et al. (Amer. J. Physiol., 1976) in which they tried dopamine after stopping blood flow. Under both circumstances there was still a chemosensory inhibition induced by dopamine.

Kiwull-Schöne: I believe rabbits would be very useful for your studies because it is known from comparisons between cat and rabbit that the chemosensitive activity is not very sensitive to low blood pressure in the rabbit. One is able to reduce the blood pressure to very low values and not get an increase in chemosensitive activity (Ott et al.: Pflüg. Arch. 325, 28-29, 1975).

McDonald: It is my impression that the observations of Mitchell in his studies of the cat carotid body support the comments made by Willshaw. In Mitchell's experiments phenoxybenzamine was infused intravenously over a period of 30 min. It is a slow infusion rather than an injection. When administered that way, a dose of 10 mg/kg did not change blood pressure.

Zapata: I found similar changes in spontaneous rate of discharge to those described previously by Mitchell and McDonald (In: The Peripheral Arterial Chemoreceptors, 1975) with the use of phenoxybenzamine (10-20 mg/kg). However, these changes in the resting rate of discharge were also well correlated with the decrease in blood pressure, in agreement with observations by Torrance (Lee et al. Quart. J. exper. Physiol., 1964) on the effect of blood pressure on aortic nerve chemosensory discharge. However, phenoxybenzamine (3 mg/kg), while antagonizing the vascular effects of norepinephrine injections, did not result in a great fall in blood pressure. Under these circumstances there were no significant changes in the resting rate of discharge of chemosensory fibers or in the inhibitory response to dopmaine injections. In contrast, doses of dopaminergic blockers that did not produce real changes in blood pressure produced a tremendous increase of resting rate of chemosensory discharge.

Torrance: In several of your dose-response curves of the blockers the response to dopamine was reduced, but your curve seems to remain at the same level of about 50% response over a 10-fold range of dose. This result is puzzling to me. Would you like to comment on it, perhaps in terms of two types of action of dopamine?

Paintal: Did you find that a small dose of dopamine had an excitatory effect as Nishi showed? That is a very important observation.

Zapata: In our experiments in situ the minimal doses of dopamine that were producing chemosensory inhibition did not produce any effect if they were assayed again after dopaminergic blockers. The dose had to be increased usually about 10 times to produce a chemosensory excitatory effect.

Variable Influences of the Sympathetic Nervous System Upon Carotid Body Chemoreceptor Activity

R. G. O'Regan

Summary

In anesthetized cats two distinct patterns of excitation of carotid chemosensory discharges by the sympathetic were noted. The characteristic features of the more usual pattern were marked excitatory changes in the initial 10-20 s of stimulation, resistance to alpha-adrenoceptor blockade, exaggeration of effect after administering haloperidol, and an involvement of a nicotinic receptor site in the carotid body. The other type demonstrated excitatory changes that appeared 5-10 s after stimulation began and that became more marked as stimulation progressed. This latter excitation was eliminated after administration of alpha-blockers, and presumably a vasoconstrictor mechanism was involved in its production. The mechanisms responsible for the excitation resistant to alpha-blockade are obscure and may not depend on blood flow changes. Inhibition mediated by the sympathetic was occasionally noted.

Introduction

The well-known excitation of carotid chemoreceptor activity during sympathetic activation is considered to be the consequence of an associated reduction of carotid body blood flow. Recent investigations by McDonald and Mitchell (2) indicate that the sympathetic not only supplies the blood vessels of the carotid body but also innervates the glomus cells of the organ. They postulate that the effect of this innervation is to engender release of dopamine from these cells. This agent is known to suppress the discharge of the chemosensory endings (4). However, a possibility exists that this nonvascular innervation could contribute to the excitation of chemoreceptor discharges by the sympathetic. The investigation reported here reexamines the influences exerted by the sympathetic upon chemoreceptor activity.

Methods

In cats anesthetized with pentobarbitone, chemosensory activities were recorded from 112 filaments peeled off the cut sinus nerve. Any alterations of these activities were examined both during and after bouts of electrical stimulation (10-20 Hz, 0.6 ms) of the preganglionic cervical sympathetic trunks for periods of between 30 and 60 s. The voltages employed (8-12 V) were 50% greater than those needed to cause maximal dilation of the ipsilateral pupil. The ganglioglomerular nerves containing the postganglionic fibers supplying the carotid body were also stimulated and gave essentially the same findings as those obtained when the preganglionic trunk was used. Drugs were administered intravenously (i.v.) or into the carotid artery (i.c.).

Results

Certain criteria were used to distinguish genuine sympathetic effects from changes that could have arisen from random variations of chemoreceptor activity. Excitation or inhibition of chemosensory activity during or after electrical stimulation of the sympathetic was considered to be genuine and significant (a) if the increase or decrease of activity exceeded 10% of the prestimulation values and (b) if repeated bouts of sympathetic stimulation were associated with similar patterns of excitatory or inhibitory changes. Employing these criteria the discharges of 49 chemosensory units were enhanced by sympathetic activation, whereas in a further 15 preparations inhibition occurred.

Excitatory Effects. The most prominent features of the two excitatory patterns that were observed can be seen in Fig. 1. In the pattern depicted by the squares chemosensory discharges increased soon after commencing stimulation, but the increase was poorly sustained, discharge rates returning toward control values as stimulation proceeded. While the most prominent feature of this excitatory pattern was the marked elevation of discharge in the initial 10-20 s of stimulation, it was usual for a another increase of activity to appear in the latter periods of stimulation, this accentuation then being carried on into the poststimulatory periods. This pattern of excitation, which was noted in 28 preparations, was never affected following i.v. or i.c. administration of the alpha-adrenoceptor antagonists dibenzyline and phentolamine, and it can be suitably referred to as non-alpha-adrenergic. In the pattern of excitation depicted by the circles chemosensory activity was unaffected or showed slight depressive changes in early stimulation. Discharge built up 10 s after stimulation began,

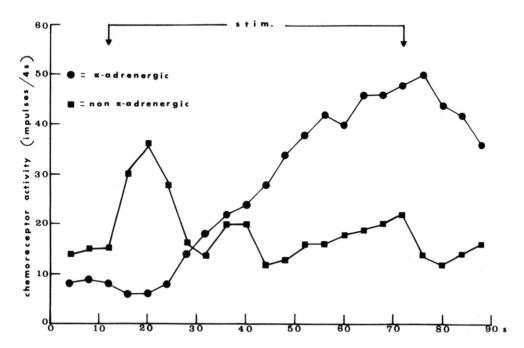

Fig. 1. Patterns of excitation of carotid chemoreceptor activity during or after electrical stimulation of preganglionic sympathetic trunks. Sympathetic stimulation carried out between arrows. (See text for further explanation)

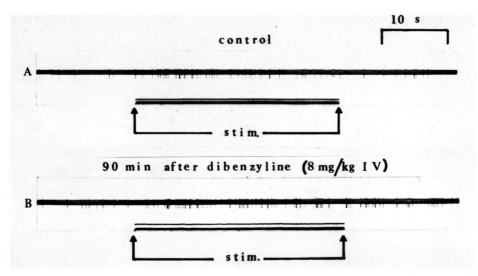

Fig. 2. Inability of alpha-adrenoceptor antagonists to affect the excitatory effect of sympathetic upon chemoreceptor discharge. Chemosensory activity recorded from filament of sinus nerve with stimulus artifacts gated out. Stimulation of preganglionic sympathetic trunk carried out between arrows and also shown by tops of stimulus artifacts recorded on other beam of oscilloscope. (A) Before dibenzyline administration (cat breathing air); (B) after dibenzyline administration (cat breathing oxygen)

became more marked as stimulation proceeded, and was present in the poststimulatory period. Excitatory changes of this pattern were usually of much smaller magnitude than that shown in Fig. 1 and were no longer elicited after administering alpha-antagonists. They can be conveniently designated as alpha-adrenergic.

Fig. 2 shows the inability of alpha-blockers to affect the excitatory changes with marked increases in early stimulation. Chemoreceptor discharge was increased in the initial period of stimulation (Fig. 2A). Excitation was less marked in the late stimulatory and poststimulatory periods. After recording A was taken, dibenzyline was injected in graded doses (2 mg/kg i.v. at 15 min intervals), which were designed to minimize hypotensive effects of the drug. A cumulative dose of 8 mg/kg did not abolish or reduce the excitatory effect of sympathetic stimulation (Fig. 2B), although much smaller doses of this drug (0.5-3 mg/kg) irreversibly abolished the changes of the ipsilateral pupil to preganglionic cervical sympathetic stimulation.

A feature of the non-alpha-adrenergic effect was its appearance or exaggeration following administration of the dopamine antagonist haloperidol in doses adequate to abolish the inhibitory changes of chemosensory discharges to i.c. injections of dopamine. Non-alpha-adrenergic excitations by the sympathetic were abolished, however, following i.c. injections of hexamethonium (0.25-0.75 mg). This effect of hexamethonium need not have involved the superior cervical as similar findings using this drug were recorded when the ganglioglomerular nerve was stimulated. This suggests that this drug can exert its influence within the carotid body. These doses of hexamethonium eliminated the excitations of chemosensory activity to i.c. injections of acetylcholine.

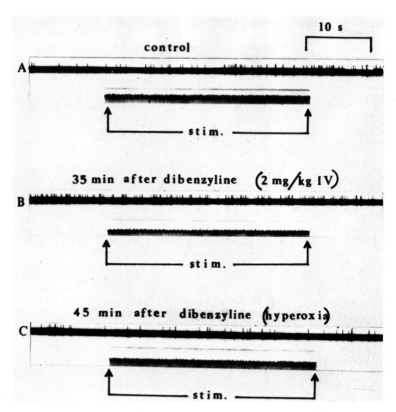

Fig. 3. Effect of dibenzyline on excitatory change of chemosensory discharge to sympathetic activation. Chemosensory activity recorded from filament of sinus nerve with stimulus artifacts gated out. Stimulation of preganglionic sympathetic trunk carried out between arrows and also shown by tops of stimulus artifacts recorded on other beam of oscilloscope. (A) Before dibenzyline administration (cat breathing air) air); (B) after dibenzyline administration (cat breathing air); (C) after dibenzyline administration (cat breathing oxygen)

Fig. 3 shows an excitatory effect of the sympathetic that proved to be alpha-adrenergic. Discharge rates increased 10 s after the onset of stimulation, and the accentuation was marked in late stimulation and in the poststimulatory period (Fig. 3A). After dibenzyline sympathetic stimulation had no effect (Fig. 3B). Prestimulatory discharge rates before (Fig. 3A) and after (Fig. 3B) injecting dibenzyline were vastly different, and this disparity may have affected the result. After addition of oxygen to the inspired air, discharge rates decreased to levels prior to injecting the drug, yet sympathetic stimulation did not excite the discharge (Fig. 3C). The response in Fig. 3A would, therefore, appear to be a genuine alpha-adrenergic effect. Excitations of this type were noted in 21 chemoreceptor preparations and were no longer elicited after i.v. (1-3 mg/kg) or i.c. (0.1-0.3 mg/kg) injections of small doses of alpha-blockers.

Inhibitory Effects. Inhibition of discharge also displayed two main patterns. In 10 preparations depressive changes during sympathetic stimulation were limited to the initial 10-15 s of stimulation. Inhibitory effects in 5 preparations appeared 5-10 s after stimulation started, became more pronounced as stimulation progressed, and were present in the poststimulatory periods. During sympathetic stimulation depressive effects occurred, which resulted from the spread of current. While true inhibitory influences of the sympathetic were abolished after applying local anesthetic onto the preganglionic trunk distal to the stimulating electrodes, adventitious effects were not. Depressive effects, in general, were little affected after administration of haloperidol, dibenzyline, or phentolamine.

Effects of 6-Hydroxydopamine (6-OHDA). 6-OHDA is an agent used to produce a chemical sympathectomy (5). The most sensitive catecholamine-containing structures affected by this drug are the noradrenergic endings and SIF cells, with glomus cells probably being less vulnerable (3). 6-OHDA does destroy the norandrenergic endings in the carotid body (2), but it is also accumulated by the glomus cells (1). Conduction is rapidly lost at the nerve ending.

In nine cats 6-OHDA was used to abolish the function of the noradrenergic terminals in the carotid body without affecting adversely the glomus cells and so possibly unmask a non-alpha-adrenergic influence of the sympathetic. Initially, in two cats i.v. administration of the drug was employed, but marked deterioration of the animals' condition within 3-4 h precluded any useful study. It was decided to give i.c. injections of 6-OHDA (15-25 mg) and to examine the changes of chemoreceptor discharges to sympathetic stimulation prior to and after injections. The results were inconclusive. In four cats excitatory changes to sympathetic activation were eliminated within 30 min of injecting the drug, and in two of these animals inhibitory changes appeared. However, inhibitory effects were transient, the sympathetic being ineffective within $1\frac{1}{2}$ to 2 h of injection. In the remaining cats sympathetic activation did not alter the discharge before or after the injection.

Discussion

The investigation reported here indicates that the sympathetic influences carotid chemoreceptor activity in a complex manner in that two mechanisms of excitation appear to exist as well as occasional inhibitory effects. However, 43% of the chemoreceptor preparations studied did not respond to sympathetic activation and those that were excited usually showed small increases of discharge. Indeed, in a substantial number of cats, sympathetic stimulation sufficient to cause maximal dilation of the ipsilateral pupil failed to alter the discharges of any of the chemoreceptor preparations studied. After administration of haloperidol, the number of responsive units increased markedly, suggesting that dopamine is, in some obscure manner, antagonizing excitatory influences mediated by the sympathetic.

Excitatory effects abolished by alpha-blockers presumably depend on vasoconstrictor mechanisms. However, the mechanism involved in the production of the excitatory changes that were unaffected by large doses of alpha-antagonists is obscure. Without further investigations involving the monitoring of carotid body blood flow, a vascular effect cannot be ruled out. Nevertheless, the results using hexamethonium show that nicotinic receptor sites are implicated, and if this is so, the findings could be explained by involvement of the preganglionic sympathetic fibers described by McDonald and Mitchell (2), which terminate on the type I cells. By some as yet unknown mechanism, such a neurally mediated effect could alter chemoreceptor discharge. Exaggeration of the excitatory effects after haloperidol again suggests that dopamine within the carotid body is responsible for lessening the magnitude of these effects. As the role of this catecholamine in chemoreception is oscure at present, it would be mere speculation to attempt to explain the results obtained with haloperidol. Further investigations are needed if the mechanism and significance of this non-alpha-adrenergic effect of the sympathetic is to be elucidated.

Genuine inhibitory effects mediated by the sympathetic were unusual and their mechanism of production and significance difficult to explain. They may represent, as postulated by McDonald and Mitchell (2), a release of dopamine from the glomus cells, which in turn suppresses the discharge of the sensory terminals. But the resistance of the inhibitory effects to haloperidol noted in the present investigation does not support this interpretation.

The experiments using 6-OHDA gave inconclusive results. On superficial examination abolition of excitatory changes after administering this agent suggests that these changes depend on a normal behavior of the noradrenergic endings and that a true nonnoradrenergic influence is inhibitory. However, the replacement of excitatory effects to sympathetic stimulation by inhibitory changes was noted in only a small number of experiments, and since there is no evidence as to whether the glomus cells were involved, the significance of the findings with 6-OHDA is open to question.

References

1. Hess, A.: in: The Peripheral Arterial Chemoreceptors. Purves, M.J. (ed.). New York: Cambridge U. Pr. 1975, pp. 51-68
2. McDonald, D.S., Mitchell, R.A.: J. Neurocytol. *4*, 177-230 (1975)
3. Sachs, C., Jonnson, G.: Biochem. Pharmacol. *24*, 1-8 (1975)
4. Sampson, S.R.: Brain Res. *45*, 261-263 (1972)
5. Thoenen, H., Tranzer, J.: Naunyn Schmiedebergs Arch. Pharmacol. *261*, 271-288 (1968)

DISCUSSION

Wiemer: A striking feature of your recordings was that the alpha-excitatory activity needed more than 1 min to build up during stimulation. This is the longest time for any stimulus I know. Was this regular?

O'Regan: The effects that proved pharmacologically to be alpha-adrenergic appeared within 5-15 s of stimulation and kept building up as stimulation proceeded. I do not know when it would have reached a maximum value as I did not stimulate for longer than a minute.

Trzebski: Together with Majcherzyk I did experiments about the sympathetic influence on chemoreceptor discharge. By giving intracarotid injection of CO_2 in Locke's solution we could see a very strong sympathetic response of the ganglioglomerular nerve of the contralateral side, while there was not so marked a response in the other sympathetic outflow. So the question arises whether it is possible that the chemoreceptor sympathetic reflex has some modulatory effect on the chemoreceptors themselves. By cutting the sympathetic supply of the contralateral carotid body and stimulation of the opposite carotid area there was a very clear excitatory response in the chemoreceptor discharge of the denervated carotid body followed by an inhibitory period. So it may be that there is some kind of a tonic inhibitory influence from the sympathetic fibers.

O'Regan: By what mechanism did you get the excitation of discharge?

Trzebski: Perhaps these fibers are sensitive to the blood pressure rise, because in these experiments there is an immediate increase of the blood pressure.

O'Regan: I think that the inhibitory effects could explain your results. It is possible that the inhibitory effects occur more often, but I had to eliminate several experiments because I did not think the effects were genuine.

Belmonte: In a superfused preparation we could not see any effect of sympathetic stimulation. So perhaps the effects you have seen were vascular.

O'Regan: I do not dispute that they could be vascular, and I have tried additional experiments on blood flow. After giving haloperidol the flow was greatly reduced, and the sympathetic was inaffective in altering the flow.

Torrance: Do you imagine that the variety of effects arises because there are several groups of fibers coming from the cervical sympathetic ganglion to the carotid body? Have you done anything to try to call in reflexly one group rather than another so you might get your responses more regularly?

O'Regan: I cannot answer your question. It would be interesting to peel away filaments of the sympathetic and stimulate them independently, but I have not done it.

Kobayashi: I think McDonald demonstrated that the density of the small synaptic vesicles in the type I cell increases after 6-hydroxydopamine injections. What are your comments on the relation between McDonald's morphologic findings and your pharmacologic findings on the effect of 6-hydroxydopamine on the chemoreceptor discharge?

O'Regan: You are probably correct, it probably does affect the type I cells. I thought by using small doses I could get the effect limited to the adrenergic terminals.

McDonald: Actually we did not use 6-hydroxydopamine. We used a different isomer, 5-hydroxydopamine, and this substance increased the electron density of small vesicles in glomus cells but did not cause degeneration of noradrenergic nerve endings.

Zapata: When you gave hexamethonium, there is an increase of the resting discharge to a level that was very similar to the discharge during sympathetic stimulation previous to hexamethonium. Under those circumstances you did not find any effect of sympathetic stimulation. Did you try to return the resting discharge to a lower level similar to control level?

O'Regan: If you bring the discharge down to control levels, you get the same result as when the discharge was high. The levels of discharge did not seem to make much difference in this type of experiment. But the one I have shown has a sparse discharge and it is easy to reproduce these levels. If you have a very high rate of discharge, you get the same effects of the sympathetic and the same effects of hexamethonium.

Wiemer: What is in your opinion the mechanism for the slow rise of the alpha-excitatory activity, and is some substance produced that can be blocked by alpha-blocking agents?

O'Regan: I assume that this is a vasomotor effect. Why it builds up so slowly I cannot answer. But if you clamp the carotid arteries, which are more or less producing a partial ischemia of the carotid body, you also get a slow build up of discharge. I think that the oxygen levels are possibly dropping slowly.

Wiemer: The problem is that a vasomotor is slower than anything when you excite an efferent motor nerve.

O'Regan: If you clamp the carotid arteries, you can get an immediate increase of discharge or a gradual build up depending on the chemoreceptor preparation used.

Mechanism of Inhibition of Chemoreceptor Activity by Sinus Nerve Efferents

P. Willshaw

Although activity arising from the central end of the sinus nerve had been described by Joels and Neil at the Wates Symposium held at Oxford in 1966, it was not until 1968 that the first comprehensive account of this phenomenon was published by Biscoe and Sampson (2). They described two patterns of neuronal activity. The first type arose from sympathetic fibers, originating in the superior cervical ganglion, which joined the sinus nerve near its junction with the glossopharyngeal nerve and coursed back to the carotid body region within the sinus nerve itself. The second type of activity that could be recorded from the central end of the sinus nerve was not affected by preganglionic Sect. of the cervical sympathetic or by Sect. of the other buffer nerves and vagi in the neck. This type of activity will be referred to as sinus nerve efferent activity. Sinus nerve efferent activity is positively identified by its response to an intravenous injection of epinephrine, when a marked increase in firing rate is seen. Other stimuli for sinus nerve efferent activity are shown in Fig. 1. Efferent activity is enhanced by hypoxia and hypercapnia (2), by alkaline artificial cerebrospinal fluid (6), and by increased afferent activity in the same sinus nerve (8). Majcherczyk and Willshaw (in preparation) have found that hypocapnia may stimulate sinus nerve efferents.

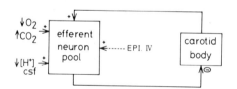

Fig. 1. Schematic diagram summarizing known stimuli to sinus nerve efferent fibers. EPI, epinephrine. Stimulus provided by CO_2 may be more complex than shown (see Willshaw, discussion Sect. of various papers)

It is now generally agreed that sinus nerve efferent activity travels to the carotid body chemoreceptor where it is capable of causing an inhibition of chemoreceptor afferent activity. The experimental evidence for such a conclusion is based upon two types of preparation: first, electrical stimulation of the main sinus nerve trunk while recording from a separated, unstimulated few-fiber slip of the nerve (3,7). Second, it is possible to record chemoreceptor activity from the otherwise intact sinus nerve and not a difference in response to a given stimulus such as hypoxia after cutting the nerve central to the recording site (8,10).

Although agreement on the inhibitory nature of sinus nerve efferents has been reached, there is a sharp division of opinion on the mechanism of efferent inhibition. This problem has been compounded by the recent publication of papers by Belmonte and Eyzaguirre (1) and McCloskey (5) and appears to be the result of the method of efferent stimulation used. The question at issue is whether the inhibition is mediated by a vasomotor action of the efferents or by their direct action upon the chemoreceptor specific cells.

At this point it is necessary to reconsider the results of Neil and O'Regan (7), Fidone and Sato (3), and Goodman (4). In these three papers chemoreceptor inhibition was induced by electrical stimulation of the sinus nerve trunk. Neil and O'Regan's basic finding was that stimulation of the nerve trunk resulted in the inhibition of chemoreceptor afferent activity recorded from a separate, unstimulated strand of the same nerve. They also noted a marked hyperemia of the carotid body, its blood flow almost doubling during some periods of stimulation. They found that local injection of atropine blocked the hyperemic response without affecting the electrically induced inhibition and concluded that the mechanism of inhibition was independent of blood flow effects.

Fidone and Sato (3) have shown that strong electrical stimulation of the sinus nerve could result in antidromic depression due to stimulus current spreading to and depolarizing the fibers on which chemoreceptor afferent activity was recorded. At lower stimulus strengths they could still observe inhibition of chemoreceptor activity in the absence of antidromic depression and termed this effect efferent inhibiton. They devised a method for testing which type of inhibition was taking place.

Goodman (4) criticized the results of Neil and O'Regan, basing his criticism on Fidone and Sato's work. Goodman argued that, because Neil and O'Regan had not monitored their preparation for antidromic depression, their observation of the persistence of chemoreceptor inhibition after blocking the hyperemic response with atropine was almost certainly a result of antidromic depression. Goodman repeated Neil and O'Regan's experiment, carefully avoiding antidromic depression, and found that atropine blocked the efferent inhibition. Although Goodman did not monitor carotid body flow, he came to the conclusion that efferent inhibition was mediated by the hyperemic response to electrical stimulation. Neil and O'Regan (7) presented cogent arguments in their paper to support the idea that stimulus current was not spreading to the carotid body itself. For example, application of procaine between the stimulating and recording sites blocked the effects of electrical stimulation. This would not have resulted if the effects were due to current spread. Also, examination of their published records does not reveal a poststimulatory rebound of chemoreceptor afferent activity, which is described as being a characteristic of antidromic depression by Fidone and Sato.

It seems obvious that experiments designed to investigate this problem further must fall into two classes. First, is it possible to obtain efferent inhibition when no blood is flowing, i.e., in the ischemic carotid body? Second, is it possible to devise an experiment in which blood flow is measured during inhibition but in the absence of antidromic depression? O'Regan (9) was the first to show that the ischemic discharge of the carotid body could be inhibited by intravenous injection of epinephrine. In this preparation the carotid body is arterially isolated for the common and external carotid arteries. Chemoreceptor afferent activity is recorded from the otherwise intact sinus nerve and the carotid body rendered ischemic by clamping the common carotid artery and opening the external carotid artery to the atmosphere. Chemoreceptor afferent discharge builds up an intense rate over a period

169

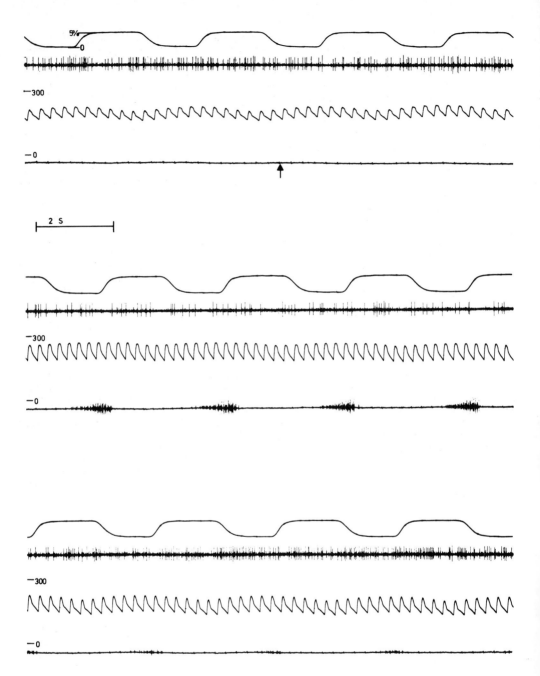

Fig. 2. Three records, originally continuous, are shown. *Top trace*: airway CO_2; *second trace*: chemoreceptor afferent activity recorded from otherwise intact sinus nerve; *third trace*: femoral arterial pressure (mmHg); *Fourth trace*: diaphragmatic electromyogram. Carotid body was ischemic throughout recording period. Ipsilateral ganglioglomerular nerve was cut. (*arrow*) 20 µg epinephrine injected intravenously. Cat. Sodium pentobarbitone anesthesia. Artificial ventilation

of 1 or 2 min. Intravenous injection of epinephrine at this point quickly evokes a depression of chemoreceptor discharge. This is illustrated in Fig. 2 (11). Epinephrine was shown by Biscoe and Sampson to evoke sinus nerve efferent activity, and because the inhibition of ischemic discharge is lost after sinus nerve Sect., the inference that sinus nerve efferent activity is capable of inhibiting chemoreceptor afferent activity in the absence of blood flow through the carotid body is not unreasonable. However, the argument can be put forward that under natural circumstances the direct inhibitory effect of sinus nerve efferents may complement a hyperemic action.

Willshaw (12) investigated this point by studying efferent-induced inhibition at the same time as measuring total carotid body blood flow. In this preparation the carotid body was arterially isolated so that carotid body blood flow could be measured from the rate of inflow of blood contained in a graduated pipette under pressure. Sinus nerve efferent activity was increased by superfusion of the ventral brainstem surface with alkaline cerebrospinal fluid. The sinus nerve was left untouched during this procedure. Later, a few-fiber slip was peeled from the otherwise intact nerve, and the changes in chemoreceptor afferent activity observed during the superfusion period. The result from such an experiment is shown in Fig. 3. An inhibition to about 80% of control afferent activity is seen during the superfusion period. The previous tests showed that blood flow did not vary by more than 5% during the superfusion period. These results contrast strongly with the supposition of Goodman (4) that total carotid body flow should show a very marked increase to induce this degree of inhibition.

In conclusion, it can be stated that chemoreceptor afferent activity is undoubtedly inhibited by sinus nerve efferent in the absence of carotid body blood flow and is not accompanied by significant changes in total blood flow when vascularly perfused. In both these experiments sinus nerve efferents were stimulated by nonelectrical methods. In the papers of Belmonte and Eyzaguirre (1) and McCloskey (5) electrical stimulation was used, and inhibition was not found in the absence of carotid body perfusion. It seems that two distinct inhibitory mechanisms exist in the carotid body, one of which is the sinus nerve efferent pathway. This pathway does not require the dilation of blood vessels to achieve its effects.

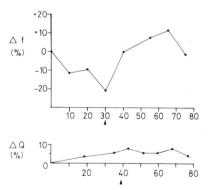

Fig. 3. Abscissa in both graphs shows time (s). Brain-stem superfusion with artificial cerebrospinal fluid at pH 7.8 started at time zero and ended at the points marked by the triangles. *Upper graph*: percent change in chemoreceptor afferent activity samples from otherwise intact sinus nerve. Ipsilateral ganglioglomerular nerve was cut. *Lower graph*: percent change in carotid body blood flow. Control blood flow was 65 µl/min. Cat. Chloralose-urethane anesthesia

References

1. Belmonte, C., Eyzaguirre, C.: J. Neurophysiol. *37*, 1131-1143 (1974)
2. Biscoe, T.J., Sampson, S.R.: J. Physiol. *196*, 327-388 (1968)
3. Fidone, S.J., Sato, A.: Brain Res. *22*, 181-193 (1970)
4. Goodman, N.W.: J. Physiol. *230*, 295-311 (1973)
5. McCloskey, D.I.: Respir. Physiol. *25*, 53-61 (1975)
6. Majchercyk, S., Willshaw, P.: J. Physiol. *231*, 26-27 (1973)
7. Neil, E., O'Regan, R.G.: J. Physiol. *215*, 15-32 (1971a)
8. Neil, E., O'Regan, R.G.: J. Physiol. *215*, 33-47 (1971b)
9. O'Regan, R.G.: in: The Peripheral Arterial Chemoreceptors. Purves, M.J. (ed.). New York: Cambridge University Press 1975, pp. 221-240
10. Sampson, S.R., Biscoe, T.J.: Experientia *26*, 261-262 (1970)
11. Willshaw, P.: Ph. D. thesis, University of London 1973
12. Willshaw, P.: in: The Peripheral Arterial Chemoreceptors. Purves, M.J. (ed.). New York: Cambridge University Press 1975, pp. 253-267

DISCUSSION

Belmonte: How do you measure blood flow? Even without having big changes in total blood flow, you can have a redistribution of the local blood flow.

Willshaw: Yes. But we can also get profound inhibition on the ischemic carotid body. O'Regan and I have developed methods to stimulate only efferent fibers. By stimulating electrically you are stimulating sinus nerve efferents, sympathetic fibers running in the sinus nerve, and, antidromically, chemoreceptor afferent fibers. So the big difference between us is that you in your preparations ascribe inhibition to vasomotor effects, which I certainly would not do because I use an entirely different method of stimulation.

Belmonte: Of course, I accept that. But I think the inhibition is produced by vasomotor changes. Your evidence is that the total blood flow changes are about 5%. We do not know how much that flow has to change to obtain an inhibition. Second, you can have flow redistribution in the carotid body during stimulation.

Willshaw: What I am saying is that the original postulate that 15% inhibition is caused by 200% change in total blood flow must be wrong.

Belmonte: You do not think that with epinephrine you can change the critical closing pressure, the venous pressure, and all the circulatory circumstances in the carotid body?

Willshaw: You can keep your carotid body ischemic and can repeatedly give injections of epinephrine. If there is any small amount of blood or plasma left in the carotid body whose flowing causes inhibition, then after repeated tests you should not have any flow left at all inside the carotid body.

Zapata: This physiological method of evoking efferent activity is not an evidence of efferent neurons but only of efferent activity. There is still the possibility that these effects would be mediated by some kind of primary afferent depolarization of central terminals.

Willshaw: There is evidence to suggest that efferent activity does not arise from primary afferent depolarization. The voltage of the action potentials of the efferent fibers is often very much smaller than the voltage recorded from primary afferents. If I were recording from the central end of the primary afferent, I would expect to get far bigger voltages.

O'Regan: Neil and I did not say that the vasomotor effects cannot affect the discharge. We said that there is a possibility a second mechanism may exist that is difficult to elicit. We did truly ischemic preparations, and all that remained between the carotid body and the rest of the animal was external carotid artery and common carotid artery.

Trzebski: It puzzles me that over a range of pH there is not a continuum of efferent activity of the chemosensitive medullary area when using artificial cerebrospinal fluid. The activity starts only when you enter the alkaline side.

Loeschcke: In these areas neurons exist that react positively to increases of hydrogen ion concentration. But these neurons lead into the central nervous system. The way in which the efferent fibers of the sinus nerve are influenced is unknown.

Torrance: You have the carotid body connected to the nervous system only by the sinus nerve. Have you any suggestion about what histologic structure may be associated with the response?

Willshaw: I am terribly sorry, but I cannot help.

Eyzaguirre: I wonder if there is an inhibitory pathway. First of all, we should have considered Verna's work with the rabbit - apposition of nerves to nerves.

Verna: Yes, I have seen some synaptic contacts between such cells that looked like nerve fibers. But the big problem is to distinguish between a nerve fiber and a cytoplasmic projection of a type I cell.

Willshaw: We should also consider the space constant of the fibers within the carotid body. I do not know whether the morphologists have investigated this. Some respiratory motor axons have a space constant in the order of centimeters. If there is some tight packing within the carotid body, we may also have large space constants.

McDonald: Ten days after cutting the glossopharyngeal nerve central to the petrosal ganglion, we found no change in the response produced by antidromic electrical stimulation of the carotid sinus nerve. In other words, we could evoke efferent inhibition of chemoreceptors even though this operation presumably caused the degeneration of efferent fibers in the sinus nerve. We have no evidence of a synaptic input to the chemoreceptive nerve endings from cells other than glomus cells. The question is, what is the mechanism of efferent inhibition after this operation has been done?

Willshaw: Maybe the so-called efferents are in fact primary afferents that are in some way depolarized. But I think they do actually operate in physiological conditions. We, in fact, no longer use alkaline cerebrospinal fluid; we hyperventilate, resulting in a very marked increase in efferent activity.

Purves: The efferent fibers appear to have at least two origins of which one is presumably sympathetic because the discharge disappears after the superior cervical ganglion is removed. Have you been able to distinguish between the efferent discharge, which responds to epinephrine, and the discharge that responds to superfusion of alkaline cerebrospinal fluid?

Willshaw: Yes, in all studies we made on true sinus nerve efferents, all these fibers do respond to alkali in the cerebrospinal fluid and also to epinephrine.

Bingmann: In histologic pictures of the sinus nerve you can see that up to 10-20 fibers are enveloped in a Schwann cell. It is possible that the efferent mechanism does not take place in the carotid body but in the sinus nerve by short circuit?

Willshaw: To prove this we have to get some microstimulating and microrecording electrodes right in the axons running closely together. I favor the idea that within the carotid body there are generator regions before reaching the myelinated part of the axon. There may be some interaction.

Further Studies on the Fluctuation of Chemoreceptor Discharge in the Cat

M. J. Purves and J. Ponte

The response of carotid body chemoreceptors to differing but steady levels of arterial blood gas tensions and pH has been extensively documented (e.g., 3), and further studies have shown that some form of interaction between hypoxic and hypercapnic stimuli occurs at chemoreceptor level (9,11). These observations are of value but of limited physiological importance because it is clear that these stimuli in arterial blood are rarely, if ever, steady. In addition to breath-to-breath variations in tidal volume, breatholding as in phonation, drinking etc., there is clear evidence of within-breath fluctuations of blood gas tension (16) and pH (1) to which the chemoreceptors respond with a fluctuation of discharge of similar period (2).

The question arises as to whether these fluctuations of chemoreceptor discharge have any physiological significance in addition to changes in the mean level of discharge. Such significance has been proposed under two conditions. First, it has been shown that a burst of chemoreceptor discharge caused chemically or electrically is maximally effective in enhancing the next breath if it coincides with the inspiratory phase of respiration (4,7). Second, such fluctuations of chemoreceptor discharge could be an important additional drive to ventilation if they are of the sufficiently large amplitude that could be predicted for active exercise where a relatively large gradient exists between mean mixed venous and alveolar levels of CO_2 and O_2 (6). A third possibility is that fluctuations of chemoreceptor discharge or a combination of the amplitude of the fluctuation and the mean discharge could provide the respiratory controller with information about the deviation of blood gas tensions and pH from optimum levels, a topic discussed in some detail by Priban and Fincham (15). It should be emphasized that most of the studies referred to above have been carried out in experiments in which the amplitude of fluctuations of the chemoreceptor discharge has been artificially increased and/or the animals were made to breathe at frequencies significantly lower than normal. The question referred to above cannot be answered until the experiments have been repeated under tolerably normal conditions.

The first attempt to provide an answer was that by Ponte and Purves (13) who formally measured the frequency response of chemoreceptors in the cat to P_aO_2, P_aCO_2, and arterial pH and found that, compared to the response to different steady levels, the response to sinusoidal stimuli was rapidly reduced for stimulus frequencies at or above the normal respiratory frequency (ca. 30-35/min). Only with CO_2 did the chemoreceptors respond more than to different steady-state levels, but only at a frequency substantially less than normal respiration, i.e., 10-15/min. These results, together with the responses to step functions in which the time to peak response to changes in CO_2 was ca. 1 s, those to changes in PO_2 and pH being rather longer, suggested that the response of the chemoreceptor complex, including the vascular component, was sluggish relative to most mechanoreceptors. The results also suggested that the chemoreceptors could probably respond phasically to slow transients associated with forms of breatholding, but that at normal respiratory frequencies the phasic component of their discharge pattern was likely to be extremely small and even smaller if respiratory frequency increased.

This study has been criticized on two grounds. First, the frequency response was expressed to individual chemical stimuli only and thus excluded possible interactions between O_2 and CO_2 that might have enhanced the magnitude of the fluctuations in the receptor discharge This certainly was possible though in our view unlikely, because the receptor response to hypoxia was substantially slower than to hypercapnia. The second criticism was that the use of valves and tubing could have caused microemboli and the release of vasoactive substances, which could have affected the dynamic sensitivity of the receptors. Accordingly, we have looked at the problem again during the course of experiments in which the respiratory response of cats, decerebrate or anesthetized with pentobarbitone sodium or a chloralose-urethane mixture, was measured when CO_2 was administered either by infusion via the inferior vena cava, thus simulating increases in metabolic rate, or by inhalation (14). A particular reason for looking at this point was the proposal that a very substantial, if not major proportion of the drive to ventilation when CO_2 was thus infused was derived from fluctuations in receptor discharge (5,12).

Two groups of experiments were carried out. In the first the cats breathed spontaneously, and responses to CO_2, given either by infusion or inhalation, were measured before and after bilateral cervical vagatomy. In the second group the cats were paralyzed and ventilated mechanically, the frequency of the pump being varied over, above and below the physiological range, while stroke volume was altered reciprocally to yield an approximately constant level of P_aCO_2.

Chemoreceptor discharge was measured from a single-fiber preparation of the sinus nerve, amplified, monitored on an oscilloscope face, and displayed as a ratemeter output in parallel with tidal volume, end-tidal CO_2, and, in some experiments, the output of a rapidly responding pH electrode placed in an ipsilateral external carotid artery-to-jugular vein loop. In each experiment the discharge was also stored on magnetic tape for subsequent computer analysis. Each respiratory cycle was divided into 20 time bins, and the discharge over number of respiratory cycles sufficient to yield 1000 to 1500 potentials was accumulated. The resulting histogram of the action potential distribution within the respiratory cycle showed varying degrees of respiratory modulation of the receptor discharge. The simplest way of describing this effect was to measure the difference between maximum and minimum discharge within the cycle and to express the modulation as ± percent of the mean discharge.

Fig. 1 illustrates the effect of cutting the second vagus nerve (the other having been cut 5 min previously) in a cat being perfused with CO_2 via the inferior vena cava and breathing rather faster than normal (42 breaths/min). Prior to the second vagotomy a relatively modulation of arterial pH could just be detected, and averaging the chemoreceptor discharge during this period yielded a modulation of ± 3.2% of the mean discharge, which is barely distinguishable from noise. Following the second vagatomy the respiratory modulation of the arterial pH signal was not quite obvious, and the modulation of chemoreceptor discharge now increased to ± 12.4% of mean discharge. This observation suggested a close inverse relation between respiratory frequency and amplitude of chemoreceptor discharge fluctuation, even though this fluctuation was observed to be substantially increased during CO_2 perfusion.

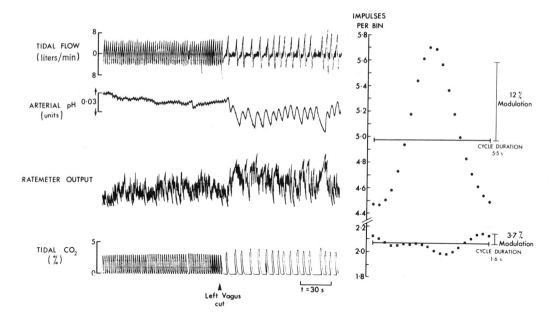

Fig. 1. Tracings of tidal flow, arterial pH, ratemeter output of a few-fiber chemoreceptor afferent discharge, and tidal CO_2. Right vagus had been previously cut and left vagus was cut during experiment (*arrow*). On the right, demodulated average discharge of chemoreceptor with left vagus intact (below) and with left vagus cut (above)

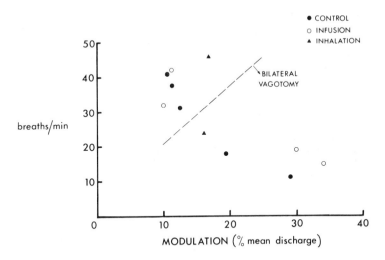

Fig. 2. Relation between frequency of respiration (breaths/min) and modulation percent mean discharge for one cat breathing spontaneously

Fig. 2 shows for one cat the expected relation between the modulation of chemoreceptor discharge and respiratory frequency. It should be noted that modulation of any size was observed only at the lower frequencies after vagatomy; at normal or higher frequencies the modulation

177

was very small indeed. More importantly, the responses of the chemoreceptor regardless of whether the cat inhaled or was infused with CO_2 at normal respiratory frequency were not distinguishable. From this result we may reasonably conclude that any increase in the magnitude of the discharge fluctuation with increased $\dot{V}CO_2$ is essentially offset by any increase in respiratory frequency.

Fig. 3 shows essentially the same relation in a typical experiment in a paralized, artificially ventilated cat. At constant PCO_2 (30 ± 2 mmHg) and at constant average chemoreceptor discharge, there is a clear inverse relation between the modulation of discharge and respiratory frequency. The point is made again that a normal respiratory frequencies or above the modulation of discharge is very small and, in many instances, indistinguishable from noise.

These findings are similar to those reported by Goodman et al. (10), although these authors did not formally relate the amplitude of discharge fluctuation with respiratory frequency. The conclusion is that, *for the cat* breathing normally at rest, fluctuations of chemoreceptor discharge, having the same period as respiration, are so small that they are unlikely to contribute significantly to respiratory drive. With respect to the question of respiratory control in exercise, our experiments similarly do not support the view that such fluctuations are important, because a simulated increase in metabolic rate sufficient to cause a four- to fivefold increase in ventilation was not accompanied by a significantly increased amplitude of modulation of the receptor afferent discharge.

But even assuming that with changes in metabolic rate beyond those that we have explored and with modulation of the discharge increasing significantly, it would still have to be shown that, to be effective in stimulating respiration, the peak of the fluctuation would have to be synchronized with the excitable phase of the respiratory neurons during inspiration. At rest the variability of the lung/carotid transit

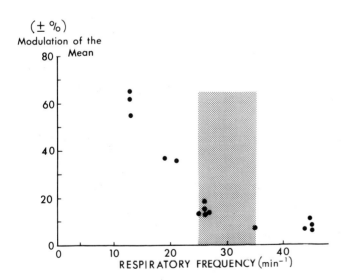

Fig. 3. Relation between modulation ± percent of mean and respiratory frequency for one cat paralized and artificially ventilated. *Hatched area*: range of normal respiratory frequency

time and of the respiratory cycle duration makes it unlikely that any such synchronization occurs. In exercise it is even less likely, because with the decrease in cycle duration it would be required that there was (a) an exactly proportional reduction in lung/carotid transit time, (b) a greater than proportional reduction in the variability of between-breath transit times, (c) a reduction in the phase lag of the chemoreceptor itself, and (d) some mechanism ensuring that the peak of the receptor discharge modulation coincided with the phase of increased excitability of the respiratory neurons. To prove that such exact matching exists and contributes to respiratory drive in exercise would require simultaneous measurement of a number of variables in a naturally exercising animal. This has not yet been attempted.

Until such time as these experiments are carried out and this particular question is resolved, our present view may be summarized as follows. Within-breath fluctuations of blood gases and chemoreceptor discharge occur and indeed are inevitable in animals that breathe tidally. Similarly, cyclic variations in excitability of respiratory medullary neurons occur (8,17) and are inevitable in an oscillatory neuronal system. However, it does not follow that because these phenomena exist they are related, that is, that the chemoreceptors feed back phasic information (even though a respiratory cycle or two delayed) in the same way as pulmonary stretch receptors and intercostal muscle spindles undoubtedly do. It is probable that the importance of the phasic responses of chemoreceptors has been exaggerated from experiments in which animals have been made to breathe at low frequencies. The results of our experiments persuade us, on the contrary, that in a naturally breathing animal and with moderate CO_2 loading, the phasic information from peripheral chemoreceptors is slight and unlikely to contribute significantly to respiratory drive.

References

1. Band, D.M., Cameron, I.R., Semple, S.J.G.: J. Appl. Physiol. *26*, 261-267 (1969)
2. Biscoe, T.J., Purves, M.J.: J. Physiol. *190*, 389-412 (1967)
3. Biscoe, T.J., Purves, M.J., Sampson, S.R.: J. Physiol. *208*, 121-131 (1970)
4. Black, A.M.S., Torrance, R.W.: Respir. Physiol. *13*, 221-237 (1971)
5. Cameron, I.R., Linton, R.A.F., Miller, R.: J. Physiol. *256*, 21-22 (1975)
6. Chilton, A.B., Stacy, R.W.: Bull. Math. Biol. *14*, 1-18 (1952)
7. Eldridge, F.L.: J. Physiol. *222*, 319-334 (1972)
8. Euler, C. von, Trippenbach, T.: Acta Physiol. Scand. *97*, 175-188 (1976)
9. Fitzgerald, R.S., Parks, D.C.: Respir. Physiol. *12*, 218-219 (1971)
10. Goodman, N.W., Neil, B.S., Torrance, R.W.: Respir. Physiol. *20*, 251-269 (1974)
11. Lahiri, S., DeLaney, R.G.: Respir. Physiol. *24*, 249-266 (1975)
12. Linton, R.A.F., Miller, R., Cameron, I.R.: Respir. Physiol. *26*, 383-394 (1976)
13. Ponte, J., Purves, M.J.: J. Appl. Physiol. *37*, 635-647 (1974)
14. Ponte, J., Purves, M.J.: Commun. Phys. Soc. (1976)
15. Priban, I.P., Fincham, W.F.: Nature, *208*, 339-343 (1965)
16. Purves, M.J.: Respir. Physiol. *1*, 281-296 (1966)
17. Trzebski, A., Lipski, J., McAllen, R.M.: This workshop (1976)

DISCUSSION

O'Regan: Your results fit in very well with what Wassermann showed for carotid body receptors in humans. These receptors contribute to exercise responses only when lactic acidemia is present.

Loeschcke: The amplitude of the oscillations for O_2 and CO_2 depends very much on the frequency of breathing. Plaas-Link and Müller made artificial oscillations of CO_2 by using a mixing system of equilibrated blood. They tried to trigger the artificial oscillations with the actual breath by the pneumotachograph. The result is that the oscillation is most effective as a ventilatory drive if it is perfectly superimposed on the breath.

Torrance: I do not think that your observations differ much from other observations on chemoreceptors. You do not get much of an oscillation in the discharge if respiration is at a normal frequency of 20-30/min at rest. However, in exercise you can increase the O_2 uptake 10-fold or more, and you increase the CO_2 production similarly. You would then expect much greater oscillations in the discharge of chemoreceptors, and so you would really have to consider their phase relation. We mimicked exercise, so far as metabolism is concerned, by giving dinitrophenol and found that there was a marked increase in the amplitude of the oscillations in discharge. What are the oscillations in P_aO_2 and P_aCO_2 after increasing the O_2 and CO_2 transport by dinitrophenol or by exercise?

Purves: I do not know if anybody can answer that question; I cannot. However, if CO_2 production is increased by infusing CO_2 via venous blood and if the respiratory pump frequency is held constant, the amplitude of of the pH fluctuation within each respiratory cycle increases, but not by as much as one would expect, probably because of the accompanying rise in end-tidal CO_2, which effectively diminishes the CO_2 difference between alveolar gas mixed venous blood. If the maneuver is repeated with the animal breathing spontaneously, no such rise in the amplitude of the pH fluctuation within breath is observed, and indeed, it either remains the same or is reduced. This is almost certainly the result of the increased frequency of respiration brought about by the CO_2 stimulus. Almost exactly the same changes are seen in the modulation of chemoreceptor discharge within each breath, and at normal or higher than normal respiratory frequencies it is very difficult to distinguish any modulation of the discharge at all. Furthermore, if the modulation of receptor discharge is plotted against frequency of respiration, it is not possible to distinguish between the values obtained when CO_2 was inhaled or when it was infused.

Willshaw: The various groups working on the problem of the oscillations in chemoreceptor afferent discharge are evenly divided as to whether a venous load of CO_2 causes an increase in ventilation. Cameron, Linton, and Miller have shown quite clearly that a load of CO_2 given intravenously is a far more effective ventilatory stimulus than CO_2 breathed in. Of course, Cunningham has shown that the pattern and timing of alveolar CO_2 can influence the way you breathe. There is an extremely important secondary effect, which is the cardiac output increases with CO_2 load.

Bingmann: In Bristol we have demonstrated that the venous blood pressure can effect chemoreceptor discharge. We have recorded oscillations of venous blood pressure in parallel with the discharge of chemoreceptors. Did you try to find such a correlation?

Purves: I have not measured carotid body venous pressure, so I cannot exclude that if influenced our results.

Willshaw: Band et al. were investigating these oscillations and found a better correlation between the chemoreceptor discharge and the pH cycle in the blood than with the respiratory cycle measured by the air flow. I think Bingmann's venous pressure changes would be more closely allied to the air flow changes.

Bingmann: These chemical agents have a marked effect, which exceeds the effects of venous pressure. But the question is still, what is the influence of the venous pressure?

Willshaw: By using a mixing chamber in the carotid artery the chemical oscillations can be mixed. Oscillations in chemoreceptor discharge can no longer be recorded, but you do not interfere with the venous system.

Torrance: Goodman found that, if he clipped the external carotid and so delayed the arrival of blood at the carotid body, he delayed the oscillations, but surely the venous pressure changes would have retained the same relation to ventilation. This would suggest chemistry rather than pressure.

Bingmann: Mitchell demonstrated that the venous blood pressure effect was a short-lasting effect, so the mechanism may be a very sensitive one that depends on the frequency and the oscillation. But to return to the initial question, has the experimental arrangement any effect on venous blood pressure that could be detected?

Purves: If the air way into the cat is obstructed, the changes in chemoreceptor discharge are delayed by about 2-3 s, in other words about the transit time from lungs to carotid body. If these changes were primarily the result of changes in venous pressure, they should be seen virtually instantaneously. This finding is an important reason for thinking that the effect is mainly chemical.

Torrance: Has anybody measured the pH oscillations of arterial blood and the changes during venous CO_2 loading?

Purves: Yes. We have found that the amplitude of the pH oscillations increases during CO_2 loading if respiratory frequency id held constant, even though end-tidal CO_2 rises and pH falls. However, in the spontaneously breathing cat, CO_2 loading causes a substantial increase in respiratory frequency, which largely offsets this increase in the amplitude of both pH and chemoreceptor oscillation.

Carotid Body Chemoreceptor Afferent Neurons in the Solitary Tract Nucleus Area of the Cat*

J. Lipski, R. M. McAllen, and A. Trzebski

Afferent input from the carotid body chemoreceptors to single brainstem neurons has been studied very little. Trzebski and Peterson (19) demonstrated that stimulation of the carotid chemoreceptors by close intra-arterial injection of lobeline influenced the pattern of discharge of inspiratory and expiratory medullary neurons. Similar results were recently obtained by Davies and Edwards (3). However, for the analysis of the chemoreceptor-induced sympathetic and cardiovagal reflexes, the input to those medullary neurons that are silent or do not show any obvious respiratory modulation of their spontaneous activity seems to be of more interest.

Afferent fibers from the carotid chemoreceptors (and baroreceptors) terminate centrally in the middle part of the nucleus of the solitary tract (NTS) at the level of the obex (2,11,12). Electrical stimulation of the carotid sinus nerve (CSN) excites single neurons in the NTS area and in other parts of the lateral medulla (1,3,8). In the experiments in which the intact CSN was stimulated, lateral baroreceptors could also be excited by carotid distension, and it was shown that some of the neurons excited by electrical shocks to the nerve did not respond to baroreceptor stimulation. Consequently, the suggestion was made that these neurons receive a projection from the carotid chemoreceptor afferents present in the CSN (10,15). Miura and Reis (18) reported that intracarotid injection of lobeline excited single nonrespiratory neurons within the dorsal medulla. However, the rather long latency neuronal responses, which overlapped in their records with the reflex blood pressure rise, might indicate that the responses were not evoked primarily by chemoreceptor stimulation of the CSN and that the neurons could be excited subsequently by stimulation of the carotid chemoreceptors with a latency shorter than that of the blood pressure rise. We have assumed that such neurons represent the primary projection of the carotid chemoreceptor afferents. Our experimental approach was similar to that proposed by McAllen and Spyer (1972) for identification of baroreceptor neurons in the medulla. Their procedure was also applied in our experiments, in addition to chemoreceptor stimulation, so the response of the same neuron to baroreceptor stimulation could be checked. In this way a possible interaction and convergence of both inputs from the carotid chemo- and baroreceptors could be analyzed on single medullary neurons. A short communication on the results has already been published (13).

---

*This study was supported by the Polish Academy of Sciences grant 10.4.

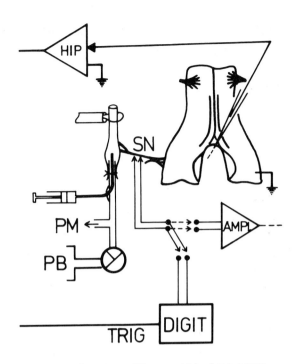

Fig. 1. Schematic representation of experimental setup. SN, carotid sinus nerve; PM, pressure monitor; PB, pressure bottle; AMPL and HIP, AC amplifiers; DIGIT, stimulator (Digitimer D4030); TRIG, trigger source of oscilloscope (Tektronix 5103) and computer (Anops-3). In upper right-hand corner exposed medulla oblongata and microelectrode are shown

Methods

Experiments were performed on 15 adult cats of both sexes (2.6-3.8 kg) anesthetized i.v. with chloralose-urethane (40 and 400 mg/kg, respectively). The animals were bilaterally vagotomized, subjected to pneumothorax, paralyzed with gallamine (Flaxedil 3 mg/kg), and artificially ventilated. End-tidal CO_2 was continuously recorded with Beckman LB-2 CO_2 gas analyzer and ventilation adjusted to keep expired CO_2 at 3.5-4%. Fig. 1 shows the schematic setup of the experimental arrangement (for more details, see (13)). In preliminary experiments we compared the effects of chemoreceptor stimulation by Locke-CO_2, lobeline (50 µg), and cyanide (1-5 µg) solutions on CSN and phrenic nerve activity. It was shown that the responses to Locke-CO_2 were much more prompt and abrupt, so this stimulus was used in most of the following experiments.

Results

Out of several hundred units tested, 37 responded to CSN stimulation with the latency below 20 ms. Among them 10 units were spontaneously active with a clear inspiratory (8 cells) or expiratory rhythm (2 cells). The respiratory units were not a subject of analysis here because the chemoreceptor input to the respiratory neurons was shown earlier (19) and will be subjected to more detailed analysis later (Lipski, McAllen, Spyer; in preparation). Out of the remaining 27 neu-

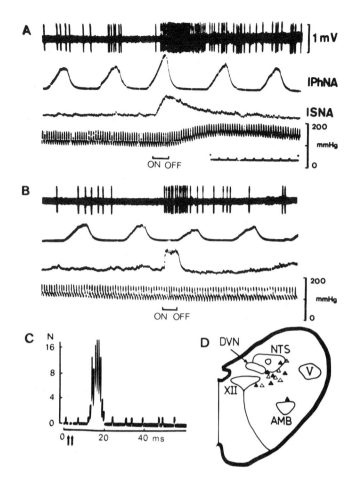

Fig. 2. Response of a unit to local chemoreceptor stimulation by Locke-CO_2 solution (A), to carotid baroreceptor stimulation (B), and to electrical CSN stimulation (C). (A and B) from above, single-unit discharge, integrated phrenic nerve activity (IPhNA), integrated sinus nerve activity (ISNA), and femoral artery blood pressure. (C) Poststimulus time histogram to a pair of stimuli (*arrow*). Time markers for A and B, 1 s. (D) Anatomic localization of 13 out of 16 chemoreceptor-sensitive neurons found. Neurons with baroreceptor convergence (*closed triangles*), neurons responding to chemoreceptors only (*open triangles*). NTS, nucleus of tractus solitarius; DVN, nucleus dorsalis nervi vagi; XII, nucleus nervi hypoglossi; AMB, nucleus ambiguus; V, nucleus nervi trigemini

rons, 19 fired irregularly and 8 were silent. We assumed that these neurons were not a part of respiratory complex. Eight of them tested with interruption of the ventilation and asphyxia for 30-40 s still did not reveal any respiratory rhythm. Nine of the 27 nonrespiratory neurons, which responded to electrical CSN stimulation, did not respond to either chemoreceptor or baroreceptor stimulation. Both stimuli evoked typical reflex response: increase of the arterial blood pressure and of the phrenic burst following a properly timed chemoreceptor stimulation, and opposite effects following longer-lasting (5-8 s) baroreceptor stimulation. The lack of response of these neurons to the physiological stimuli raises some doubt as to the physiological meaning of the technique of CSN stimulation alone. We can presumably exclude the possibility of the spread of the current to adjacent nerves, because the strength of the stimulus was kept at voltages below the threshold for the twitch of the surrounding muscles measured before injection of galamine (12).

Out of 18 remaining neurons 2 responded both to electrical stimulation of the CSN and to baroreceptor stimulation but not to chemoreceptor stimulation. The remaining 16 neurons responded both to CSN stimulation and to chemoreceptor stimulation (Fig. 2A and 2C; Fig. 3A). The latency to CSN stimulation was short and varied between 3 and 11 ms (mean 7.5 ms). The response to chemoreceptor stimulation had the form of a high-frequency burst of activity and started with a latency of less than 0.5 s measured from the beginning of the afferent barrage in the CSN. Thus, the neuronal response preceded the reflex rise in blood pressure, which appeared 2-3 s later. This excluded the possibility of secondary

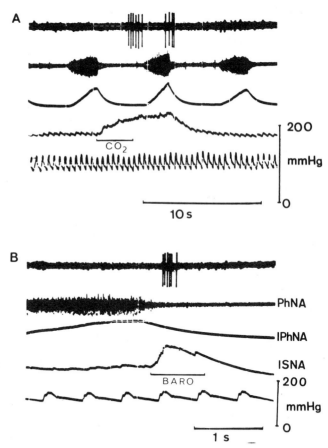

Fig. 3. Response of a unit to chemoreceptor stimulation by Locke-CO_2 solution (A) and to carotid baroreceptor stimulation (B). Unit was localized in vicinity of nucleus ambiguus (see Fig. 2D). From above, single unit discharge, phrenic nerve activity, integrated phrenic nerve activity, integrated carotid sinus nerve activity, femoral artery blood pressure. Excitatory response could be evoked only when stimulus was applied in expiratory phase. Prolonged burst of activity was interrupted during subsequent phrenic nerve burst and resumed again in following expiration

unit response to the blood pressure changes. Surprisingly, when tested with carotid sinus distension, 9 out of 16 chemoreceptor-sensitive neurons were excited by both baro- and chemoreceptor stimulation (Figs. 2 and 3). The localization of the neurons is shown in Fig. 2D. One neuron with convergence was identified within the nucleus ambiguus (Figs. 2D and 3). The unit could be excited only during the expiratory cycle (Fig. 3). Following chemoreceptor stimulation during expiration a prolonged burst of activity started, which was inhibited during inspiration but resumed again in the next inspiration.

Discussion

We identified the cells in the dorsal medulla that are functionally close to the chemoreceptor afferent input. They do not seem to be a part of the respiratory complex of the brain-stem, as no respiratory modulation of their spontaneous activity was present even in the conditions of severe asphyxia. It can not be excluded, however, that they relay to respiratory neurons.

Whatever the final projection of these neurons may be to the respiratory, sympathetic, or vagal preganglionic neurons, the most surprising finding is the synergistic convergence of the baroreceptor and chemoreceptor inputs onto about half of the identified medullary neurons. The site of this convergence seems to be mainly central. We have tried to exclude the possibility of peripheral interaction of the baroreceptor and chemoreceptor stimuli at the receptor level. Great care was taken to avoid any intracarotid pressure change during chemoreceptor stimulation and to remove entirely the chemical agent before baroreceptor stimulation was applied. The experimental procedure did not result in simultaneous stimulation of both kinds of receptors with the stimuli specific for one of them. However, the possibility should be considered of a nonspecific stimulation either of baroreceptors by cyanide or Locke-CO_2 solution, or vice versa, a stimulation of chemoreceptors by mechanical stimuli. However, there are indications against the possibility that cyanide excites baroreceptor fibers, both myelinated and nonmyelinated (4). Our own unpublished observations did not show any excitatory influence of Locke-CO_2 on the myelinated carotid baroreceptor fibers (see also (6)). If such a nonspecific effect exists, it is hardly possible that it could appear with a latency as short as that observed in our experiments. On the other hand, we cannot ignore the possibility that sudden mechanical stimuli may influence chemoreceptor discharge (17). Recent observations on the proximity of baroreceptor endings within the carotid body vessels (5) are an indirect hint at the feasibility. Functional significance of synergistic convergence of baro- and chemoreceptor input onto single medullary neurons could be explained as a pathway for cardiac vagal motoneurons, which are excited both by baroreceptor and by chemoreceptor stimulation. One neuron found in the nucleus ambiguus, where cardiac vagal motoneurons are located (16), may be example of this population.

Another possible function of these neurons may be a multisynaptic pathway to the sympathetic neurons, which ultimately supply skin vessels and which are inhibited by systemic hypoxia, presumably through a chemoreceptor reflex (9). In view of the unexpectedly large proportion of neurons that show synergistic convergence, one cannot exclude the possibility that they are involved in both reflexes. They could also function as a kind of antagonistic, self-limiting, negative feedback system, which, on one hand, limits excessive excitatory effects of chemoreceptor stimulation and, on the other hand keeps in check baroreceptor-induced inhibition.

Expiration-locked response to chemoreceptor stimulation of the single neuron within the nucleus ambiguus corresponds well with the recent results suggesting a direct inhibitory input from the inspiratory neurons to the cardiac vagal motoneurons within the nucleus (14). We were unable to find in the present study a clear antagonistic interaction of both baroreceptor and chemoreceptor inputs. Antagonistic interaction of both reflexes shown at the preganglionic sympathetic fibers (20) presumably takes place at a later stage of the central reflex pathways, elsewhere in the brain stem, or at the spinal cord level.

References

1. Biscoe, T.J., Sampson, S.R.: J. Physiol. *209*, 359-374 (1970)
2. Cottle, M.K.: J. Comp. Neurol. *122*, 329-343 (1964)
3. Davies, R.O., Edwards, M.W.: Respir. Physiol. *24*, 69-79 (1975)
4. Fidone, S.J., Sato, A.: J. Physiol. *205*, 527-548 (1969)
5. Gorgas, K., Böck, P.: This workshop
6. Haymet, B., McCloskey, D.I.: J. Physiol. *245*, 699-712 (1975)
7. Hellon, R.F.: J. Physiol. *214*, 12 (1971)
8. Humphrey, D.R.: in: Baroreceptors and Hypertension. Kezdi, P. (ed.). New York: Pergamon Press 1967, pp. 131-168
9. Jänig, W., Kümmel, H.: Pflügers Arch. *365*, R37,1 (1976)
10. Lipski, J., Trzebski, A.: Pflügers Arch. *356*, 181-192 (1975)
11. Lipski, J., McAllen, R.M., Spyer, K.M.: J. Physiol. *225*, 30-31 (1972)
12. Lipski, J., McAllen, R.M., Spyer, K.M.: J. Physiol. *251*, 61-78 (1975)
13. Lipski, J., McAllen, R.M., Trzebski, A.: Brain Res. *107*, 132-136 (1976)
14. Lipski, J., McAllen, R.M., Spyer, K.M.: J. Physiol. (1977) (in press)
15. McAllen, R.M., Spyer, K.M.: J. Physiol. *222*, 68-69 (1972)
16. McAllen, R.M., Spyer, K.M.: J. Physiol. *244*, 82-83 (1975)
17. Majcherczyk, S., Chruścielewski, L., Trzebski, A.: Brain Res. *76*, 167-170 (1974)
18. Miura, M., Reis, D.: J. Physiol. *223*, 525-548 (1972)
19. Trzebski, A., Peterson, L.H.: in: Drugs and Respir.. New York: Pergamon Press 1964, p. 59
20. Trzebski, A., Lipski, J., Majcherczyk, S., Szulczyk, P., Chruścielewski, L.: Brain Res. *87*, 227-237 (1975)

DISCUSSION

Nishi: By injection of a CO_2 saturated solution into the carotid body region in the cat, bradycardia can usually be evoked. But you did not show anything about bradycardia. If you inject sodium cyanide 15-17 s later, you can see an increase in the baroreceptor activity of the sinus nerve. Even though you stimulated the specific chemoreceptor, the input to the central nervous system is complicated by the stimulation of the chemoreceptor activity together with the baroreceptor activity.

Trzebski: If you have a vagatomized cat, you never see a bradycardia when stimulating the carotid body chemoreceptors provided you prevent a blood pressure increase, which secondarily stimulates baroreceptors and is responsible for indirect inhibition of the caridiac sympathetic fibers activity (Trzebski et al.: Brain Res. 87, 227-237, 1975). In this response a stabilized ventilation plays an important role because of the pulmonary stretch receptors. But it is also very important to have a centrally modulated vagal spontaneous activity. You can very easily inhibit the vagal tone, and then even with the chemoreceptor stimulation you have tachycardia. This inhibition can be done by some kinds of anesthetics. To the second problem, perhaps Fidone will respond because he has shown that cyanides do not excite the baroreceptor.

Fidone: Yes, that is usually the case, but we also recorded a small number of fibers that had a very regular pattern of discharge and a chemosesitivity to sodium cyanide (J. Physiol. 205, 527-548, 1969). We could not be sure whether these were chemoreceptors or baroreceptors, although at that time we concluded that they most probably were chemoreceptors since they did not respond to mechanical stimulation of the sinus area.

Paintal: Sodium cyanide is, of course, a very dangerous drug. Fidone showed in 1969 that the baroreceptor is stimulated by cyanide; we have also found that the gastric stretch receptors are stimulated by cyanide.

Trzebski: Most of the results presented here are results of using CO_2 as a stimulus. The problem is whether CO_2 or hydrogen ions excite baroreceptors. Only in a very early stage of our experiments did we use cyanide, which we then abandoned because its effects are smaller and smaller with repeated applications, showing a kind of tachyphylaxis. On the other hand, it is possible to have very short and very rapid responses with CO_2, which can be very nicely timed to the respiratory cycle. So most of the experiments here are done with CO_2, and cyanide is not important enough to be included.

Willshaw: You should not be too worried about this nonspecific effect of CO_2 on mechanoreceptors. In the integrated phrenic activity the chemoreceptor stimulus gave an increased firing in the phrenic nerve, and the baroreceptor stimulus gave a decrease in the firing of the phrenic nerve.

Trzebski: It is our opinion that there is a predominance of the baroreceptor with mechanical stimulation and a predominance of the chemoreceptor with chemical stimulation. The problem is whether there is some kind of mutual contribution that is not manifested in the phrenic response.

Zapata: With respect to the barosensory nerve terminals that Böck has shown, in some experiments we did with Eyzaguirre with both normal and regenerated carotid nerves, we performed a crush between the carotid body and sinus to avoid contamination from barosensory fibers of the carotid sinus. In both cases we recorded a few barosensory fibers that originated in the carotid body. The potentials of those fibers were of higher amplitude than those of most chemosensory fibers.

Trzebski: May I ask you if these particular fibers responded to chemical stimuli or did you observe any spontaneous activity of these fibers?

Zapata: If sodium cyanide was given in large doses, the fibers did not respond at the same time as chemosensory fibers. But when the blood pressure went up, their discharges were very well correlated with these vascular changes, but not with the discharge of chemosensory fibers.

Trzebski: They cannot be taken into account in these experiments, because in all our responses we were very careful to include only very early response, earlier than the pressure response. We selected CO_2 because it always gave these immediate, fast effects. The response was so early that it could be clearly distinguished in time from the secondary effect of blood pressure changes.

O'Regan: Neil and I years ago were recording efferent activity in the otherwise intact sinus nerve, and we utilized the natural branches of the carotid body nerve near the carotid body. We found that you can record baroreceptor activity, which goes up the sinus nerve and back down to the carotid body. Perhaps there could be some interaction by this mechanism.

Trzebski: May be.

Fidone: Just a delayed reply to Paintal's earlier comment about dangerous drugs: it was acetylcholine that was the dangerous drug in 1969, not sodium cyanide. We obtained the response of the baroreceptors only to acetylcholine and not to sodium cyanide.

Multifactor Influences on the Functional Relationship Between Ventilation and Arterial Oxygen Pressure*

H. Kiwull-Schöne and P. Kiwull

The functional significance of the peripheral chemoreflex in the hypoxic drive of pulmonary ventilation has been demonstrated for most laboratory animals, but experimental data are not sufficient to analize the functional relationship between ventilation ($\dot{V}$) and arterial oxygen pressure (P_aO_2) by means of mathematical statistics. The mathematical treatment of human steady-state P_aO_2-$\dot{V}$ responses (3,9) was developed by Cunningham's and Severinghaus's groups (4,8). Since in human subjects experimental procedures, i.e., application of too low oxygen concentrations, arterial measurements, and denervations, are limited, a more extensive, quantitative study on the peripheral respiratory control system was attempted in anesthetized animals.

Methods

The experiments were carried out on 50 rabbits weighing 3.03 ± 0.54 kg (± SD). An initial i.v. dose of pentobarbital sodium of 40.9 ± 8.7 mg/kg was given for the 1-2 h operation. A subsequent continuous infusion of 7.8 ± 2.7 mg/kg/h was sufficient to maintain a light anesthesia. $\dot{V}$ was determined by pneumotachography, entidal CO_2 pressure (P_aCO_2) by infrared absorption, and (P_aO_2) by polarography. In addition, blood pressure and heart rate were continuously recorded, and from arterial samples PCO_2 and acid-base values were determined. Pure O_2 and O_2 at different fractions with N_2 or N_2/CO_2 (1.0, 0.12, 0.21, 0.07, and again 1.0) was applied for inhalation for 10-12 min. Steady-state values of $\dot{V}$ were evaluated at different levels of P_aO_2 with {1} hypocapnia resulting from hypoxic hyperventilation, {2} constant P_aCO_2 kept at the hyperoxic control level, or {3} both carotid sinus nerves cut. Conditions 1-3 were investigated further after vagal reflexes has been eliminated, usually by reversible cold block.

For evaluation of the average P_aO_2-$\dot{V}$ relationship in the whole population, group mean values of $\dot{V}$ were calculated from the individual data at iso-P_aO_2 levels of 600, 100, 80, 60, 40, 30, and (25) mmHg. When directly measured values were not available at a desired P_aO_2, they were interpolated. The significance of experimentally induced effects on the resulting mean values (Fig. 1) was studied by a paired Students t-test of within animal comparisons. The best-fitting equation for the description of the group mean values as a function of P_aO_2 was tested for the rectangular hyperbola $\Delta\dot{V}=\dot{V}-\dot{V}_o= a(P_aO_2)^{-1}$, the power function $\Delta\dot{V}=\dot{V}-\dot{V} = a(P_aO_2)^{-b}$, and the exponential function $\Delta\dot{V}=\dot{V}-\dot{V}_o= a\exp^{-b}(P_aO_2)$, where $\dot{V}$ signifies the control level of ventilation during hypoxia. The parameters a and b were evaluated by the method of least squares. The

---

*This work was supported by the Deutsche Forschungsgemeinschaft, SFB 114.

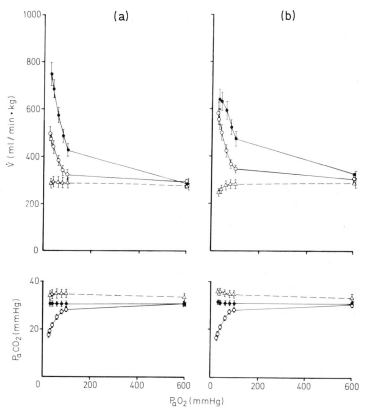

Fig. 1A and B. P_aO_2 response of ventilation (*upper diagrams*) and end-tidal CO_2 pressure (*lower diagrams*) with (A) intact and (B) eliminated vagus nerves. Group mean values (±SE) at iso-P_aO_2 levels for 19 rabbits with free running P_aCO_2 (open circles), 23 rabbits with P_aCO_2 kept constant at the hyperoxic control (closed circles), and 23 rabbits with cut carotid sinus nerves (open triangles)

three proposed functions were subjected to an F-test (10): The hypothesis that H_o, the variance of group mean values about the regression line, is equal to the sum of variances of single values within each group was rejected with an alpha risk (α) calculated from the F-distribution.

Results

1. Mean values of $\dot{V}$ and P_aCO_2 (±SE) at different levels of P_aO_2 (Fig. 1): During *hyperoxia* control values of ventilation - those in parentheses during vagal block - were in the hypocapnic group 292±15.0 (307±14.8)ml/min/kg, in the normocapnic group 285±15.8 (326±18.2)ml/min/kg, and in the chemodenervated group 279±20.3 (289±19.4)ml/min/kg. Correspondingly, control values of P_aCO_2 were 30.7±0.8 (30.6±0.8) mmHg, 30.7±1.1 (31.0±1.1) mmHg and 33.7±1.6 (33.4±1.6) mmHg. During *normoxia* an increase of $\dot{V}$ by 9.7±1.7% (13.1±2.2%) above the control value was found in the hypocapnic group, resulting in a decrease of 2.5±0.4 (2.6±0.3) mmHg of P_aCO_2. Correspondingly, an increase of $\dot{V}$ by 51.1±6.8% (43.3±5.0%) without change of P_aCO_2 was found in the normocapnic group, and no significant change in $\dot{V}$, 3.2±1.8% (-1.7±1.3%), but a significant

rise in P_aCO_2 of 1.2±0.4 (1.2±0.3) mmHg in the chemodenervated group. During *hypoxia* the maximum increase of $\dot{V}$ in the hypocapnic group was 171.7±7.3% (187.3±10.4%) of the control, resulting in a fall of P_aCO_2 of 13.2±1.2 (14.2±0.9) mmHg; in the normocapnic group this was 268.4± 16.6% (198.0±11.2%) of $\dot{V}_o$, and in the chemodenervated group there was a significant change neither in $\dot{V}$, 2.7±2.5%, nor in P_aCO_2, 0.4±0.5 mmHg, provided the vagus nerves were left intact. Without vagal reflexes a marked hypoxic depression of $\dot{V}$ (-13.6±3.1% of $\dot{V}_o$) was followed by a distinct increase of P_aCO_2 (2.2±0.5 mmHg).

2. Test for the probability of three different functional relationships between $\dot{V}$ and P_aO_2 (Fig. 2): Group mean values of $\Delta\dot{V}=\dot{V}-\dot{V}_o$ and P_aO_2 were plotted for comparison in different systems of coordinates with either the ordinate or both ordinate and abscissa transformed into logarithmic scales. In order to find evidence of the linearity of the transformed data, the best fitting straight lines with either variable slopes (Fig. 2B and C) or a slope of -1 (Fig. 2A) were evaluated. The plot was most closely related to the straight line representing the logarithmic transformation of an exponential function and least of all to that of a rectangular hyperbola. In order to substantiate this evidence by statistical means, the curvilinear regression analysis of the three proposed functions was subjected to the F-test described above. The exponential statement was the most probable one, being highly significant at the 0.5 and 1.0% level of α, except for the normocapnic hypoxia

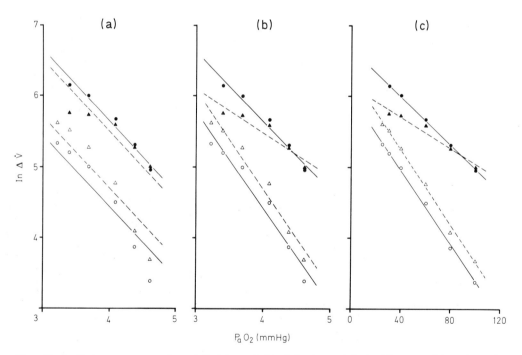

Fig. 2A-C. P_aO_2 response of ventilation plotted in logarithmically transformed systems of coordinates. Group mean values of $\Delta\dot{V}$ with intact vagus nerves, P_aCO_2 free running (open circles) or constant (closed circles), as well as with eliminated vagus nerves under same conditions (open triangles, closed triangles). Linear regression lines correspond to logarithmic transformation of : (A) rectangular hyperbola, (B) power function, and (C) exponential function (same population as Fig. 1)

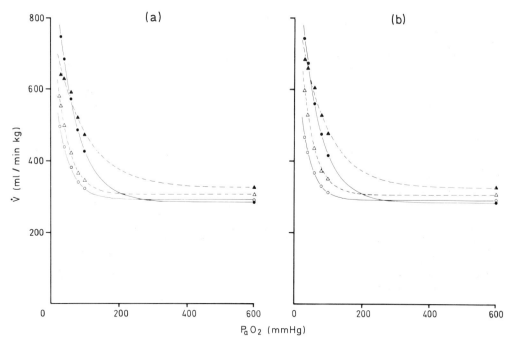

Fig. 3A and B. Curvilinear fit of P_aO_2 response of ventilation by exponential function $\dot{V}-\dot{V}_o=a\exp^{-b(P_aO_2)}$ under different experimental conditions. $\dot{V}_o$ refers to hyperoxic control value. The parameters a and b are listed in Table 1. (A) Group mean values of P_aO_2 response of ventilation with intact vagus nerves, P_aCO_2 free running (open circles) or constant (closed circles), as well as with eliminated vagus nerves under same conditions (open triangles, closed triangles). (B) Calculated chemoreflex portion of P_aO_2 response shown in A, considering change of ventilation after carotid body denervation (same population as Fig. 1)

response with eliminated vagi. Under all experimental conditions among the other two equations, the rectangular hyperbola was less probable (α 60-99%) than the power function (α 50-60%).

3. Ventilation as an exponential function of P_aO_2 under different experimental conditions (Fig. 3): The best-fitting parameters of the function $\dot{V}-\dot{V}_o=a\exp^{-b(P_aO_2)}$ are listed in Table 1 and illustrated by the regression lines in Fig. 3. $\dot{V}_o$ corresponds to the asymptote of the curves toward high oxygen tensions; a refers to the overall sensitivity factor represented by the point of intersection with the y-axis, b intensifies or weakens the slope differences of the steeper hypoxic and the flatter hyperoxic part, and k is inversely related to b indicating the change of P_aO_2 required for a change of $\dot{V}$ by the exponential factor. The experimental conditions are reflected by parameter changes tested for significance by a paired Student's t-test. Since the total hypoxic response includes both the chemoreflex ventilatory drive and a direct effect of hypoxia upon the respiratory system, the pure chemoreflex was evaluated while considering the hypoxic change of $\dot{V}$ after chemodenervation. There were no significant parameter differences between the total hypoxic and the chemoreflex response with intact vagus nerves. Without vagal reflexes a significant reduction in the overall sensitivity of the hypoxic response was indicated by a pronounced decrease

Table 1. Best-fitting parameters of the Estimated Function $V-V_o = a\exp^{-b(P_aO_2)}$
[a] (A) Total hypoxic response and (B) chemoreflex response of ventilation with intact and eliminated (*) vagus nerves

	Experimental conditions[a]	a (ml/min/kg)		b (mmHg)		$k=1/b$ (mmHg)	
A	Hypocapnia	413.5	559.8*	0.0265	0.0269*	37.8	37.1*
	Normocapnia	785.7	469.9*	0.0170	0.0111*	58.7	89.8*
B	Hypocapnia	457.5	653.3*	0.0308	0.0272*	32.4	36.7*
	Normocapnia	785.3	544.8*	0.0177	0.0123*	56.6	80.8*

in parameter a, whereas b remained unchanged. This emplies that blocking of the vagi leads to an overall sensitivity gain of the hypocapnic chemoreflex drive. With intact vagi the normocapnic differed from the hypocapnic chemoreflex drive by a significant increase in a but not in b, whereas after removal of vagal reflexes only b decreased significantly. This implies that blocking of the vagi causes a smoothing over of the normocapnic chemoreflex drive without sensitivity gain.

Discussion and Summary

Anesthesia was kept as light as possible so that an average normoxic $\dot{V}$ within the range reported for conscious rabbits (2,5) was achieved. Moreover, in order to avoid systematic errors due to anesthesia, the initial and final control values of $\dot{V}$ were not allowed to deviate by more than 10%. Nevertheless, about 20% reduction of the hypocapnic P_aO_2-$\dot{V}$ response had to be considered in the range of 25-30 mmHg P_aO_2 when compared with that of conscious rabbits (2) or humans (9).

As regards the functional relationship between P_aO_2 and $\dot{V}$, the exponential equation yielded without exception the smallest variances of values about the best-fitting repression line as compared to the power function or the rectangular hyperbola. This, in principle, agrees with the analysis carried out in humans by Kronenberg et al. (8) as far as the normocapnic condition with intact vagus nerves is concerned, but not with Cunningham's mathematical model (4) based on the rectangular hyperbola, which requires additional assumptions for the severe hypoxic range and the interaction with CO_2. The two parameters of the exponential function used in this paper were sufficient to describe all investigated conditions. The overall sensitivity parameter a was generally influenced by processes proportional to the stimulus (P_aO_2) or to the respiratory reaction. This was true first for the direct, not reflexogenic action of hypoxia on ventilation, provided vagal reflexes were eliminated. With intact vagi the hypoxic depression of $\dot{V}$ was balanced out by a changed pattern of breathing (7). Whether this is a central effect requiring vagal input or a hypoxic activation of vagal afferents remains uncertain. The second mechanism responsible for changing the sensitivity factor could be a volume-related inhibitory reflex, because vagal cold block significantly increased a in the hypocapnic chemoreflex drive. Since, in contrast, under normocapnic blood gas conditions vagal block lowered a, other vagal mechanism (possibly dependent on the inspiratory CO_2 (1)) should also be taken into account. The third effect, determinative for parameter a, is an increasing loss of CO_2 drive in proportion to the hypoxic hyperventilation when not kept normocapnic. The interpretation of parameter differences between the normocapnic and the hypocapnic response curve is complicated by interaction phenomena of CO_2 and O_2 (6), reflected by parameter b. No significant difference in b could be demonstrated with intact vagi, indicating that there was no considerable change in CO_2 sensitivity of $\dot{V}$ at any O_2 level. In contrast, the parameter b was significantly decreased in the normocapnic response with eliminated vagi, whereas parameter a remained unchanged. This indicated a decreasing CO_2 sensitivity with the drop in P_aO_2, which agrees with the hypothesis that the mode of central processing of different respiratory drives is primarily occlusive and independent of blood gas conditions (6).

To summarize, the functional relationship between ventilation and P_aO_2, investigated in anesthetized rabbits by statistical methods, was found to be of exponential mode. The functional parameters showed a variability dependent on experimental conditions rather than indicating of a unique function. This indicated a complex interrelationship between the peripheral chemoreflex, directly acting effects of hypoxia, inhibitory as well as facilatory vagal reflexes, and finally the central CO_2 drive of respiration.

References

1. Bartoli, A., Cross, B.A., Guz, A., Jain, S.K., Noble, M.I.M., Trenchard, D.W.: J. Physiol. *240*, 91-109 (1974)
2. Chalmers, I.P., Korner, P.I., White, S.W.: J. Physiol. *188*, 435-450 (1967)
3. Cormack, R.S., Cunningham, D.J.C., Gee, J.B.L.: J. Exp. Physiol. *42*, 303-319 (1957)
4. Cunningham, D.J.C.: in: Respiratory Physiology. Physiology. Guyton, A.C., Widdicombe, J.G. (eds.). London: Butterworths University Part Press 1974, Ser. 1, Vol. II, pp. 303-369
5. Honda, Y.: Respir. Physiol. *5*, 279-287 (1968)
6. Kiwull, P., Kiwull-Schöne, H., Klatt, H.: in: Acid Base Homeostasis of Brain Extracellular Fluid. Loeschcke, H.H. (ed.). Stuttgart: Thieme 1976, pp. 146-158
7. Kiwull-Schöne, H.F., Kiwull, P.J.: Fed. Proc. *35*, 553 (1976)
8. Kronenberg, R., Hamilton, F.N., Gabel, R., Hickey, R., Read, D.J.C., Severinghaus, J.: Respir. Physiol. *16*, 109-125 (1972)
9. Loeschcke, H.H., Gertz, K.H.: Pflügers Arch. *267*, 460-477 (1958)
10. Sachs, L.: Angewandte Statistik 4th ed. Berlin-Heidelberg-New York: Springer-Verlag 1974

DISCUSSION

Willshaw: This is another paper in which no mention of sinus nerve efferents is made. The sinus nerve efferent activity does change during the development of hypoxia. Sampson and Biscoe showed that the oxygen response curve of the cut sinus nerve is significantly different from that of the intact sinus nerve. Also we found that if we cause the arterial CO_2 to fall by hyperventilation, we produce a very marked increase in the sinus nerve efferent drive.

Kiwull-Schöne: Sinus nerve efferents could play a role when there is hypocapnia resulting from hypoxic hyperventilation. According to Trzebski these efferents are activated by an alkaline shift in the cerebrospinal fluid. We have measured only the arterial pH, but we never observed such an alkaline shift, which was reported to produce efferent activity. Furthermore, one has to consider that the pH values in the cerebrospinal fluid must be even more acid than in the arterial blood.

Willshaw: I disagree with the last statement. Yes, the shifts in the cerebrospinal fluid toward alkalinity were very gross when artificial cerebrospinal fluid was used. But in our latest experiments we effectively repeated that work by hyperventilation of the animal to bring the CO_2 down from a control level of 4% end-tidal to 2.5%. This is possibly what is going to happen in hyperventilating animals. We got an extremely marked increase in efferent drive from that drop in CO_2.

Kiwull-Schöne: What was the pH value in the arterial blood?

Willshaw: We did not measure the pH in the arterial blood, because what would that have told us?

Kiwull-Schöne: It would be very interesting for comparison of actual alkalinity during artificial ventilation and the hypoxic hypercapnia in our experiments.

Willshaw: Yes, but the pH of the arterial blood has got very little to do with the pH of the cerebrospinal fluid.

Kiwull-Schöne: Considering the direction, we have to expect more acidic values in the cerebrospinal fluid than in the arterial blood.

Willshaw: In absolute terms, but we have been talking about a shift in cerebrospinal fluid pH. When the CO_2 goes down, the cerebrospinal fluid pH shifts alkaline and we get efferent activity.

Paintal: Just a small point that I am sure you must have considered. The addition of CO_2 has an inhibitory effect on pulmonary stretch receptors. This has been shown for the dog and the cat. I do not know whether it has been shown for the rabbit.

Kiwull-Schöne: No, not directly but there is some evidence for it.

Paintal: I think you might consider this after you have proved it in the rabbit.

O'Regan: We do not have to involve cerebrospinal fluid changes because Neil and I showed that the input from the carotid body can effect the efferents that come down and suppress the chemoreceptor activity. But what I want to ask is, how do these studies relate to experiments carried out in a nonanesthetized animal? It seems to me that your ventilation increased at about the same level as the chemoreceptor discharge increased. I would be different, I would suppose, if the animals were unanesthetized. The ventilation increased at about 100 mmHg in the arterial blood, whereas if you do studies on unanesthetized animals, you do not get increased ventilation until 60 mmHg in the arterial blood. Have you compared the animals with unanesthetized ones?

Kiwull-Schöne: We did not try. We only compared our results for anesthetized animals with the results obtained by other authors for unanesthetized animals, and we were in the same range. The rabbits were only lightly anesthetized by permanent infusion of pentobarbital, and we tried to keep the amount very low. Some of the animals had strong motor reactions to hypoxia, which indicates that our anesthesia was very light. We also tried to keep the anesthesia level as constant as possible, and only those animals were taken for final evaluation in which the control values were reproducible. But to answer your question, we have no evidence that there might be great differences between the reaction of our anesthetized animals and the reaction of conscious animals as far as I can see from the literature. Our control values are exactly in the range of those for ewake animals, and with respect to the hypoxic response there might be a reduction of more than 20%.

Session V

Morphometric Analysis of Ultrastructural Changes in the Carotid Body Tissue

Histofluorescent and Ultrastructural Studies on the Effects of Reserpine and Calcium on Dense-Cored Vesicles in Glomus Cells of the Rat Carotid Body

A. Hess*

The dense-cored vesicles of the glomus cells of the carotid body are presumably the storage site of neurotransmitter substances because of the similarity in appearance of these granules to those found in catecholamine-storing sites of other cells. In addition, the glomus cells fluoresce strongly after exposure to hot paraformaldehyde vapor, and this formaldehyde-induced fluorescence (2) is a specific histochemical test indicating the presence of monoamines. The greenish yellow color of the fluorescent glomus cells, in contrast to the more brilliant yellow of the serotonin-containing surrounding mast cells, indicates that the glomus cells contain catecholamines.

Reserpine is an agent that specifically causes depletion of catecholamines from cells by preventing uptake of these substances into vesicles, resulting in cytoplasmic accumulation and leading to deamination of the catecholamines intracellularly without postsynaptic effects (1). Three hours after administration of 15 mg/kg reserpine intraperitoneally or 1 mg/kg subcutaneously or overnight after 5 mg/kg intraperitoneally, the formaldehyde-induced fluorescence of the rat carotid body is virtually absent (3,4). This indicates that the catecholamines are in the glomus cells. However, perhaps surprisingly, the dense-cored vesicles appear unaffected and as numerous, and the cores are still present and as dense as after reserpine.

We then resorted to morphometric methods (11) to attempt to determine if any changes were reduced in the numerical or volume densities of dense-cored vesicles by reserpine. One hundred random electron micrographs (10 levels spaced 5 μm apart) were taken of each of 4 normal and 4 reserpinized (15 mg/kg for 3 h) rat carotid bodies. Volume and numerical densities of dense-cored vesicles expressed as percentages of glomus cell cytoplasm were determined by a point counting method; measurements were made of the diameters of 800 dense-cored vesicles from the normal and reserpinized groups. It was found that the vesicles after reserpine are smaller since the volume density is slightly reduced, but there was no significant change in the numerical density of dense-cored vesicles. The diameter measurements reveal that vesicles of 60-90 nm (vesicles vary from 75-135 nm in diameter) are increased in frequency of occurrence in reserpinized over normal rats by 43% (Fig. 1). Since increase in number of vesicles does not occur and new, smaller vesicles are not induced to form after reserpine, the dense-cored vesicles must have undergone shrinkage in diameter after reserpine. It is thus entirely possible that this reduction in diameter is caused by the vesicles being depleted of their stores of catecholamines by reser-

---

*The author's investigations are supported by research grant NS-07662 from the National Institutes of Health, U.S. Public Health Service. Dr. E.S. Hearney collaborated on the morphometric studies, to be published in detail elsewhere.

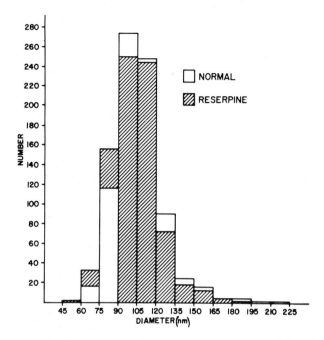

Fig. 1. Histogram of diameter measurements of 800 dense-cored vesicles in reserpinized (15 mg/kg i.p. for 3 h) and normal rats

pine. The persistence of the dense core may indicate that the dense core itself is not the catecholamine, but may represent some form of electron-dense binding substance that serves to store and sequester catecholamine within the vesicle (5).

The glomus cell, therefore, appears to be catecholaminergic and to react to reserpine like other catecholaminergic cells. It was decided to see if the glomus cell reacts in other situations like other catecholaminergic cells. Calcium pretreatment prevents reserpine from depleting the catecholamine stores of the brain (10). To study the effects of pretreatment with calcium on the effects of reserpine on the carotid body, rats received three subcutaneous injections of 100 mg/kg calcium chloride 15 min apart followed by 1 mg/kg reserpine subcutaneously or 15 mg/kg intraperitoneally. Quantitative microfluorimetric measurements were made of the subsequent histofluorescence preparations (4). The pretreatment with calcium inhibited the depleting action of reserpine on catecholamines of the carotid body. If the intensity of fluorescence of a calcium-injected control is considered as 100%, rats treated with reserpine only had a fluorescence intensity of about 10% of controls, those pretreated with calcium before 1 mg/kg reserpine had an intensity of fluorescence about 80% of the control animals, and rats pretreated with calcium before 15 mg/kg reserpine revealed essentially the same intensity of fluorescence as rats receiving only reserpine without calcium. Hence, the calcium injections were successful in inhibiting lighter, but not heavier doses of reserpine. However, the depletion of fluorescence in the carotid body caused by a light dose of reserpine (1 mg/kg) is just as severe as that of a heavy dose (15 mg/kg).

Subsequent experiments have shown that injections of calcium intraperitoneally, in a site different from that of subcutaneously administered reserpine, are also successful in inhibiting the depletion of catecholamines from the carotid body by reserpine. It was suggested that, to prevent the depleting action of reserpine, calcium might occupy the attachment loci of the dense-cored vesicles in the glomus cells and thereby prevent reserpine from gaining access to the vesicular membrane; the action of reserpine on hindering uptake of catecholamines into vesicles and causing their intracellular degradation would thus be somewhat negated, resulting in the preservation of catecholamines seen in histofluorescent preparations.

Support for this suggestion is perhaps provided by the finding that calcium-binding sites are located on the vesicular membranes of dense-cored vesicles of glomus cells (6). A method of fixing cells in the presence of calcium has been used to localize calcium along the membranes of squid giant axons and in synaptic vesicles of the frog neuromuscular junction (7,8,9). It is believed that this localization of calcium represents the calcium-binding sites in these structures. Rat carotid bodies were fixed in 2.5% glutaraldehyde in 0.1M collidine buffer at pH 7.3 overnight at 4°C, washed in collidine buffer, and postfixed in 1% osmium tetroxide in collidine buffer. For tissues in which calcium-binding sites were to be identified, 50mM CaCl$_2$ was added to the fixative, the buffer wash, and the osmium tetroxide solution. The carotid bodies were examined in the electron microscope. Fixatives containing calcium reveal a dense dot or particle (5-20 nm) in about half the dense-cored vesicles (Fig. 2B); cells fixed without calcium do not have such a dot on their vesicles (Fig. 2A). The dot is eccentrically located, more dense than the dense core, is restricted in number to one dot per vesicle, and is found only in vesicles and in mitochondria. The presence of a calcium-binding site on the catecholaminergic vesicle of the glomus cells might have importance in the process of stimulus-secretion coupling and release, presumably by exocytosis, of the neurotransmitter substances from within the vesicle (6).

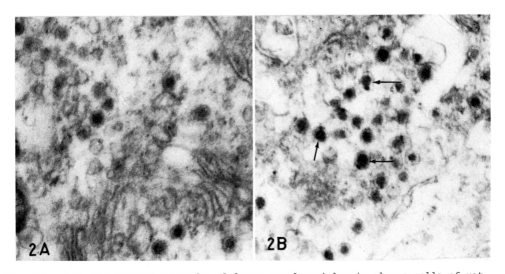

Fig. 2A and B. Electron micrographs of dense-cored vesicles in glomus cells of rat carotid body. (A) Fixed in calcium-free solution. No dense particles in dense-cored vesicles. (B) Fixed in presence of calcium. (*arrows*) Dense-cored vesicles with dense dot or particle. 60,000×

The above studies, then, yield some information on the nature of the dense-cored vesicles in glomus cells of the carotid body and contribute further to the concept that these cells are like catecholaminergic cells elsewhere. Considerations of glomus cells in this light should enable future meaningful experiments to be conceived and performed to determine the role of these cells and their organelles in carotid body function.

References

1. Carlsson, A.: Pharmacol. Ther. B. *1*, 393-400 (1975)
2. Fuxe, K., Hökfelt, T., Jonsson, G., Ungerstedt, U.: in: Contemporary Research Methods in Neuroanatomy. Nauta, W.J.H., Ebbesson, S.O.E. (eds.). New York: Springer-Verlag 1970, pp. 275-314
3. Hess, A.: Brain Res. *98*, 348-353 (1975)
4. Hess, A.: Brain Res. Bull. *1*, 359-362 (1976)
5. Hess, A.: in: Tissue Hypoxia and Ischemia. Reivich, M., Coburn, R., Lahiri, S., Chance, B. (eds.). New York: Plenum Press 1977 (in press)
6. Hess, A.: The calcium binding sites of dense-cored vesicles in the catecholaminergic glomus cells of the rat carotid body (1977) (in preparation)
7. Hillman, D.E., Llinás, R.: J. Cell Biol. *61*, 146-155 (1974)
8. Oschman, J.L., Hall, T.A., Peters, P.D., Wall, B.J.: J. Cell Biol. *61*, 156-165 (1974)
9. Politoff, A.L., Rose, S., Pappas, G.D.: J. Cell Biol. *61*, 818-823 (1974)
10. Radouco-Thomas, S.: Int. Z. Klin. Pharmak. Ther. Toxikol. *5*, 271-278 (1971)
11. Weibel, E.R.: Int. Rev. Cytol. *26*, 235-302 (1969)

DISCUSSION

Hellström: I agree with Dr. Hess: You need 5 mg/kg reserpine to completely deplete the norepinephrine and dopamine. However, if you look at the micrographs the dense-cored vesicles remain. Do you think that the depletion has the same mechanism as that caused by natural stimuli?

Hess: No, reserpine acts to prevent reuptake of the catecholamine into the vesicle. There is constant turnover of catecholamines from the vesicle membrane into the cytoplasm of the cell, hence the catecholamine is left and deaminated intracytoplasmically.

Kobayashi: I agree with you that the membrane-bound vesicles contain catecholamine, but I cannot agree that the small dots are calcium binding sites. In the adrenal medulla, calcium is important for the formation of a storage complex with ATP and adrenaline or noradrenaline, but the chromaffin granules have no such binding sites.

Hess: Did you fix the tissue with calcium chloride and so on?

Kobayashi: No.

Hess: Then why say there are no binding sites?

Böck: I will clarify this. I have tested whether there are calcium binding sites in the adrenal medulla, and there are not.

Hess: There are in the neuromuscular junction. If you stimulate the frog neuromuscular junction and then fix it with 19 mM $CaCl_2$ you find these dense dots. If you then stimulate the junction intensely and fix with calcium, there are no more binding sites. I'm using that as an analogy.

Acker: At the Kashmere symposium, we presented a paper which showed that Ca^{2+} decreased the nervous activity and increased the oxygen consumption of the carotid body. There is a calcium effect on the membrane, of course, but perhaps it also decreases or abolishes catecholamine release.

Hess: Calcium hyperpolarizes cells in general, so that could be true. However, I did not get a consistent increase in histofluorescence with the calcium injections alone; sometimes there was a 20% increase in fluorescence, sometimes none, and sometimes 20% less. In addition, reserpine does not cause discharge of catecholamines from cells, but has an intracellular effect.

Blessing: I confirm your observations that you cannot see any exocytosis of dense cored vesicles. I don't think that if it occurs at all, it is a major way of releasing catecholamines. Did you find any change in the location of dense cored vesicles? Were they located near the cellular border after reserpine or near the nucleus? We have seen that after prolonged hypoxia the vesicles are situated predominantly near the cell border.

Hess: We noticed no change in the distribution of the vesicles and purposely did look very carefully. Hypoxia may be the more natural stimulus causing release of vesicles, while reserpine produces intracellular depletion.

McDonald: What is the evidence that indicates the dots were calcium and indicates a calcium binding site?

Hess: The evidence for that is from Llinas is squid axon and Pappas in the neuromuscular junction using ion probe X-ray analysis in the electron microscope.

McDonald: Yes, but those granules have been shown in the dense cored vesicles in SIF cells in sympathetic ganglia; and in carotid body glomus cells without using extremely high concentrations of calcium we see them with calcium concentrations of 2 mM. Is there any evidence to suggest that calcium is preserving something that is normally extracted, and in your preparations does that dot have its own electron density?

Hess: Yes. There were not lead stained.

McDonald: You said that you used osmium, I believe?

Hess: Yes. Osmium was used but no lead staining. Llinas and Pappas showed the dots had the same X-ray spectrum as calcium and assumed, therefore, that it was calcium and represented the calcium binding site.

Eyzaguirre: I notice in the size distribution of your vesicles there was a unimodal curve. Do you want to comment on that?

Hess: Why not?

Eyzaguirre: Well, we have heard here that there is a bimodal distribution.

Hess: We treated the vesicles as one population and counted eight hundred of them. We tried various formulas and procedures in deference to McDonald and Hellström to make two populations without success. I treat the vesicles as one population; when that is done I do not know how easy it is to get a bimodal distribution. If you look first and divide the cells into two classes then it might be possible to come up with large and small diameter vesicles.

Hellström: I think there are differences in the way we measure the vesicles. I measure all vesicle profiles in each cell; if you do that and do not sample vesicles from different cells you may possibly see differences between cells.

Zierold: I have a question about the microanalytical methods for determining calcium. Do you know what the concentration of calcium is in the vesicles? Any speculation about the amount?

Hess: No, I just think, as these other papers stated, that this is somehow revealing calcium binding sites.

Zierold: Another question: Have these measurements been made on thick specimens or on Sects.?

Hess: They can do them on Sects. in the electron microscope.

Korkala: Do you think that your failure to observe a decrease in number of density of the dense cored vesicles after reserpine might be explained partly by the presence of an extragranular storage site for catecholamines which is invisible in electron micrographs?

Hess: Well, I did not assume that. I just assumed that more small vesicles were present and that the vesicles in general were shrinking.

Pallot: The finding that you do not get the depletion after you have denervated the carotid body is interesting. Have you done the same experiment after denervating the adrenal gland?

Hess: No, I have not done that.

Torrance: What is the evidence that the granules did contain catecholamine rather than, for example, some polypeptide which has not been removed?

Hess: Circumstantial. Whatever a catecholamine has been found in the periphery, dense cored vesicles occur. The histofluorescence procedure is specific for catecholamines. If one finds an area that fluoresces and then looks in the electron microscope, electron dense cored vesicles are found. About Pearse's cells, what other cells does he put into this class?

Torrance: The cells in the thyroid which secrete calcitonin, those in the gut which secrete secretin and pancreozymin, those in the stomach which secrete gastrin and a whole array of cells that secrete a small polypeptide hormone.

Hess: Are not those cells argertaffin in most cases?

Torrance: Yes but they contain and secrete a small polypeptide hormone which is stored in these vesicles.

Ultrastructural Changes in Sensory Nerve Endings Accompanying Increased Chemoreceptor Activity: A Morphometric Study of the Rat Carotid Body

D. McDonald*

Most nerve endings on glomus cells of the carotid body are part of sensory neurons whose cell bodies are located in the petrosal ganglion of the glossopharyngeal nerve (5,6,9,11,13,15,17). Ultrastructural analyses of the synaptic connections in the carotid body of the rat (11), cat (13,17), duck (16), and chicken (10) have shown that some sensory nerve endings and glomus cells are interconnected by reciprocal synapses. At such synapses a sensory nerve ending is *presynaptic* to a glomus cell at one synapse and *postsynaptic* to the same cell at an adjacent synapse. Sensory nerve endings presynaptic to glomus cells exhibit the morphologic characteristics of nerve endings that release a neurotransmitter: small (synaptic) vesicles in nerve terminals are present near presynaptic dense projections, which are part of synaptic junctions. Ultrastructural studies also have shown that nerve endings on glomus cells are remarkably variable in size, shape, and synaptic vesicle content (3,15). In the present study we used morphometric methods to measure changes in the synaptic vesicle content of sensory nerve endings that accompany increased chemoreceptor activity. In addition, using quantitative methods we assessed the variability of sensory nerve ending morphology.

Female rats of the Long-Evans strain (200-250 g body weight) were anesthetized with sodium methohexital (70 mg/kg). They were ventilated via a tracheal cannula by a Harvard rodent respirator for 10 min with 100% O_2 (60 strokes/min at a stroke volume of 3 ml) before being ventilated with a test gas (see Table 1 and 2). Rats in which the carotid sinus nerve was stimulated electrically toward the carotid body (10 V pulse of 0.5 ms duration at 20 Hz) were ventilated with 100% O_2. Immediately after the 10-min period af electrical stimulation or test gas ventilation, the carotid bodies were fixed by vascular perfusion with 4% glutaraldehyde and 0.075% hydrogen peroxide in 0.075 M sodium cacodylate buffer. Details of our processing technique have been published elsewhere (11).

For morphometric studies *all* sensory nerve endings next to glomus (type I) cells or sheath (type II) cells visible in a Sect. of each carotid body supported by a 200 mesh grid (40% open area) were photographed at 12,500× with a Zeiss EM 10 electron microscope. For analysis of regional differences in sensory nerve ending morphology, the location of each nerve ending was marked on a montage of low magnification electron micrographs (2500×), which showed all of the tissue in the

---

*Dr. Robert A. Mitchell collaborated on the neurophysiological studies. I want to thank Ms. Barbara Sternitzke and Ms. Sheila Trumble for technical assistance and Dr. Jon Goerke for his help in writing the computer program. This research was supported in part by NIH Grants HL-06285 and Pulmonary SCOR HL-14201 from the U.S. Public Health Sercive.

Table 1. Effect of various stimuli on concentration of synaptic vesicles in Sects. of sensory nerve endings

Stimulus (for 10 min)	Mean vesicle concentration[a] (mean ± SEM)	No. of nerve endings	Difference from 100% O_2 (%)
100% O_2	18.2 ± 0.6	403	–
10% O_2	15.2 ± 0.9[b]	80	–17
5% O_2	13.5 ± 0.7[b]	129	–26
10% CO_2	13.3 ± 1.1[b]	64	–27
Electrical stimulation	10.1 ± 0.6[b]	109	–45

[a] Vesicles/μm^2 of Sect..
[b] Statistically different from value for 100% O_2 ($P<0.05$).

Table 2. Effect of various stimuli on packing density of synaptic vesicles in Sects. of sensory nerve endings

Stimulus (for 10 min)	Maximal vesicle packing density[a] (mean ± SEM)	No. of nerve endings	Difference from 100% O_2 (%)
100% O_2	50.9 ± 1.6	403	–
5% O_2	36.7 ± 1.6[b]	129	–28
Electrical stimulation	30.8 ± 1.7[b]	109	–40

[a] Vesicles/μm^2 of Sects..
[b] Statistically different from value for 100% O_2 ($P<0.05$).

Sect. of carotid body that was visible. We calculated the concentration of synaptic vesicles in photographs of Sects. (70 nm) of sensory nerve endings at a total magnification of 35,000×. We made this calculation directly from the number of synaptic vesicles in nerve endings and from the area of the endings measured with a Lasico computing rolling disc planimeter with a digital readout. To assess the tendency of synaptic vesicles to aggregate, the maximal packing density of synaptic vesicles, we determined the concentration of vesicles in the circular region (0.25 μm^2 in area) in which vesicles were most abundant in each sensory nerve ending. We used a Data General Corporation NOVA-2 computer to facilitate our analysis.

Concentration and Packing Density of Synaptic Versicles in Sensory Nerve Endings at Various Levels of Chemoreceptor Activity

Rats Breathing Oxygen. The mean concentration and the maximal packing density of synaptic vesicles in sensory nerve endings were higher in rats ventilated with 100% O_2 than in rats exposed to any of the conditions studied that increase chemoreceptor activity (Table 1 and 2). The comparatively high packing density of synaptic vesicles in sensory nerve endings of rats breathing oxygen reflects the tendency of vesicles to cluster (Fig. 1). Clustering was conspicuous in more than 20% of sensory nerve endings under conditions of high P_aO_2 but was not present in most sensory nerve endings under conditions of increased chemoreceptor activity (Fig. 2). For example, in rats ventilated with 5% O_2 for 10 min, clustering of synaptic vesicles was present in less than 5% of sensory nerve endings.

Rats exposed to Conditions That Increase Chemoreceptor Activity. Hypoxia produced by ventilating rats for 10 min with either 10 or 5% O_2 in nitrogen reduced both the concentration and the packing density of synaptic vesicles in sensory nerve endings (Fig. 2; Table 1 and 2). The lower concentration of oxygen produced greater reductions then did the higher one.

Hypercapnia produced by ventilating rats for 10 min with a gas mixture containing 10% CO_2, 20% O_2, and 70% N_2 also reduced the concentration of synaptic vesicles in sensory nerve endings (Table 1 and 2). These rats had a P_aCO_2 of about 70 mmHg and a P_aO_2 of about 110 mmHg (12).

Antidromic electrical Stimulation of the carotid sinus nerve for 10 min produced the greatest reduction in the concentration and packing density vesicles of synaptic vesicles in sensory nerve endings (Table 1 and 2). Although rats were ventilated with oxygen during nerve stimulation, the concentration and packing density of synaptic vesicles were reduced by 40% or more compared to values from control rats breathing oxygen.

This series of experiments showed that hypoxia, hypercapnia, and antidromic electrical stimulation of the carotid sinus nerve, which increase the activity of chemoreceptive nerves, produce changes in the ultrastructure of the sensory nerve endings. Each reduced the number of synaptic vesicles in sensory nerve endings and also caused synaptic vesicles of sensory nerve endings to become more dispersed. We postulate that ultrastructural changes measured in our experiments are morphologic manifestations of increased activity of chemoreceptive nerve endings. Presumably the reduction in the number of synaptic vesicles reflects a state in which transmitter release by exocytosis exceeds the rate of synaptic vesicle formation (12).

The identity of the neurotransmitter released by presynaptic sensory nerve endings is not known. The nerve endings apparently are not cholinergic because destruction of the sensory nerves by cutting the carotid sinus nerve does not change the amount of acetylcholine in the carotid body (7,8). Whatever the chemical nature of the transmitter, however, data from our neurophysiological studies are consistent with the concept that it has an excitatory action on glomus cells (13,14). We propose from these studies that antidromic electrical stimulation of chemoreceptive nerves can increase the release of dopamine from glomus cells.

Variability in Synaptic Vesicle Content of Sensory Nerve Endings. To assess the variability in the concentration of synaptic vesicles in sensory nerve endings under conditions of low chemoreceptor activity, we analyzed 268 sensory nerve endings on glomus cells in a carotid body of a rat ventilated with oxygen. This population of sensory nerve endings had a synaptic vesicle concentration of 19.1±0.8 vesicles/μm^2 (mean ±SEM), but values for individual nerve endings ranged from 1.0 to 70.9 vesicles/μm^2 of Sect.. About 50% of the values fell outside the range of 11.0-25.0, and 20% were outside the range of 6.0-34.0 vesicles/μm^2.

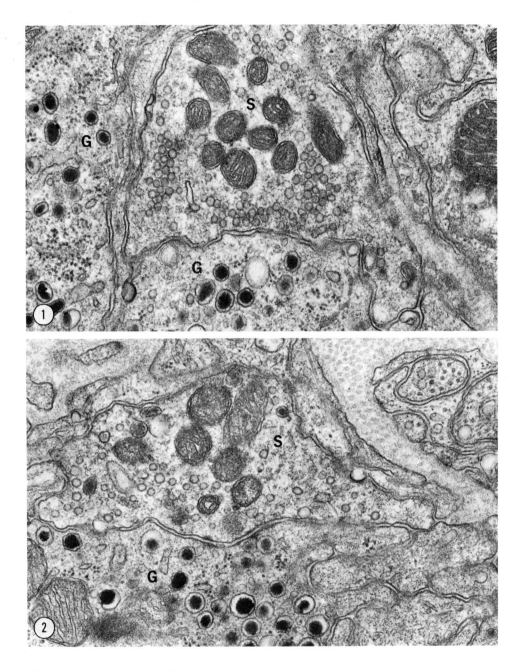

Fig. 1. Sensory nerve endings (S) next to glomus cells (G) of rat ventilated with 100% O_2 for 20 min before fixation. Synaptic vesicles of sensory nerve endings are abundant and clustered in some regions. In this Sect. of nerve ending, synaptic vesicle concentration is 38.1 vesicles/μm^2 of Sect. and maximal vesicle packing density is 108 vesicles/μm^2. 46,000×

Regional Differences in Synaptic Vesicle Content of Sensory Nerve Endings. We examined regional differences in the morphology of sensory nerve endings under conditions of low chemoreceptor activity by calculating the mean synaptic vesicle concentration of groups of sensory nerve endings in various parts of the carotid body. In a Sect. cut through the long axis of the carotid body, we established the location of each of the 268 sensory nerve endings described above. The concentration of synaptic vesicles in sensory nerve endings in the central half of the carotid body was not statistically different from that of sensory nerve endings in the peripheral region of equal area that surrounded the central zone. However, the 83 sensory nerve endings in regions at the poles of this eggshaped carotid body (at the base near the entrance of its arterial blood supply, and at the apex near the junction with the carotid sinus nerve) had 15.8±1.1 (mean±SEM) synaptic vesicles/μm^2 of Sect.. In comparison, the 185 sensory nerve endings located elsewhere in the Sect. had as a group 20.6±1.0 synaptic vesicles/μm^2 of Sect.. This difference in synaptic vesicle concentration was significant ($P<0.005$). Thus the mean concentration of synaptic vesicles of sensory nerve endings in some regions of a carotid body of a rat breathing oxygen resembled that of sensory nerve endings of rats made hypoxic by breathing 10% O_2 (see Table 1). Such regional differences in synaptic vesicle concentration contribute to the marked variability in ultrastructure of sensory nerve endings present in a carotid body.

Because of our evidence that the concentration of synaptic vesicles of chemoreceptive nerve endings reflects their physiological activity, we believe that regional differences in synaptic vesicle concentration may be morphologic manifestations of regional differences in chemoreceptor activity *within* a carotid body. Our data are consistent with the concepts that at a high PO_2 of arterial blood, chemoreceptive nerves in some regions of the carotid body either are more sensitive than other chemoreceptive nerves or are exposed to an environment that differs from that of nerves in other regions of the carotid body.

Acker et al. (2) found marked regional differences in tissue PO_2 in the cat carotid body. Furthermore, the range of tissue PO_2 values obtained for a carotid body increased as the P_aO_2 increased (1). Acker and Lübbers postulated that regions of low tissue PO_2 in carotid bodies of animals with a high P_aO_2 resulted from the high oxygen consumption of carotid body cells coupled with the perfusion of blood vessels of such regions with plasma containing few red blood cells (plasma skimming) (1). Consistent with our morphologic data and with the measurements of tissue PO_2 in the carotid body is the report by Biscoe et al. (4) that the frequency of discharge of different chemoreceptive axons in cats with a P_aO_2 of 200 mmHg ranged from 0.2 to 2.3 impulses/s. While each chemoreceptor exhibited a characteristic response to hypoxia, the firing rate of some fibers at high P_aO_2 equalled that of other fibers at a much lower P_aO_2.

Fig. 2. Sensory nerve ending (S) next to glomus cell (G) of rat ventilated with 5% O_2 for 10 min before fixation. Note that synaptic vesicles in nerve ending are not clumped, even though concentration of synaptic vesicles in this Sect. of nerve ending exceeds mean concentration for population of nerve endings sampled in this carotid body. Nerve ending has synaptic vesicle concentration of 23.2 vesicles/μm^2 of Sect. and maximal vesicle packing density of 52 vesicles/μm^2. 46,000×

References

1. Acker, H., Lübbers, D.W.: in: The Peripheral Arterial Chemoreceptors. Purves, M.J. (ed.). New York: Cambridge U. Pr. 1975, pp. 325-343
2. Acker, H., Lübbers, D.W., Purves, M.J.: Pflügers Arch. *329*, 136-155 (1971)
3. Biscoe, T.J., Pallot, D.: Experientia *28*, 33-34 (1972)
4. Biscoe, T.J., Bradley, G.W., Purves, M.J.: J. Physiol. *208*, 99-120 (1970)
5. Castro, F. de: Trav. Lab. Rech. Biol. Univ. Madrid *25*, 330-380 (1928)
6. Fidone, S.J., Stensaas, L.J., Zapata, P.: J. Neurobiol. *6*, 423-427 (1975)
7. Fidone, S.J., Weintraub, S.T., Stavinoha, W.B.: J. Neurochem. *26*, 1047-1049 (1976)
8. Hellström, S.: in: Nonstratal Dopaminergic Neurons. Costa, E., Gessa, G.L. (eds.). New York: Raven Press 1977 (in press)
9. Hess, A., Zapata, P.: Fed. Proc. *31*, 1365-1382 (1972)
10. King, A.S., King, D.Z., Hodges, R.D., Henry, J.: Cell Tissue Res. *162*, 459-473 (1975)
11. McDonald, D.M., Mitchell, R.A.: J. Neurocytol. *4*, 177-230 (1975)
12. McDonald, D.M., Mitchell, R.A.: in: Morphology and Mechanisms of Chemoreceptors. Paintal, A.S. (ed.). Delhi: Vallabhai Patel Chest Institute 1976, pp. 248-266
13. McDonald, D.M., Mitchell, R.A.: Brain Res. 1977 (in press)
14. Mitchell, R.A., McDonald, D.M.: The Peripheral Arterial Chemoreceptors. Purves, M.J. (ed.). New York: Cambridge U. Pr. 1975, pp. 269-291
15. Nishi, K., Stensaas, L.J.: Cell Tissue Res. *154*, 303-319 (1974)
16. Osborne, M.P., Butler, P.J.: Nature *254*, 701-703 (1975)
17. Smith, P.G., Mills, E.: Brain Res. *113*, 178-194 (1976)

DISCUSSION

Hess: Hellström has said that the small granules are probably found in cells that are noradrenergic and large granules in cells containing dopamine. Your small granules cells have no sensory terminals on them, so do you think that there are different neurotransmitters involved?

McDonald: We both agree that there is no direct evidence linking a given neurotransmitter with a given cell type, although the most abundant catecholamine in the rat carotid body is dopamine, as Hellström has shown. I have considered glomus cells to be interneurons, and just because one type of interneuron is not related to a sensory nerve ending but only to other types of interneurons seems reasonably logical.

Hess: Is the small granule cell related to other glomus cells?

McDonald: It forms synapses with other glomus cells but not with sensory nerve endings.

Hess: What is presynaptic to the small vesicle cell?

McDonald: The sensory nerve endinjs lead onto the large vesicle-containing cells, and the small vesicle cells are related to these cells. The small vesicle cell can form reciprocal synapses with the large vesicle cell. We have not identified nerves having a synaptic input to the small vesicle cell.

Torrance: Would you suppose that processes of type II cells would limit the interaction between type I cells, that a single type II cell encloses a functional group of type I cells?

McDonald: In the rat carotid body there are about three or four type I cells to every type II cell. The type II cell has a much different configuration from the type I cell. Type II cells have very long processes, which envelop a large part of the type I cells, but they do not form a barrier to the diffusion of macromolecules, as the work of Woods has shown.

Nishi: I thought there were many fundamental differences in the morphology of the carotid body of the rat and the cat. In the cat I have shown by serial reconstruction that only three out of 21 glomus cells were not innervated. In the rat this figure is higher. Most physiological studies are done on the cat, so why not do a similar presentation for the cat?

McDonald: Mitchell and I have done a number of physiological and morphologic studies on the cat. There are probably several types of glomus cells in the cat carotid body, but there does not seem to be a simple distinction as there is in the rat. The fact that 3 of 21 glomus cells, as you have shown, were not innervated in the cat is a very important point. Maybe it is 50% in one species and 10 or 20% in another. With regard to the reciprocal synapses in the cat, the proportion is different from that of the rat, but they do exist.

Paintal: I would like to know the criteria for defining a reciprocal synapse.

McDonald: A synapse is defined morphologically by the collection of vesicles next to an asymmetrical membranous junction involving dense projections on the presynaptic side and a fuzzlike density on the postsynaptic side. In a reciprocal pair of synapses the presynaptic dense projections are on one side at one synapse and on the other side at an adjacent synapse. They have been studied most in the retina and olfactory buib.

O'Regan: You have shown very clearly what Eyzaguirre calls a sensory unit, which contains up to 20 cells. The sensory axons must branch at least 10 times. From my calculations you could have quite a number of efferents in the sinus nerve if they do not branch and are in synaptic contact with only a few glomus cells. Through gap junctions between glomus cells, a few fibers could affect the activity of the entire sensory unit. It would be very easy to miss one such fiber among 10-20 branches of the afferents. Fidone's findings that only 92% of the synaptic terminals on the glomus cells contained the radioactive label confirm this.

McDonald: The studies we have done using axonal degeneration techniques were evaluated statistically. The amount of variability found in the studies was such that if a small number of efferent axons were present, say, a few percent, we would not be able to detect them. That is, we would not be able to detect a 5% difference from the control state after cutting the glossopharyngeal nerve above the petrosal ganglion. Nonetheless, it is perhaps relevant that the synaptic vesicle concentrations of nerve endings on glomus cells from axons in the carotid sinus nerve have as a group a unimodel distribution. Thus, they seem to be a homogeneous population. However, the efferent nerve endings on glomus cells from preganglionic sympathetic fibers have a vesicle concentration different from that of sensory nerve endings. The mean vesicle concentration in such efferent nerve endings is over 60 compared to around

20 for sensory nerve endings. To me the most intriguing part about the presumptive efferent system is that we can induce efferent inhibition electrically even after the efferent axons have been eliminated from the carotid sinus nerve by cutting the glossopharyngeal nerve central to the petrosal ganglion.

Bingmann: When you stimulate a nerve, do you ever get a spike back up the same nerve? In the retina where reciprocal synapses occur, backfiring also occurs.

McDonald: We have no evidence of that.

Fidone: We observed a rebound phenomenon following efferent inhibition. When recording from one slip and stimulating the other, following the inhibition there was a marked rebound. I do not know how it was mediated, through the cells or synaptically, but it may involve a backfiring phenomenon.

Kobayashi: The subcellular membrane-bound particles in the type I cell should be classified into two categories, large dense-cored vesicles and small synaptic vesicles. Your morphometric classification and that of Hellström were done based upon the dense-cored vesicles. What do you think is the function of the so-called small synaptic vesicles?

McDonald: In the large vesicle-containing glomus cells, there are two populations of vesicles: the mean diameter of the larger ones is about 120 nm, and that of the smaller ones is about 50 nm. I do not know the function of the smaller ones.

Eyzaguirre: It is remarkable that an electrical stimulus can produce a morphologic change in the nerve ending. In the early phases of electron microscopy people tried without success to find this. I would also like to comment on these reciprocal synapses. There are a few micrometers at best between the afferent and the efferent synapses. Unless we postulate some barriers rectifying from one side to the other, substances could diffuse from one synapse to another.

McDonald: Heuser and Reese and several others have shown that electrical stimulation of the neuromuscular junction results in a decrease in the number of synaptic vesicles within the nerve endings and that this decrease coincides with neurotransmitter release. There is apparently a recycling process by which synaptic vesicle membrane moves to the plasma membrane and then back into the synaptic terminal as coated vesicles.

Fidone: In the animal exposed for 60 s to nitrogen, the mitochondria were demolished. I do not know how powerful a stimulus is, but that animal must be spending a lot of time remanufacturing nerve endings. Can you rule out the possibility that the gas mixture these animals receive just prior to perfusion does not alter the effect of the fixative on the endings? The fixative rather than the gases themselves may be reponsible for the condition of the mitochondria and vesicles in the ending.

McDonald: We consider this unlikely. The preganglionic sympathetic endings on the adrenal medullary cells do not show these changes. Furthermore, not only are the changes reversible, but also they are restricted to sensory nerve endings. If exposure to nitrogen is continued for 5 min or so, the entire carotid body is damaged, and these changes are irreversible. The swelling of mitochondria is much less with such other stimuli as electrical stimulation, elevated P_aCO_2, and less severe degrees of hypoxemia.

Lübbers: Our tissue PO_2 measurements also show quite large regional differences in the carotid body. Usually there is a zone in the periphery that has a low tissue PO_2 and toward the center a higher one. But unfortunately, the anatomical structures are not always regular, particularly in the rabbit, so there may be a zone with a low tissue PO_2 in the center of one carotid body and in the periphery in another carotid body.

McDonald: We compared the central with the peripheral region as a simplistic way of looking at regional differences. That obviously did not work. Nonetheless, it was clear to us that there were regional differences in sensory nerve ending morphology within the carotid body, and we tried to find some systematic way of characterizing them. In the rat regions of low tissue PO_2 may be scattered throughout the carotid body.

Pallot: Of 100 sensory nerve endings, what proportion would exhibit reciprocal synapses?

McDonald: In the rat carotid body about 35% of sensory nerve endings are postsynaptic to the glomus cells, 20% are presynaptic, and 5% form reciprocal synapses. In the cat about 5% of nerve endings are postsynaptic to glomus cells, 15% are presynaptic, and about 1% form reciprocal synapses.

Dense-Cored Vesicles and Cell Types in the Rabbit Carotid Body

A. Verna

The histologic structure of the carotid body is characterized by the presence of cell islets in which it is easy to distinguish two types of cells: type I (or glomus cells) and type II (or sustantacular cells). The type I cell cytoplasm contains a variable number of dense-cored vesicles, whereas the type II cells never contain such vesicles.

The diameter of the dense-cored vesicles has been used by several authors to differentiate different kinds of type I cells. Morita et al. (4) have described three groups of type I cells in the cat carotid body, which differ in the mean diameter of their dense-cored vesicles. Unfortunately, these authors do not indicate if the calculated mean diameters are derived from one or several cells of each kind, and they do not give the number of measured vesicles. In a more precise study two subclasses of type I cells have been described in the rat carotid body by Hellström (1) and McDonald et al. (3). The cells of these subclasses differ significantly in the mean diameter of their dense-cored vesicles.

If the presence of different kinds of type I cells is functionally important, we can suppose that in all mammalian species the carotid body would contain differing type I cells. We have therefore attempted to distinguish different kinds of type I cells in the rabbit carotid body by measuring their dense-cored vesicles.

The study was made on a female rabbit about 2.5 kg, anesthetized with intraperitoneal sodium pentobarbital and artificially ventilated with oxygen for 30 min. The carotid body was perfused with a glutaraldehyde-formaldehyde mixture in sodium cacodylate buffer, rapidly dissected out, and then cut into three blocks. These blocks were postfixed in osmium tetroxide and embedded in araldite. The stained Sects. (about 70 nm thick) were observed in a Philips EM 300 electron microscope, and 30 type I cells, taken at random from each of the three blocks, were photographed. The magnification of the microscope was calibrated with a cross-grating replica (2160 lines/mm), and the final magnification of the photographic prints was 50,000×. The measurements were made with graded circles drawn on a transparent plastic stencil (5), and the results were distributed into classes of 10 nm width. An average number of about 250 dense-cored vesicles (140-434) were measured for each cell, and the frequency curves were traced.

The results are characterized by a great variability from cell to cell. For a majority of cells the frequency curve is unimodal (Fig. 1a). For the other cells the frequency curve is more difficult to interpret, showing sometimes two or even three peaks (Fig. 1b). These observations may have different interpretations:

1. The type I cells, or at least some of them, contain different kinds of dense-cored vesicles, which differ in their mean diameter. However, the positions of the different peaks vary from cell to cell, and the

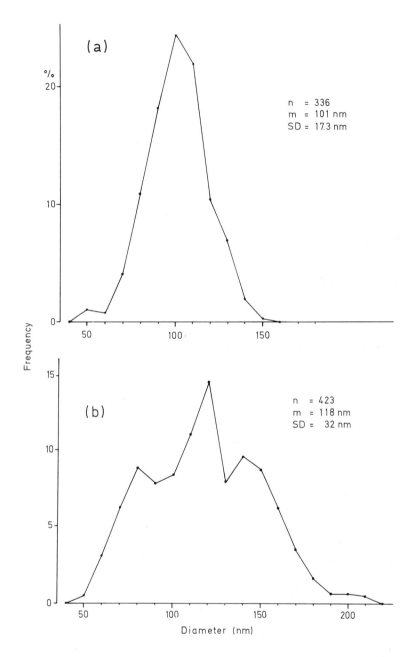

Fig. 1a and b. Frequency curves of dense-cored vesicle diameters in two different cells of rabbit carotid body. n, number of measured vesicles; m, mean diameter; SD, standard deviation. (a) Frequency curve shows only one peak. (b) Frequency curve shows three peaks

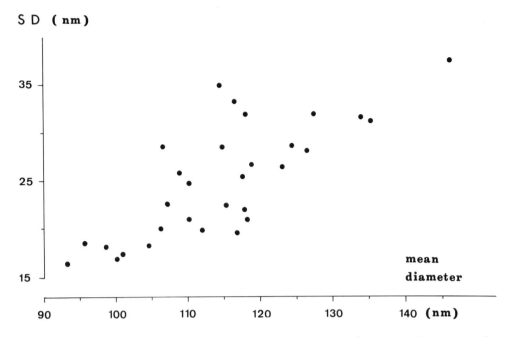

Fig. 2. Mean diameter-standard deviation relationship for 30 cells: the larger the mean diameter, the larger the standard deviation

relative proportions of the different populations of vesicles (if they really exist) also vary continuously from cell to cell. It is therefore impossible to distinguish clear-cut categories of type I cells.

2. On the other hand, it is possible that type I cells contain a single population of vesicles but that their repartition into large and small is perhaps not uniform in the cytoplasm. In this case the different peaks of the frequency curve would be artifacts, and it would be necessary to measure a great number of vesicles in different areas of the cell before having a good representation of the distribution. However, in the case of Fig. 1b, for example, this interpretation is somewhat doubtful because the number of measured vesicles ($n=423$) is relatively high. The cells that contain large dense-cored vesicles also show a particularly large range of vesicle diameters: the larger the mean diameter, the larger the standard deviation (Fig. 2). It was still impossible to separate two groups of cells, as is shown in the unimodal frequency histogram of the 30 mean diameters (Fig. 3a).

The cumulate frequency curve for all the measured vesicles ($n=7530$) is dissymmetric (Fig. 3b) and not Gaussian by the square test, but it is impossible to decide if this distribution is homogeneous or not. On the other hand, it must be added that many rabbit carotid bodies show some type I cells clearly degenerating. These cells exhibit a pycnotic nucleus and a disorganized cytoplasm. The membrane of the dense-cored vesicles is often disrupted, and the dense cores are always very large in size. We can suppose therefore that there is a morphologic evolution of type I cells, which may be reflected by the size of the dense-cored vesicles. The observation of degenerating type I cells suggests that there is some renewal of these cells, which is confirmed by the occurrence of mitotic figures in type I cells of the adult rabbit carotid body. A similar observation in the guinea pig has been made by Kondo (2)

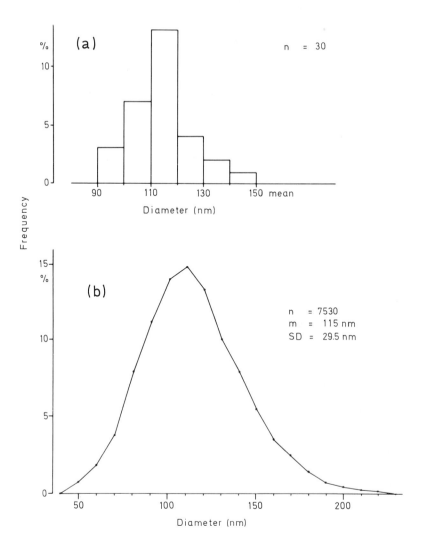

Fig. 3. (a) Frequency histogram of mean diameters for 30 cells; distribution does not allow to separate two groups of type I cells. (b) Frequency curve of all vesicle diameters measured. This distribution is not Gaussian, and it is hard to decide if there is more than one kind of dense-cored vesicles

In summary,
- The size of the dense-cored vesicles is characterized by a great variability not only from cell to cell, but also within a given cell.
- It is impossible to distinguish two kinds of type I cells by comparing the mean diameter of their dense-cored vesicles. However, some cells seem to contain more than one kind of dense-cored vesicles; a more comprehensive analysis of this point is required.
- There is a renewal of the type I cells, and the evolution of these cells may be accompanied by variations in size of their dense-cored vesicles.

These results differ from the findings of Hellström (1) and McDonald et al. (3) in the rat carotid body. It would be necessary now to determine whether these discrepancies are the result of methodological factors or of species differences.

References

1. Hellström, S.: J. Neurocytol. *4*, 77-86 (1975)
2. Kondo, H.: J. Ultrastruct. Res. *37*, 544-562 (1971)
3. McDonald, D.M., Mitchell, R.A.: J. Neurocytol. *4*, 177-230 (1975)
4. Morita, E., Chiocchio, S.R., Tramezzani, J.H.: J. Ultrastruct. Res. *28*, 399-410 (1969)
5. Weibel, E.R.: Int. Rev. Cytol. *26*, 235-302 (1969)

DISCUSSION

Hess: Just one point. I did not realize that Verna was considering mitosis in terms of turnover. I think this is extremely important biologically because it is well known that the taste buds have a very rapid turnover of receptor cells. This probably shows again a common trait between carotid body glomus cells and other receptors.

Verna: I would like to add that the type I cells in mitosis are very rare in the adult carotid body.

Hellström: I have done some morphometric studies on newborn rat, and it is possible to divide the cells into two populations in respect to the size of the vesicles.

Eyzaguirre: Is there a relationship between the occurrence of mitosis and the age of the animal?

Verna: I do not know because I have seen insufficient mitoses.

Blessing: There must be several mitoses in those carotid bodies, and yet we see few. I have done some studies where I could achieve large numbers of degenerating cells. Although I could see no mitoses, the carotid body was restored nearly to normal after a few months, so there must have been quite a few mitoses.

The Carotid Body Chief Cell as a Paraneuron*

S. Kobayashi

What is a Paraneuron?

The term paraneuron was proposed by Fujita (9) in a symposium on chromaffin, enterochromaffin, and related cells. Fujita listed the following characteristics as common to all the varieties of paraneurons:{1} they secrete special peptides with hormone actions; {2} they contain substances like neurotransmitters; {3} they possess membrane-bound particles that are formed through the granular endoplasmic reticulum-Golgi complex system and released by the process of exocytosis; {4} they are receptosecretory in function, receiving adequate stimuli at a specific site and in response releasing secretory substances; and {5} they are of neuroectodermal origin.

Paraneuronic Nature of the Chief Cell

The vascularization of the carotid body can be demonstrated by scanning electron microscopy of a resin cast of blood vessels. Fig. 1a illustrates a scanning electron micrograph showing the fine three dimensional distribution of the blood vasculature in the monkey carotid body (17).

In all the vertebrate species studied so far, the parenchyma of the carotid body contains a glomus complex consisting of chief cells, sustentacular cells, and nerve terminals with connective tissue between and around the glomus complexes. Myelinated and nonmyelinated nerve fibers are also present. The morphology and cytochemistry of the chief cell generally resembles that of other paraneurons that synthesize peptide hormones and/or monoamines.

The particular features of the chief cell as a paraneuron are as follows: {1} The presence of characteristic basophilic peptides was suggested by Capella and Solcia (4) from histochemical studies of the human carotid body chief cell and chemodectoma cells. Pearse (21) proposed the name glomin for the presumptive low-molecular weight peptide hormone. {2} Controversy regarding the presence of monoamines in the chief cell has been resolved (3,12). It is now well established that the chief cell contains dopamine and other catecholamines and/or indolamines (3,12). A high level of acetylcholine in the cat carotid body has been reported (8). However, there is still no direct evidence that the chief cell is the storage site of this transmitter substance. {3} The chief cells have subcellular dense-cored vesicles in all the species examined so far (2,3,13). Characteristics of these cytoplasmic vesicles will be discussed below. {4} The receptosecretory nature of the carotid body chief cell has been established. If the chief cell is

---

*This work was supported by a grant from the Ministry of Education of Japan.

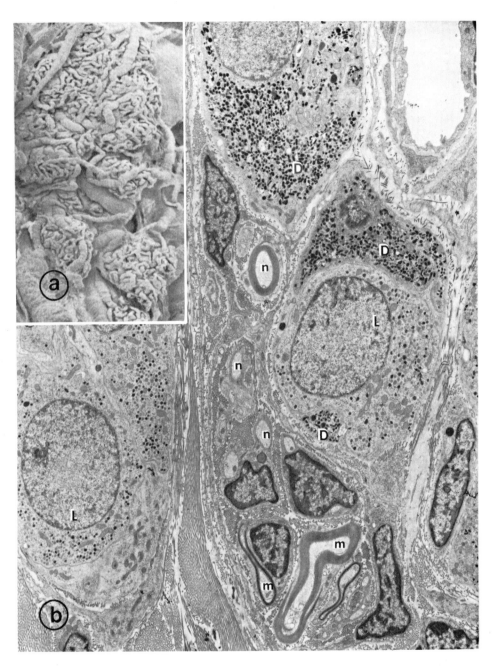

Fig. 1. (a) Scanning electron micrograph of resin cast of vasculature of rhesus monkey carotid body. Three-dimensional distribution of fine blood vessels in carotid body is demonstrated (by Dr. T. Murakami) 150×. (b) Survey electron micrograph of dog carotid body. Glomus complexes consisting of chief cells, sustentacular cells, and nerve terminals are illustrated. Between glomus complexes there are connective tissues containing thick collagen fibers and myelinated (m) and nonmyelinated (n) nerve fibers. There are two types of chief cells. One (L) has subcellular particles of moderate electron opacity, and the other (D) is characterized by extremely electron-dense particles. 4000×

an endocrine paraneuron, the major source of the stimuli is neural. If it is an internuncial paraneuron, the neural stimulus is modified and amplified in the cell and then transmitted to a chemoreceptor. The chief cell possesses a special receptor site for the blood PO_2, PCO_2, and pH (for classification of paraneurons, see (16)). {5} Earlier debates on the origin of the carotid body chief cell have been reviewed by Adams (1). Pearse (22) classified the chief cell into the category APUD cells of *proven* neural crest origin.

Subcellular Membrane-bound Particles in the Chief Cell

Large Dense-cored Vesicles. Lever and Boyd (19) first described the membrane-bound dense-cored particles in the chief cell of the rabbit carotid body. Similar particles have been described in the chief cells of many other animals.

Subtypes of the Chief Cell. Several authors have investigated whether the chief cells form a single population based on the ultrastructure of the subcellular particles and on the density of the cytoplasmic matrix. Earlier electron microscopists classified the chief cells as light, which contained a few dense-cored particles, and dark, which contained many particles. However, as pointed out by Biscoe (2), most of the earlier attempts to subdivide the chief cell were fruitless because of poor fixation of the material. Based on light- and electron-microscopic studies, Kobayashi (11) subdivided the chief cell of the dog carotid body into chromaffin cells with extremely dense particles of irregular shape and nonchromaffin cells with rounded particles of moderate electron opacity (Fig. 2). The two types of chief cells reported by Kobayashi (11) cannot be explained as fixation artifact. McDonald and Mitchell (20) and Hellström (10) introduced morphometry to study whether the chief cell constitutes a homogeneous population with reagard to the subcellular dense-cored vesicles. McDonald and Mitchell (20) reported that the mean diameter of the large dense-cored vesicles of the cells called type A was nearly 30% larger than that of type B cells. Similar results were reported by Hellström (10). It is well-documented that many peptide-secreting endocrine organs consist of several populations of endocrine cells. The occurrence of heterogeneous populations in the chief cell supports the view that the carotid body is not a simple receptor organ, but rather that it has an additional role as a peptide-secreting endocrine organ.

Small Synaptic Vesicles. Kobayashi and Uehara (18) reported that the chief cell of the mouse carotid body contained two types of subcellular particles. The first one consisted of large dense-cored vesicles, which were 80-90 nm in diameter and which frequently tended to accumulate along the cell membranes. The second one was made up of small synaptic vesicles 30-40 nm in diameter, which aggregrated particularly in the region where the nerve ending made synaptic contact with the chief cell (Fig. 3). Considering the size, uniformity, and unique localization of the small synaptic vesicles, it was suggested that these small vesicles contain a substance that was responsible for the transmission of chemosensory impulses across the synapse. Similar small synaptic vesicles were later described in various species by different authors (16).

Recent Autoradiographic Study Based on the Paraneuron Hypothesis Concerning the Chief Cell

Subcellular membrane-bound particles of the adrenal chromaffin cells contain peptides, catecholamines, and ATP. These substances form a storage complex in combination with divalent cations (24). Coupland

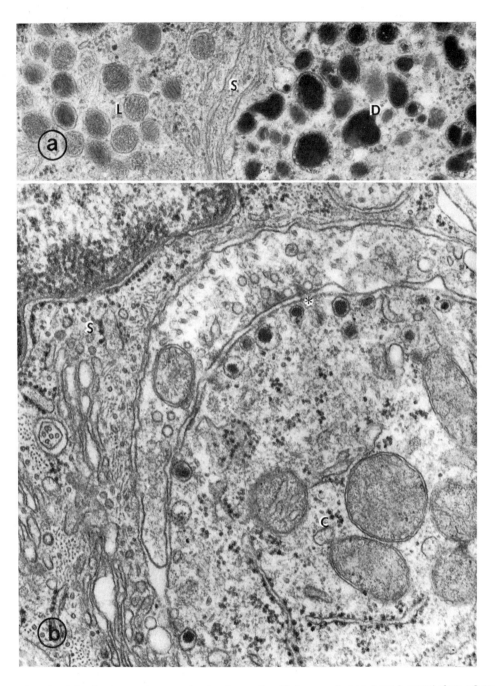

Fig. 2. (a) Electron micrograph showing subcellular membrane-bound particles of two types of chief cells (D and L) observed in dog carotid body. S, thin cytoplasmic process of sustentacular cell intercalated between chief cells. 35,000×. (b) Electron micrograph of mouse carotid body showing efferent synapse of chief cell. Region of synaptic complex is marked by an asterisk. S, sustentacular cell; C, chief cell. Large dense-cored vesicles tend to accumulate along cell membrane of chief cell. 55,000×

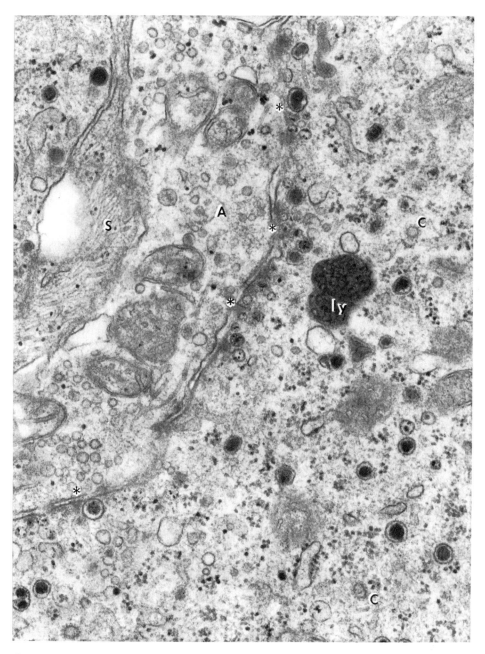

Fig. 3. Electron micrograph of mouse carotid body showing two types of subcellular membrane-bound particles in cytoplasm of chief cell. Large dense-cored vesicles are apparently scattered throughout the cytoplasm among granular endoplasmic reticulum, free polysomes, mitochondria, lysosome (ly), microtubules, and cytoplasmic microfilaments. On the other hand, small synaptic vesicles gather at synaptic membrane thickenings (*asterisk*) between chief cell and afferent nerve terminal (A). All small vesicles immediately beneath synaptic membrane thickening apparently contain dense precipitates. Coated vesicles (C) are variously scattered. S, process of sustentacular cell. 60,000×

and Kobayashi (5) examined the cellular and subcellular localization of recently synthesized peptides and catecholamines in the chromaffin cells of the mouse adrenal medulla by light- and electron-microscopic autoradiography using {^{3}H}leucine and {^{3}H}dopa, respectively. Kobayashi (14,15) has performed a similar autoradiographic study of the endocrine cells in the mouse stomach. In both the adrenal chromaffin cells and gut endocrine cells, an accumulation of {^{3}H}leucine-derived radioactivity was demonstrated in the Golgi area 15-30 min after the injection of the label, which indicates the existence of newly synthesized peptides. At longer time intervals the radioactivity was apparently highest in the cytoplasmic area, which was rich in the membrane-bound particles (5,14,15). On the other hand, at all time periods after the {^{3}H}dopa injection, the highest radioactivity resulting from newly synthesized catecholamines was associated with the membrane-bound particles (5,7). Based mainly on his studies on the adrenal chromaffin cells and gut endocrine cells, Kobayashi (15) proposed that one essential common characteristic in all the varieties of paraneurons is a uniform intracellular transport system for secretory peptides and monoamines.

In order to substantiate this hypothesis for the chief cell, a comparative autoradiographic study of the carotid body has recently been carried out in the mouse using {^{3}H}leucine, {^{3}H}dopa, {^{3}H}dopamine, and {^{3}H}ATP (6). In addition to the first three isotopes, the {^{3}H}ATP was subjected to autoradiographic study because it may be the third major component of the subcellular particles in the chief cell as well as in the adrenal chromaffin cell. It was hoped that autoradiography might help elucidate the nonmetabolic function and dynamics of ATP in the paraneurons. The combined existence of catecholamines and ATP was suggested about a decade ago by Torrance (23). The result of the recent autoradiographic study in the mouse carotid body is summarized as follows:

1. The chief cell incorporated less {^{3}H}leucine-, more {^{3}H}dopa-, and slightly more {^{3}H}ATP-derived radioactivity than the nerve cell in the superior cervical ganglion. Both kinds of cells, as well as most other cells, did not incorporate exogenous dopamine in a significant amount.

2. In the electron-microscopic autoradiography after the {^{3}H}leucine injection, no accumulation of recently synthesized peptides was demonstrated in the Golgi area at any time.

3. After the injection of {^{3}H}dopa, the highest radioactivity appeared always to be associated with the cytoplasmic area rich in the large dense-cored vesicles.

4. {^{3}H}ATP-derived radioactivity was found in all kinds of cells in the carotid body and was mostly localized in the nucleus.

Based on these observations it was concluded that, if secretory peptides in the chief cell, like those in the adrenal chromaffin cells and gut endocrine cells, are synthesized via the granular endoplasmic reticulum-Golgi complex system, the net product may not be much larger than that for the nonsecretory peptides. So accumulation of radioactive peptides could not be demonstrated in the Golgi area by the electron-microscopic autoradiography.

Summary and Comments

In this report the paraneuronic characteristics of the carotid body chief cell have been described and discussed. One of the most important morphologic features of this enigmatic cell is the small synaptic vesicles first described in the mouse. They are the principal component of the only structure that fulfills the criteria for the afferent synapse.

It seems likely that peptides, monoamines, and ATP form a storage complex in the subcellular particles of the carotid body chief cell and of the adrenal chromaffin cells and gut endocrine cells. The result of the recent autoradiographic study using {^{3}H}leucine, {^{3}H}dopa, {^{3}H}dopamine, and {^{3}H}ATP indicates that the turnover of the secretory material is much slower in the caroitd body chief cell than in the adrenal chromaffin cells and gut endocrine cells (16).

References

1. Adams, W.E.: The Comparative Morphology of the Carotid Body and Carotid Sinus. Springfield: Thomas 1958
2. Biscoe, T.J.: Physiol. Rev. *51*, 437-495 (1971)
3. Böck, P.: Ergeb. Anat. Entwickgesch. *48*, 1-82 (1973)
4. Capella, C., Solcia, E.: Virschows Arch. (Cell Pathol.) *7*, 37-53 (1971)
5. Coupland, R.E., Kobayashi, S.: in: Chromaffin, Enterochromaffin and Related Cells. Coupland, R.E., Fujita, T. (eds.). Amsterdam: Elsevier 1976, pp. 59-81
6. Coupland, R.E., Kobayashi, S., Crowe, J.: J. Anat. *122*, 403-413 (1976)
7. Coupland, R.E., Kobayashi, S., Kent, C.: J. Endocrinol. *69*, 139-148 (1976)
8. Eyzaguirre, C., Koyano, H., Taylor, J.R.: J. Physiol. *178*, 463-476 (1965)
9. Fujita, T.: in: Chromaffin, Enterochromaffin and Related Cells. Coupland, R.E., Fujita, T. (eds.). Amsterdam: Elsevier 1976, pp. 191-208
10. Hellström, S.: J. Neurocytol. *4*, 77-86 (1975)
11. Kobayashi, S.: Arch. Histol. Jpn. *30*, 95-120 (1968)
12. Kobayashi, S.: Arch. Histol. Jpn. *33*, 319-339 (1971a)
13. Kobayashi, S.: Arch. Histol. Jpn. *33*, 397-420 (1971b)
14. Kobayashi, S.: Arch. Histol. Jpn. *37*, 313-333 (1975)
15. Kobayashi, S.: in: Chromaffin, Enterochromaffin and Related Cells. Coupland, R.E., Fujita, T. (eds.). Amsterdam: Elsevier 1976a, pp. 303-316
16. Kobayashi, S.: Arch. Histol. Jpn. *39*, 295-317 (1976b)
17. Kobayashi, S., Murakami, T.: in: The Peripheral Arterial Chemoreceptors. Purves, M.J. (ed.). New York: Cambridge U. Pr. 1975, pp. 301-313
18. Kobayashi, S., Uehara, M.: Arch. Histol Jpn. *32*, 193-201 (1970)
19. Lever, J.D., Boyd, J.D.: Nature *179*, 1082-1083 (1957)
20. McDonald, S., Mitchell, R.A.: J. Neurocytol. *4*, 177-230 (1975)
21. Pearse, A.G.E.: J. Histochem. Cytochem. *17*, 303-313 (1969)
22. Pearse, A.G.E.: in: Chromaffin, Enterochromaffin and Related Cells. Coupland, R.E., Fujita, T. (eds.). Amsterdam: Elsevier 1976, pp. 147-159
23. Torrance, R.W.: in: Arterial Chemoreceptors. Oxford: Blackwell 1968, p. 113
24. Winkler, H., Hörtnagl, H.: in: Frontiers in Catecholamine Research. Usdin, E., Snyder, S. (eds.). Oxford: Pergamon 1973, pp. 415-421

DISCUSSION

Hess: What was the turnover rate for dopa and dopamine in the carotid body?

Kobayashi: I did not carry out the systematic study in the carotid body, but I did in the adrenal medulla where it is in the order of 1 week to 10 days.

Hess: A very slow turnover. Where were the radioactive dopamine and the dopa located in the carotid body?

Kobayashi: In the carotid chief cells.

Hess: I thought you said dopamine was in the nerve terminal.

Kobayashi: Yes, and in some of the chief cells.

Hess: Do you think it is usual for dopamine to be in the nerve terminals?

Kobayashi: No. I could get similar results in the stomach and other tissues containing adrenergic nerve terminals.

Hess: But most people believe that dopamine in the carotid body is contained within the glomus cells.

Kobayashi: Yes, but not all chief cells incorporate significant amounts of exogenous dopamine.

Böck: The chief cells in the carotid body rapidly take up L-dopa but not dopamine. Only the precursor is taken up.

Hess: If you give a monoamine oxidase inhibitor, then perhaps the dopamine will be taken up more rapidly by the glomus cells because it is likely that it is deaminated by the monoamine oxidase.

Eyzaguirre: Did I hear correctly that you did not find any peptides within the carotid body chief cell?

Kobayashi: Yes, I tried some fragmentary immunohistochemistry on the dog carotid body using ACTH antisera, insulin antisera, antiglucagon sera, and some others but found nothing.

Böck: The histological staining of carotid body type I cells indicates the existence or the occurrence of a specific type of protein that has a staining property in common with adrenal medullary cells and pancreatic islet cells and other cells of the APUD system. There is also indirect evidence of an unusual protein present in the chief cell.

Torrance: May I ask about this APUD group of cells. If you give reserpine to an animal, you disperse the fluorescence that they will show without dispersing the hormone they contain.

Kobayashi: Böck might like to comment.

Böck: Yes. So if you give reserpine to the pancreatic islet, the catecholamines are lost, but no insulin or glucagon is lost.

Torrance: Reserpine will abolish the fluorescence?

Böck: Yes.

Torrance: And they contain a hormone in membrane-bound granules?

Böck: Yes, and the hormone remains.

Torrance: That is fascinating.

Böck: The effect of reserpine is fascinating, but there is a specific stimulus known for the pancreatic islet. For instance, if you stimulate secretion of insulin, you lose both the hormone and the catecholamine.

O'Regan: You said that the hair cells had the characteristics of a paraneuron. How similar are they to the chief cells of the carotid body? Would you like to comment on that?

Kobayashi: No comment.

Session VI
Environmental Conditions for the Chemoreceptive Process in the Carotid Body

Factors Affecting O_2 Consumption of the Cat Carotid Body

W. J. Whalen and P. Nair

Knowledge concerning the O_2 consumption ($\dot{V}O_2$) of the carotid body is important in the understanding of the mechanism of discharge of the chemoreceptors. Previous studies have yielded some conflicting results. In the situ blood-perfused preparation Daly et al. (3) found an extremely high value for $\dot{V}O_2$ of about 9 ml/100 g/min. This study is open to criticism since, in order to obtain a measurable arteriovenous (A-V)O_2 difference, blood flow was reduced to 9 µl/min. Purves (6) using more accurate techniques, was able to measure an (A-V)O_2 difference of 0.34 ml/100 ml at the normal flow rate of 41.5 µl/min yielding a $\dot{V}O_2$ of 0.147 µl/min for the carotid body, which is not far below the value of 0.18 reported by Daly et al. (3). Purves did not give the weight of the carotid body, but assuming it was comparable to that of Daly et al., the $\dot{V}O_2$ would be about 7 ml/100 g/min.

Fay (4) measured the $\dot{V}O_2$ of the excised carotid body during perfusion with cell-free media and reported a mean value of only 1.5 ml/100 g/min based on a carotid body weight of 2 mg (like 3). On the other hand, Leitner and Liaubet (5) measured the $\dot{V}O_2$ volumetrically of the core of the cat carotid body (weight 60-450 µg) and found a high $\dot{V}O_2$ of about 1.1 ml/g/min of dried tissue. The $\dot{V}O_2$ was PO_2-dependent. At a PO_2 of about 680 mmHg the $\dot{V}O_2$ of the carotid body was roughly seven times that of brain slices treated similarly.

In the hope of shedding some light on these conflicting results, we have recorded the rate of disappearance of O_2 from the core of the carotid body upon stopping the blood flow or cell-free perfusion solution. These disappearance curves (DCs) should provide at least a relative measure of the VO_2. In the absence of red cells the DC of O_2 should give an absolute measure for $\dot{V}O_2$ provided that {1} O_2 is homogeneously distributed, {2} there is no diffusion into or out of the area being sampled, and {3} no myoglobin is present (1). Previous results indicated that the DCs during cell-free perfusion were somewhat slower than during blood perfusion (8) and that part of this slower fall could result from a reduced chemoreceptor discharge (9). Here we report some results of somewhat similar experiments but usually with cell-free perfusion at a lower PO_2 and with CO_2 added. We have also measured venous outflow from the carotid body and determined $\dot{V}O_2$ from the A-V O_2 difference. DCs and $\dot{V}O_2$ values from the dog gracilis muscle have also been obtained for comparative purposes.

Methods

The preparation and the procedures followed were similar to those we have described in detail elsewhere (8). Briefly, the cats (weight 1-2.5 kg) were anesthetized with pentobarbital, kept warm, tracheotomized, heparinized, and ventilated with moistened room air. A snare was placed on the common carotid artery, and the external carotid artery was cannulated. The carotid body was vascularly isolated except for the supplying artery and the veins draining into the internal jugular, which was cannulated. The venous cannula was inserted into a miniature, flowthrough PO_2 analyzer of our own design (to be published). Flow rate was obtained by timing the flow into a calibrated microsyringe. The cut end of the sinus nerve was drawn into a suction electrode. The amplified nerve signals were squared and averaged, which provided a measure of the total nerve activity (2). During the experiments warm saline (37.5 ± 0.5°C) equilibrated with air (air-s) or N_2 (N_2-s) flowed over the preparation.

Measurements of the PO_2 in the carotid body were made with the microelectrode developed in our laboratory (10). The electrode was inserted into the carotid body to an estimated depth of 300-500 μm and the blood flow stopped by occluding the common carotid artery with the snare and simultaneously opening the cannula in the external carotid to reduce perfusion pressure quickly to zero (see Fig. 1). One or more pairs of DCs were obtained while air-s or N_2-s, in random order, flowed over the carotid body. If the tissue PO_2 fell to zero during air-s superfusion (8), the electrode was left in place and the carotid body perfused through the cannula in the external carotid artery with warm Locke's solution at a pressure just above the animal's systolic pressure. The Locke's solution (pH 7.4) included 5.6 mM of glucose and was gassed with 10% O_2-5% CO_2. After 1-2 min one or more pairs (with air-s and N_2-s) of DCs were obtained. In the calculation of the $\dot{V}O_2$ the O_2 content was taken as 0.0236 ml O_2/100 ml per atmosphere (see 4). After 8-15 min of Locke's perfusion the carotid body was again blood perfused and the procedures repeated in a new location. In nine experiments nicotine (0.01-1.00 μg) was injected into the inflowing cell-free solution and a DC obtained during maximum nerve activity. For the experiments on the vascularly isolated in situ gracilis muscle of the dog the procedures were essentially identical to those described above. The distal end of the femoral artery was cannulated in place of the external carotid, and a drop counter measured blood flow. Also only air-equilibrated Locke's solution was used, and it was not necessary to use N_2-s superfusion.

Results

Fig. 1 illustrates the protocol followed and some typical results. The DCs are slightly faster and more nearly linear than was usual. Note that the record of sinus nerve discharge has considerable stretch receptor activity until carotid pressure is reduced to zero.

Overall there was, as shown in Fig. 1, a significant decrease in chemoreceptor discharge during DCs with Locke's perfusion. Table 1 summarizes the results from 20 pairs of DCs in 20 cats. Note that the mean rate of of O_2 disappearance was almost identical in the blood-perfused versus the Locke's-perfused carotid body and that the initial PO_2 values were not markedly different.

To find out whether the decreased discharge could be involved in the slower rate of O_2 disappearance, nicotine was injected into the inflowing Locke's solution and the perfusion stopped during the peak of the sinus nerve discharge. As shown in Table 2 the DCs during the nicotine-stimulated discharge were significantly faster than in the control periods, the postcontrol DC being obtained 5-10 min after the nicotine injection.

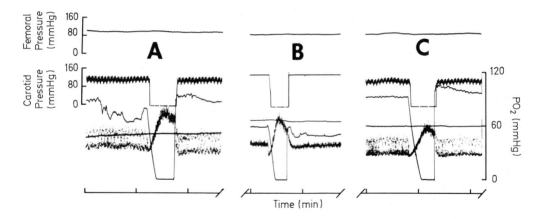

Fig. 1A-C. DCs during blood perfusion (A and C) and during Locke's-10% O_2-5% CO_2 perfusion (B) of cat carotid body. From above: systemic arterial pressure; carotid body perfusion pressure; tissue PO_2 of carotid body (calibration at right); venous PO_2 (calibration not shown); sinus nerve discharge. Approximately 6 min elapsed between segments of record. All DCs taken during superfusion with air-equilibrated, warm saline. Paired DCs during N_2-s superfusion are not shown

Table 1. Results from 20 pairs of DCs in 20 cats[a]

Parameter	Blood-perfused (mean ± SE)		Locke's-10% O_2-5% CO_2-perfused (mean ± SE)	
Initial PO_2	75 ± 5.1	70 ± 3.2	61 ± 3.9	69 ± 2.8
Time to two-thirds initial PO_2 (s)	9.4 ± 0.7	9.3 ± 0.7	7.4 ± 0.7	8.2 ± 0.8
Rate of fall to two-thirds initial PO_2 (mmHg/s)	5.5 ± 0.5	5.3 ± 0.3	5.7 ± 0.6	5.6 ± 0.5
% Maximum chemoreceptor discharge	96 ± 2.9	94 ± 2.3	75 ± 4.9	76 ± 4.8

[a] Obtained during stoppage of blood flow or Locke's perfusion while the carotid body was superfused with saline equilibrated with air or N_2.

Table 2. Effect of nicotine on DCs[a]

Parameter	Precontrol	Nicotine	Postcontrol
Initial PO_2	110 ± 13.0	98 ± 11.3	97 ± 11.9
Time to two-thirds initial initial PO_2 (s)	16.3 ± 1.4	9 ± 0.6	14.7 ± 0.9
Rate of fall to two-thirds initial PO_2 (mmHg/s)	4.5 ± 0.4	7.4 ± 0.8	4.5 ± 0.6

[a] Obtained from 11 trials in 9 cats during perfusion with Locke's solution equilibrated with air.

Table 3. Calculation of $\dot{V}O_2$ (ml O_2/100 g/min) from A-V O_2 difference and flow and from DCs[a]

Tissue	Blood		Locke's	
	A-V	DC	A-V	DC
Cat carotid body		1.09 ± 0.07	6.0 ± 1.3	1.07 ± 0.13
Dog gracilis muscle	0.26 ± 0.02	0.26 ± 0.01	0.26 ± 0.03	0.50 ± 0.1

[a] Data from 8 cats and three dogs. Initial PO_2 for cat carotid body essentially same as in Table 1; initial PO_2 for dog gracilis = 34.4 ± 3.7 (blood) and 27.2 ± 2.3 (Locke's-air). Only the results from the first 20-min period of Locke's perfusion of the carotid body are included (see text).

Our attempts to measure the A-V O_2 difference of the blood across the carotid body were unsuccessful, and they are omitted from the data shown in Table 3. In the calculation of the $\dot{V}O_2$ from the DCs it was assumed that the carotid body weighed 2 mg (3) and that the solubility of O_2 in the tissue was the same as in Locke's solution, i.e., 0.0236 ml/10 ml per atmosphere. In these experiments blood flow averaged 39.9 ± 5 µl/min and Locke's flow averaged 164 ± 33 µl/min in these first periods of 5-15 min of perfusion with Locke's. In subsequent Locke's perfusion and with time, the flow and $\dot{V}O_2$ were unusually less.

We performed a few similar experiments using the gracilis muscle of the dog. A typical result is shown in Fig. 2. In comparison with the carotid body the DCs were more curved, presumably because of myoglobin (1). Calculation of the $\dot{V}O_2$ from A-V difference and flow of blood and Locke's showed that there was no significant difference between the two values (Table 3). On the other hand, the $\dot{V}O_2$ calculated from the DCs during perfusion with Locke's solution was significantly higher than the other values. Somewhat surprisingly, the $\dot{V}O_2$ from the DC during blood perfusion is essentially the same as the value calculated from the A-V difference and flow of both blood and Locke's solution (Table 3)

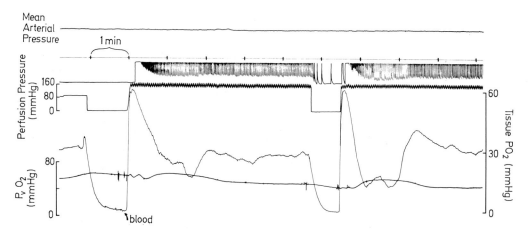

Fig. 2. DCs during Locke's-air perfusion (left) and blood perfusion (right) of dog gracilis muscle. Calibration values for venous PO_2 at left and for tissue PO_2 at right

Discussion

Our results, which show that in the carotid body the DCs are essentially the same during perfusion with Locke's solution, are difficult to explain (see also 9). The red cells should provide a store of O_2 that would substantially reduce the rate of disappearance. This appears to be the case for the soleus muscle of the cat (7) because it shows a slow disappearance curve in spite of a high A-V $\dot{V}O_2$, whereas the cat gracilis muscle, which has a much lower A-V $\dot{V}O_2$, has a faster DC. Presumably this difference in the DCs results from the much higher blood flow (and O_2 storage?) in the soleus muscle. It may be that in the carotid body the blood is extruded very rapidly. Indeed, the carotid body becomes almost white in a second or two after blood flow stops. It could be that the carotid body is perfused mainly with plasma. It is possible also that the $\dot{V}O_2$ is considerably less during cell-free perfusion even though the present preliminary results from A-V measurements do not appear to warrant such a suggestion. It seems unlikely that the small difference in discharge (Table 1) could account for the discrepancy. More work needs to be done in this connection.

Our value for $\dot{V}O_2$ of 6 ml/100 g/min calculated from A-V difference and flow of Locke's solution is not markedly different from the values reported in the literature for the blood-perfused carotid body (3,6). This value is about four times the value Fay (4) found in the cell-free perfused carotid body. However, as pointed out previously by Leitner and Liaubet (5), the estimate of the mass of the carotid body and its contribution to the $\dot{V}O_2$ of the whole preparation is crucial in these calculations. Daly et al. (3) and we, in the above calculations, assumed that the entire A-V difference was attributable to the metabolism of the carotid body alone. We, as did Daly et al. (3), recognize that this is only a guess. Fay's (4) value was obtained from the decrease in A-V O_2 difference after the carotid body was removed from the whole preparation and from an assumed weight of 2 mg for the carotid body. Fay's data, calculated as we have done so far, would give a $\dot{V}O_2$ of about 2 ml/100 g/min, still well below the A-V $\dot{V}O_2$ we obtained during Locke's perfusion. However, we have no way of knowing how much extra tissue was consuming O_2 from our experiments. Also, our values would have been lower had we used measurements made after longer periods of Locke's perfusion. It is possible that Fay's lower values stem from this effect of time. However, in our experiments the usual decrease in flow with time may have been the important factor, whereas in Fay's experiments a vasodilator was added to the perfusate, and flows were stable.

Overall, it appears to us that the $\dot{V}O_2$ of 7-9 ml/100 g/min for the carotid body based on the A-V difference across the whole preparation is an overestimation. In fact, we are inclined to give considerable weight to the $\dot{V}O_2$ values of about 1.1 (Table 3) calculated from the DCs, at least those obtained during Locke's perfusion. The rate of fall of the DCs used in these calculations was the mean value during air-s and N_2-s superfusion so that the possible influence of external PO_2 should have been neutralized. It is unlikely that edema would be a factor in these early trials. Furthermore, O_2 is distributed relatively homogeneously in the carotid body. Some additional confidence in the validity of the DC $\dot{V}O_2$ comes from the results on the dog gracilis muscle, even though the sample size is small, where there is a closer relationship between the DC $\dot{V}O_2$ and the A-V $\dot{V}O_2$. The higher DC $\dot{V}O_2$ during Locke's perfusion may result from the fact that during Locke's perfusion of the muscle few areas have PO_2 values high enough to give a good DC. Thus, the location selected probably represent peaks of PO_2.

References

1. Artigue, R.S., Hyman, W.A.: Ann. Biomed. Eng. *4*, 128-137 (1976)
2. Biro, G., Partridge, L.D.: J. Appl. Physiol. *30*, 521-526 (1971)
3. Daly, M. de Burgh, Lambertsen, C.J., Schweitzer, A.: J. Physiol. *125*, 67-68 (1954)
4. Fay, F.S.: Am. J. Physiol. *218*, 518-526 (1970)
5. Leitner, L.M., Liaubet, M.J.: Pflügers Arch. *323*, 315-322 (1971)
6. Purves, M.J.: J. Physiol. *209*, 395-416 (1970)
7. Thunin, C.A., Buerk, D.G.: Microvasc. Res. *4*, 13-25 (1972)
8. Whalen, W.J., Nair, P.: J. Appl. Physiol. *41*, 180-184 (1976)
9. Whalen, W.J., Nair, P.: Functional correlates with tissue PO_2 in the carotid body. Presented at Oxygen Electrode Colloquium: Oxygen and Physiological Function. FASEB meetings, Anaheim 1976 (in press)
10. Whalen, W.J., Riley, J., Nair, P.: J. Appl. Physiol. *23*, 798-801 (1967)

Addenda

At the time of the above writing, we, the authors, were unaware of a publication on a similar topic by Acker and Lübbers, entitled "The meaning of the tissue PO_2 of the carotid body for the chemoreceptive process" (in: The Peripheral Arterial Chemoreceptors. M. Purves, Ed. Cambridge University Press, 1975). They found, as we did, that the DCs were usually similar in blood and cell-free perfusion. $\dot{V}O_2$ calculated from the DCs was found to vary linearly with the tissue PO_2. We have not yet analyzed our results in the same way, but the fact that many of the DCs we recorded were distinctly curved is consistent with this conclusion. Their values for $\dot{V}O_2$ of about 1.2 ml/O_2/100 g/min (at very high tissue PO_2, 140-240 mmHg) is very similar to our mean value, 1.07. However, in a comparable range of tissue PO_2 their values are lower than ours.

DISCUSSION

Bingmann: How did you determine the depth of puncture of your microelectrodes in the carotid body tissue?

Whalen: The carotid body nearly always dimples so that the actual depth is uncertain. But you can remove the dimpling and actually measure the amount of movement that one must make in order to withdraw the electrode from the body.

Paintal: Could you tell me the time between the stop flow and the point when zero PO_2 is reached and does this time correspond exactly to the time for the maximum increase in discharge?

Whalen: In the first case, it went to zero in about 13 s. Discharge began at a PO_2 averaging 40-50 mmHg.

Paintal: Did the zero of the PO_2 correspond to the time of the big discharge?

Whalen: Well, it went to zero usually before the peak discharge.

O'Regan: In our experiments we found that at the beginning of saline solution perfusion the oxygen usage of the carotid body was quite high in the region as Daly et al. described. After 10-min perfusion the values had gone down to values that Fay found. After switching back onto the natural perfusion the oxygen usage did go back nearly to the high values. The chemoreceptor discharge went down and also the response to cyanide. This response came back to normal when the carotid body was re-perfused with blood. I wondered if some of the results depended on whether you have an innervated or denervated carotid body. Did you do your measurements on a fully innervated carotid body?

Whalen: We did not attempt to separate the sympathetic pathways.

Torrance: Was it in fact plasma that was important in keeping the preparation going? Neil reckoned to get this improvement of the response in a perfused preparation with a relatively small addition of blood.

O'Regan: We have found that we could not reverse this deterioration with plasma. We found a slight enhancing effect. Neither could we affect the cyanide response by adding plasma, but full concentration of plasma could not be used. We had to dilute it in the saline solution.

Whalen: We did not always see such complete recovery as you saw when we returned to blood. Usually there was some recovery but not always complete.

Eyzaguirre: Superfused in vitro preparations, in spite of early doubt, survive very well for many hours, much better than perfused preparations where there are some problems with blood pressure, extravasation of fluid, and so on. I think the trick is to keep preparations in vitro in solutions similar to the old Krebs. It is a solution I nick-name Don Carlos. It is a matter of keeping the receptor reasonably quiet and having some glutamate and other things that Krebs prescribed. It makes sense that the carotid body should be perfused with plasma.

Lübbers: How do you test that your electrode is really inside the carotid body? The carotid body is rather an inhomogeneous tissue, and if the electrode is outside, you will always find high oxygen pressure values. In the symposia in Bristol and Srinagar we described how we applied the same method and showed rather clearly that you get a PO_2-dependent respiration. I wonder if you saw these papers because you did not mention them. Did you compare your actual oxygen tension with the slope of the disappearance curves?

Whalen: I apologize for not mentioning these papers but I did not have your data. We know that the tip was 300 μm or so into the body because, after removing the dimpling, the electrode was withdrawn that distance. In the case of the superfused body we know we went through because we entered a region of higher PO_2, although we do not know exactly where the tip was.

Lübbers: My question is, have you controlled it histologically? Otherwise, you really cannot know where you are.

Whalen: When you recognize that 500 μm back from the tip the electrode is 10 μm in diameter, it is understandable that it is hard to find the track in the carotid body. I am quite sure that the electrode tip has often passed through an area where there are receptors. My basis for saying that is under low flow conditions we can get discharge by superfusing the body with nitrogen-equilibrated solution. Yet, the PO_2 that we are measuring deep in the body does not change. It is only when the electrode is raised nearer to the surface that the tissue PO_2 (and discharge) varies with the PO_2 of the superfusing solution.

Mathematical Analysis of Oxygen Partial Pressure Distribution of the Carotid Body Tissue

U. Grossmann, R. Wodick, H. Acker and D. W. Lübbers

Acker et al. (2) have shown that a decrease or an increase of the P_aO_2 induces uneven increases or decreases of the tissue PO_2 within the carotid body. After increasing the P_aO_2, the low tissue PO_2 values increase slightly, the high ones considerably. Similarly, after decreasing the P_aO_2 the low tissue PO_2 values decrease slightly, the high ones strongly. Obviously, such a reaction of tissue PO_2 cannot develop in a Krogh tissue cylinder, which is flown through by plasma in the case of constant oxygen consumption. Acker and Lübbers (1) noticed that after perfusion stops the steepness of the PO_2 decrease depends on the oxygen pressure. We attempted to simulate this response of tissue PO_2 by a mathematical model assuming a linear dependence of oxygen consumption on oxygen pressure.

We assumed a single capillary with a 6μm radius and 450μm length, flown through by plasma and surrounded by a tissue cylinder of 24μm radius. In the capillary the oxygen transport is realized by diffusion and convection; in the tissue oxygen is transported by diffusion. In both compartments longitudinal diffusion as well as transverse diffusion is being taken into account. In the tissue first-order kinetics were assumed for oxygen consumption. In a cylindrical coordinate system the equations for oxygen transport and consumption in capillary and tissue are, respectively,

$$D \cdot \left[\frac{\partial^2 p}{\partial x^2} + \frac{\partial^2 p}{\partial r^2} + \frac{1}{r} \cdot \frac{\partial p}{\partial r} \right] - V \cdot \frac{\partial p}{\partial x} = 0 \quad (1)$$

$$D \cdot \left[\frac{\partial^2 p}{\partial x^2} + \frac{\partial^2 p}{\partial r^2} + \frac{1}{r} \cdot \frac{\partial p}{\partial r} \right] - A \cdot p = 0 \quad (2)$$

where $p(r,x)$ is oxygen partial pressure, D is diffusion coefficient, V is flow velocity, and A is consumption constant.

To solve these partial differential equations boundary conditions are necessary. The outer surfaces of the tissue cylinder were assumed to be impermeable for oxygen.

$$\frac{\partial p}{\partial r}(R,x) = 0 \quad 0 \leq x \leq l \quad (3)$$

$$\frac{\partial p}{\partial x}(r,0) = 0 \quad r_c \leq r \leq R \quad (4)$$

$$\frac{\partial p}{\partial x}(r,l) = 0 \qquad 0 \leq r \leq R \qquad (5)$$

where r_c is radius of the capillary, R is outer radius of tissue cylinder, and l is length of tissue cylinder or capillary length.

A constant oxygen source was assumed to be present at the arterial inflow.

$$p(r,0) = P_aO_2 \qquad 0 \leq r \leq R_c \qquad (6)$$

The mathematical Equations (1)-(6) were solved with the aid of a finite difference method (3,4) on a digital computer (Großmann). The model allowed us to reproduce quantitatively the behavior of the measured physiological curves.

We started out with a P_aO_2 of 100 mmHg and a flow of 1000 ml/100 g/min. Oxygen consumption was 9 ml/100 g/min at 100 mmHg. Fig. 1 shows that the curve of intracapillary oxygen pressure decreases at first very sharply and that it then flattens out. After reducing the P_aO_2 to 50 mmHg or elevating it to 150 mmHg, the proportional changes in the curve are clearly visible. Low PO_2 values increase or decrease slightly, high PO_2 values increase or decrease strongly. With constant oxygen consumption the curve shifts parallely downward or upward. We also analyzed the oxygen pressure decrease along a capillary for another oxygen consumption curve. Here we assumed an oxygen consumption of 4.5 ml/100 g/min at 100 mmHg (Fig. 2). The curves show their characteristic shapes: they fall steeply, then they flatten. Low PO_2 values vary slightly, high PO_2 values vary considerably.

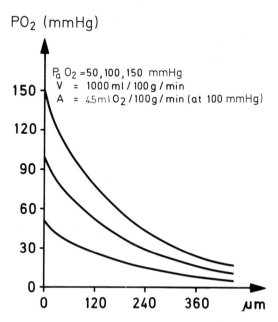

Fig. 1. Decrease of intracapillary PO_2 along a capillary for three P_aO_2 values (50, 100, and 150 mmHg). Flow: 1000 ml/100 g/min. Oxygen consumption: 9 mlO_2/100 g/min

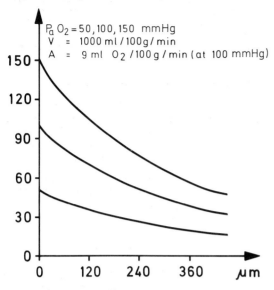

Fig. 2. Decrease of intracapillary PO_2 along a capillary for three P_aO_2 values (50, 100, and 150 mmHg). Flow: 1000 ml/100 g/min. Oxygen consumption: 4.5 mlO_2/100 g/min

Summarizing, we can say that we succeeded in simulating the behavior of measured curves assuming first-order oxygen consumption and plasma flow. We found after changing P_aO_2 and the slope of the PO_2-dependent oxygen consumption that the oxygen pressure field showed zones of relatively low PO_2 values. The transport through the tissue is strongly determined by the longitudinal diffusion, which superimposes the transverse diffusion, especially at the venous end.

References

1. Acker, H., Lübbers, D.W., Purves, M.J.: Pfluegers Arch. *329*, 136-155 (1971)
2. Acker, H., Lübbers, D.W.: Arzneim. Forsch. *23*, 1611 (1973)
3. Collartz, L.: The Numerical Treatment of Differential Equations. Berlin-Heidelberg-New York: Springer-Verlag 1966
4. Grunewald, W.: Pfluegers Arch. *309*, 266-284 (1969)

DISCUSSION

Purves: What happens if you assume that the oxygen consumption is flow-dependent? Are you assuming a constant oxygen consumption at different flows?

Lübbers: The flow-dependent oxygen consumption is not included in this analysis. The flow dependence would have similar effects as the PO_2 dependence.

Whalen: Very often we did see that the oxygen disappearance curves were almost straight lines indicating a nearly constant oxygen consumption. However, I myself favor the view that the oxygen consumption is PO_2-dependent. The second point I would like to make is that I believe that if you assume an oxygen consumption of 9 ml/100 g/min, the oxygen should disappear in 2 or 3 s. I wonder if that was your experience.

Großmann: Our model calculations concern steady-state conditions.

Purves: If there is a marked dependence of oxygen consumption upon oxygen tension - and I am really considering this in the context of the previous paper - then I wonder how valid the stop-flow method is as a measure of oxygen consumption. Presumably this method would tend to underestimate the true oxygen consumption at the initial oxygen tension.

Lübbers: That is very true and is probably one explanation of why Whalen's calculations for oxygen consumption during flow stop and the (A-V)D measurements vary so much. From other experiments we have some clues that there is a flow-dependent oxygen consumption, which was not included in this model.

Torrance: At this symposium several people have said that there are surprisingly few mitochondria in the carotid body for its high oxygen consumption, which is greater than the consumption in the heart, whose volume is about 10-30% mitochondria. In the carotid body there is a much smaller concentration of mitochondria. I wonder whether the electron microscopy is more consistent with the lower figures for oxygen uptake than with the higher ones.

Lübbers: The mitochondria agree much more with the low figures, but if there is a flow-dependent oxygen consumption, this is not necessarily confined to the mitochondria compartments. There is possibly also an oxidase around which is not bound to the mitochondria.

Loeschcke: Dis you already measure pH with these tiny pH electrodes in the carotid body? Some where there must be a source of energy such as glycolysis.

Lübbers: No, we did not.

Comparative Measurements of Tissue PO$_2$ in the Carotid Body

H. Weigelt and H. Acker

Measurements of local oxygen partial pressure (PO$_2$) performed with microneedle electrodes (5) in the cat carotid body showed results that were at first different from what was expected. Although the arterio-venous oxygen partial pressure difference is small in the carotid body, the low PO$_2$ values predominated. They were only about a third to a quarter as high as the venous PO$_2$ (1). In addition, it could be shown that the activity of the sinus nerve increases with decreasing tissue PO$_2$ and decreases with increasing tissue PO$_2$ (2). We deduced from the findings that the chemoreceptor activity depends on the low tissue PO$_2$. If the low PO$_2$ recorded in the carotid body is a basic phenomenon that accompanies the O$_2$-chemoreception, then tissue PO$_2$ values lower than the venous PO$_2$ must be present in the carotid bodies or comparable organs of other mammalian species that represent, e.g., a phylogenetic side branch of the cat, or of species such as amphibians that are at the beginning of the phylogenesis toward mammalians.

For our experiments we chose rabbits, which are lagomorphic and phylo-genetically different from the cat and represent a side branch, as well as frogs as amphibians at the beginning of the development toward mammalians. Pisces, amphibia, lagomorpha, and carnivora developed successively; the evolution line was terminated by the primates (6).

Measurements of tissue PO$_2$ were performed in the carotid bodies of rabbits of different races. The individual values of tissue PO$_2$ differed slightly but not significantly. From the total of 12 rabbit carotid bodies 331 single values were recorded. Fig. 1 shows the frequency distribution of these values as well as the frequency distribution of tissue PO$_2$ values recorded in cat carotid bodies. The latter frequency distribution was composed of 351 values. A maximum at very low PO$_2$ values is common to both frequency distributions. The maxima are between 5 and 10 mmHg for the cat and between 1 and 5 mmHg for the rabbit. The statistical comparison of the frequency distributions showed that they were significantly different. The mean tissue PO$_2$ was 25 mmHg for the cat (median 20 mmHg) and 7 mmHg for the rabbit (median 5 mmHg).

The vascularization of the carotid body, which essentially influences the distribution of tissue PO$_2$, did not show clear differences in the two species. The vascular structures of the cat and rabbit carotid bodies were made visible by the so-called Technovit cast method of injecting a synthetic resin into the arterial system near the carotid body. The synthetic resin polimerizes after some time, and then the vascular casts are exposed by corroding the vessels with 40% KOH. After drying the casts are coated with gold and examined and photographed under the scanning electron microscope. Fig. 2 shows such cast models of the carotid body. Part 1 of this figure is the model of a cat carotid body. The large arterial vessel branching from the arterial supply is striking. It is assumed to contribute essentially to the small arterio-venous oxygen partial pressure difference of the carotid body.

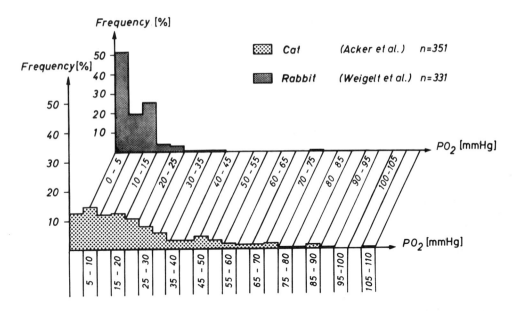

Fig. 1. Frequency distributions of tissue PO$_2$ found in carotid bodies of cat and rabbit

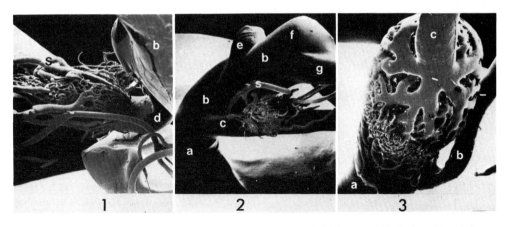

Fig. 2. Technovit cast models of carotid body of cat {1}, carotid body of rabbit {2}, and carotid labyrinth of frog {3}. a, common carotid artery; b, external carotid artery; c, internal carotid artery; d, occipital artery; e, lingual artery; f, external maxillary artery; g, internal maxillary artery; s, shunt vessel

We recognize on the surface of the carotid body big veins that join in the vena transversa, the external carotid artery, and the occipital artery. In addition, the arterial inflow is visible. The pocket-shaped carotid sinus can be seen below the carotid body. The internal carotid artery is obliterated in the cat. Part 2 of Fig. 2 shows the cast model of a rabbit carotid body. The vascular arrangement differs from that of the cat carotid body, since the rabbit carotid body is rarely supplied through the occipital artery but rather usually through the internal carotid artery. So-called arterial shunt vessels are present in the rabbit carotid body as well. They also are branching from the arterial supply. The common carotid artery, external carotid artery, internal carotid artery, occipital artery, lingual artery, and external and internal maxillary arteries are visible in the picture. In both species the carotid body is located laterally to the arterial mainstream path, which comes from the common carotid artery.

The structure of the carotid labyrinth of the frog, which corresponds to the carotid body of mammals, is different. In the frog the chemoreceptor is located in the mainstream of the arterial blood directly in the lumen of the internal carotid artery. Part 3 of Fig. 2 shows the carotid labyrinth of the brown grass frog (*Rana esculenta*), which was prepared according to the above technique and which corresponds to the findings of Kobayashi and Murakami on *R. catesbeiana* and *R. nigromaculata* (4). The common carotid artery, the external carotid artery, which envelops collarlike the lower part of the carotid labyrinth, the carotid labyrinth itself, and the internal carotid artery, which originates from the carotid labyrinth, can be seen.

Arterial vessels that could shunt a part of the arterial blood are not recognizable, and consequently, in the frog considerably higher tissue PO_2 values are to be expected than recorded in the species mentioned above. However, measurements of tissue PO_2 in the surface area of the frog carotid labyrinth showed that low tissue PO_2 values also predominate in this organ.

Fig. 3 shows the frequency distribution obtained from 174 single measurements made in the frog carotid labyrinth. Similar to the cat and rabbit, the maximum is around 5 mmHg. The mean tissue PO_2 value of 16 mmHg (median 13 mmHg) is between the mean values for the cat and rabbit. The frequency distribution is shifted to the left, as are those

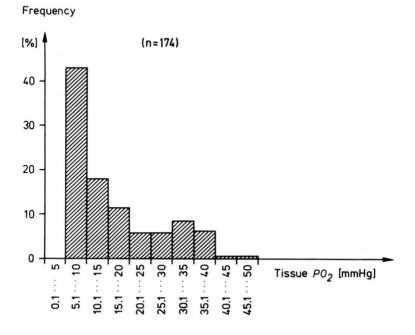

Fig. 3. Frequency distribution of tissue PO_2 in surface area of carotid labyrinth of brown grass frog (*Rana esculenta*)

values for the cat and rabbit. The left shift was found to be statistically proved. Values above 20 mmHg in the cat, above 15 mmHg in the rabbit, and above 20 mmHg in the frog were not statistically significant because of their small frequencies.

Comparison of the measurements showed low tissue PO_2 values in all the oxygen chemoreceptors, i.e., in the carotid bodies as well as in the carotid labyrinth. In addition, the left shift of the frequency distribution of tissue PO_2 values is characteristic for all carotid bodies, whereas the shape of the PO_2 frequency distributions of other organs is different. Furthermore, all the mean PO_2 values are smaller than the mean venous PO_2 values. The relationship of mean arterial PO_2 to mean tissue PO_2 (carotid body and carotid labyrinth, respectively) is 4:1 for the cat, 14:1 for the rabbit, and 7:1 for the frog under normoxic conditions. The low tissue PO_2 values can possibly be explained by the combined effects of various parameters. The strong vascularization of the organ and hence the considerably enlarged surface provide to the oxygen consuming system sufficient facilities of contacting the blood stream. It has been found in earlier investigations that under normoxic conditions practically no erythrocytes penetrate into the inner part of the carotid body. Because of this plasma skimming only physically dissolved oxygen reaches the carotid body tissue, which is rapidly consumed, and hemoglobin-bound oxygen is not available (3).

These low oxygen partial pressures lead to the question as to whether the frequency of low tissue PO_2 reflects the activation of the chemoreceptor, because this respiratory impulse differs in the individual species under hypoxic conditions. At least with regard to cat and rabbit, it can be said that under hypoxic conditions the peripheral respiratory impulse is qualitatively connected to the oxygen profile of the carotid body. The stronger left shift of the PO_2 frequency distribution recorded in the rabbit, therefore, corresponds to a stronger activation under hypoxia, i.e., to a stronger respiratory impulse than in the cat. This assumption has been substantiated in the literature (7).

The situation is more complicated in the frog because of the skin respiration, which plays a considerable role and which forms the greater part of respiration in winter. For that reason a relationship of tissue PO_2 in the carotid labyrinth to respiratory changes cannot be established by counting the respiratory frequency. The changes of tissue PO_2 in the carotid labyrinth can possibly be correlated to blood flow changes in the large cutaneous artery.

References

1. Acker, H., Lübbers, D.W., Purves, M.J.: Pfluegers Arch. *329*, 136-155 (1971)
2. Acker, H., Keller, H.P., Lübbers, D.W., Bingmann, D., Schulze, H., Caspers, H.: Pfluegers Arch. *343*, 287-296 (1973)
3. Acker, H., Weigelt, H., Lübbers, D.W., Steinhausen, M.: Pfluegers Arch. *347*, 42 (1974)
4. Kobayashi, S., Marakami, T.: in: The Peripheral Arterial Chemoreceptors. Purves, M.J. (ed.). New York: Cambridge University Press 1975, pp. 301-314
5. Lübbers, D.W., Baumgärtl, H., Fabel, H., Huch, A., Kessler, M., Kunze, K., Riemann, H., Seiler, D., Schuchhardt, S.: Prog. Resp. Res. *3*, 136-146 (1969)
6. Romer, A.S.: Vergleichende Anatomie der Wirbeltiere, 2nd.ed. Hamburg-Berlin: Paul Parey-Verlag 1966, p. 39, pp. 356-360
7. Wiemer, W., Ott, N., Winterstein, H.: Z. Biol. *114*, 230-263 (1964)

DISCUSSION

Torrance: Does the whole of the flow along the common carotid artery go to the capillary vessels of the carotid labyrinth, and do the chemoreceptor cells of the carotid body labyrinth lie in the walls of these vessels? Even if there is complete plasma skimming, should you not be able to calculate some very high oxygen consumptions for the carotid labyrinth?

Weigelt: Not at this moment. We can say only that there is plasma skimming in the cat and rabbit but not in the frog.

Torrance: But does not all the blood of the common carotid artery pass through capillary vessels and then join up again in the internal or the external carotid artery?

Weigelt: In *Rana esculenta* the common carotid artery enters into the main chamber and then spreads in small vessels. But the main chamber is also the origin of the external carotid artery. This could mean that, because of the low resistance of the external artery and the high resistance of the small vessels, the red cells pass preferentially through the external artery rather than the small vessels.

Acker: This is an assumption. I think that this is a good model to clarify the question of plasma skimming.

Eyzaguirre: Have you thought of looking at the Necturus?

Weigelt: No, we have not.

Eyzaguirre: Because that animal has very big cells, it might be a good idea to use microelectrodes in these cells because of their size.

Paintal: You said that the ratio of the P_aO_2 to tissue PO_2 is 4 in the cat. Have you determined this ratio also for a P_aO_2 of 20 mmHg?

Acker: This ratio is valid only under normoxic conditions. If you change the P_aO_2 to 60 and 200 mmHg, then you have a linear relationship. Above or below these values this relationship changes.

O'Regan: Do these measurements of PO_2 depend on whether the organ is innervated?

Weigelt: We usually measured the different species under innervated conditions.

Whalen: The distribution of our tissue PO_2 values in the cat carotid body is quite different. We find very few values below 10, and I wonder if one possibility might be that you get more dimpling than we do. If you really get severe dimpling, you get low values.

Weigelt: Yes, that would be right. If we had not avoided dimpling. We first inserted the electrode somewhat deeper and then drew it back. Under microscopic control it can be seen that there is no dimpling of the tissue in the organ. The second argument is the response time of the electrode. If we make some oxygen changes in the arterial blood and compare the response time of the catheter electrode in the arterial blood and of the microelectrode in the carotid body tissue, then we find that the response times of both are very comparable in a range of 1-3 s. This peaks against dimpled tissue.

Acker: It is very important to say that we really measured in the specific tissue of the carotid body. We have controlled this very carefully together with Seidl in the carotid body of the rabbit especially and also of the cat. There we could control the track of our electrodes and compare the course of this track with our physiological PO_2 measurements. We could show that if we were in the specific tissue we had mostly low values. When we were outside the specific tissue, then the PO_2 values increased.

Whalen: If you punctured in and then came back, you know what the track was, but that is not necessarily where your tip was.

Weigelt: We could define in these histologic Sects. the point where the tip lay in a variation of about 30 µm.

Böck: You have shown us what you called shunt vessels. Does that mean in histologic terms arteriovenous anastomoses or branching arteries?

Weigelt: In most of the cases this would mean branching arteries.

Wiemer: What happens to the respective PO_2 frequency distributions if you change the blood flow?

Weigelt: We did not change the blood flow under these experimental conditions.

A Functional Estimate of the Local PO_2 at Aortic Chemoreceptors

A. S. Paintal

The tissue PO_2 at any point in a tissue mass is determined by the distance of the point from the capillaries, its metabolic activity, and oxygen availability (i.e., oxygen content of the blood and blood flow). Variation in local blood flow and the level of metabolic activity will determine the level of tissue PO_2 at different points in the tissue mass. This should also apply to the aortic and carotid bodies. Indeed, direct measurements of tissue PO_2 have releaved that it is different in different parts of the carotid body (1,2,18,19). In addition, as was concluded earlier (12), Whalen and Nair have shown that the external environment has an obvious influence on tissue PO_2 of the exposed carotid body. This is particularly obvious if the blood flow to the carotid body is greatly reduced (18,19).

Since the tissue PO_2 differs in different parts of the chemoreceptor tissue mass, it follows that it will not be possible to determine the actual relation of the activity of a chemoreceptor to the local PO_2 at that receptor until it becomes technically possible to record neural activity with a microelectrode along with measurements of tissue PO_2 with a PO_2 electrode at close proximity to the former. Therefore, indirect estimates of local tissue PO_2 at the chemoreceptor from which impulses are recorded become of value, particularly because of contradictory values for tissue PO_2 obtained by Acker et al. (1), on the one hand, and Whalen and Nair (18), on the other.

Indirect Method of Estimating the PO_2 at a Chemoreceptor

One possible method of getting an indirect estimate is based on the assumption that, as in the case of other receptors, the stimulus-response relationship of chemoreceptors will not be altered by reduction of temperature. The reasonableness of this conclusion will be appreciated when it is realized that it must be true in the case of poikilotherms in order to ensure the normal functioning of these animals at various ambient temperatures. There is indeed no justification to assume otherwise, as the available evidence shows that this assumption is valid in the case of receptors whose responses have been examined at various temperatures, e.g., the muscle spindle of the frog (10), mammalian pulmonary stretch receptors (Fig. 1), carotid baroreceptors (Fig. 2b), and aortic baroreceptors (3) (Fig. 2a). It should be noted that in every case the response of the receptor falls with reduction in temperature because of the temperature coefficient (Q_{10}) for the activity of the receptor (Fig. 1a). However, when the responses are plotted in terms of the percentage of peak activity (for the stimulus used), then the quantitative difference in the responses of the endings at different temperatures disappears. The results of Diamond (6) on the carotid baroreceptors and of Angell James (3) on the aortic baroreceptors have been replotted in this manner in Figs. 2b and 2a, respectively.

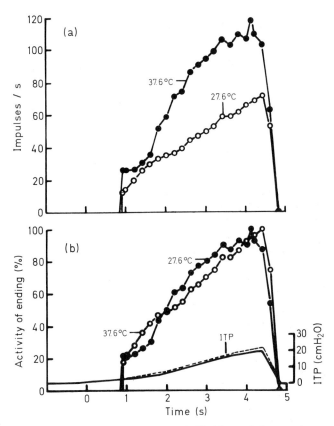

Fig. 1. Typical response of pulmonary stretch receptor to inflation of lungs at normal body temperature and after lowering temperature by 10°C. Ordinate represents frequency of discharge (reciprocal of impulse interval). (b) Same information but with ordinate showing activity of ending expressed as percentage of discharge frequency at peak of inflation so as to take into account the Q_{10} for the ending. In addition to reduction in frequency of discharge, curves also show that lowering temperature by 10°C does not alter stimulus-response relationship. Lowest curves in (b) show intratracheal pressures (ITP) (ordinate on right) at the two temperatures; ITP increased slightly at lower temperature (*dashed line*)

As shown in Fig. 1b, the responses at normal and at lower temperatures look identical. However, the available evidence suggests that lowering the temperature alters the threshold of the receptor. In some experiments it seems to have been raised (Fig. 2b), and in others it seems to have been reduced (Fig. 2a). These variations can be attributed to likely differences in experimental conditions relating to in vitro experiments. However, under in vivo conditions no change in threshold was found (Fig. 1).

Stimulus-Response Relationship of Chemoreceptor

In view of the above results it can be assumed that, in the case of chemoreceptors also, lowering the temperature will not alter their stimulus-response relationship, i.e., the relation of tissue PO_2 at the receptor to the activity of the chemoreceptor. However, as shown

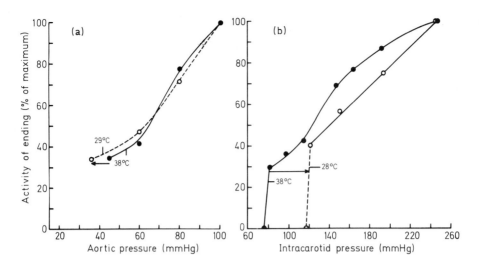

Fig. 2. (a) Stimulus-response relation of rabbit aortic baroreceptor at normal and reduced temperatures. Response measured as impulses/s has been expressed on ordinate as percentage of maximum response of ending at intra-aortic pressure of 100 mmHg (abscissa). Data were obtained from Fig. 6 in Angell James (3). Lowering temparature reduced threshold of ending but did not otherwise alter stimulus-response relationship. (b) Stimulus-response relation of rabbit carotid baroreceptor at two temperatures, 38 and 28°C. Data obtained from Fig. 2 of Diamond (6) have been plotted as in (a). Here lowering temperature raised threshold of ending in contrast to effect shown in (a)

in Fig. 3, this is apparently not the case because lowering the temperature shifts the stimulus-response relation to the left. The stimulus shown on the abscissa (i.e., P_aO_2) is not the actual stimulus at the receptor (this is the main reason for the apparent shift in the stimulus-response relationship), because the actual tissue PO_2 is less than the P_aO_2 shown on the abscissa.

The activity of the two chemoreceptors shown in Fig. 3 was recorded with the cat's chest intact. First, the response of the chemoreceptor was recorded at normal body temparature at various levels of P_aO_2 down to around 18 mmHg. The activity at a tissue PO_2 of 0 mmHg was then extrapolated from the value at a P_aO_2 of about 18 mmHg by multiplying this value by 1.47 (13). Thereafter the whole cat was cooled so that the temperature of the aortic chemoreceptor (regarded as being equivalent to the measured right artrial temperature) fell to about 27°C. The response of the chemoreceptor at various levels of P_aO_2 was then recorded, and finally the activity of the chemoreceptor was recorded at a tissue PO_2 of 0 mmHg by producing circulatory arrest (see (13)). The responses at both temperatures have been expressed as a percentage of maximum activity at 0 mmHg tissue PO_2. The curves at the two temperatures coincide at this value (Fig. 3). Thereafter at lower stimulus strengths the curves deviate from one another. This deviation can be attributed mainly to the reduction in the metabolic activity of the glomus cells such that the tissue PO_2 at about 27°C obtaining at any given P_aO_2 is higher than the tissue PO_2 at normal temperature for the same level of P_aO_2. In both experiments the PCO_2 fell and the pH rose somewhat, but neither of these factors consistently influence the activity of chemoreceptors (8,14,17).

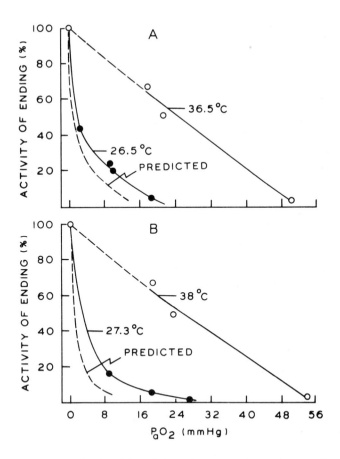

Fig. 3. The apparent stimulus-response relationship of two chemoreceptors at normal temperatures and at about 27°C. The abscissa for these two sets of curves in A and B represents the measured PO_2 of arterial blood. The interrupted line represents the predicted relation between the activity of the chemoreceptor (ordinate) and the near-real stimulus to it i.e. the local PO_2; the abscissa therefore represents PO_2 for this curve. The derivation of this curve is described in the text. In B the pH was 7.43 at 38°C and it rose to 7.59 at 27.4°C

The shift in the apparent stimulus-response curve to the left with fall in temperature (Fig. 3a and 3b) must be due to certain factors such as a fall in the metabolism of the cells of the glomus leading to a rise in the local PO_2 for a given value of P_aO_2. If for argument's sake the metabolic activity of the glomus cells is assumed to be zero at about 27°C, then the curves at about this temperature in Fig. 3 would represent the actual relation between PO_2 at the PO_2 sensor and the activity of the ending. However, since metabolic activity cannot be zero at 27°C, it follows that the true stimulus-response relation must be to the left of the curves at about 27°C in Fig. 3. The question to be answered now is, by how much should these curves at about 27°C be shifted further to the left so as to represent the relation between the actual PO_2 at the PO_2 sensor and the response of the ending? From such curves one could then use the response of the ending to provide a functional estimate of the local PO_2. A rough assessment of the needed shift can be made if it is assumed that the Q_{10} for the metabolism of the glomus is similar to that of the brain, i.e., about 3.3

(see Table 1 in (16)). (Unfortunately, there is no information about the Q_{10} for metabolism of the aortic or carotid chemoreceptors.) A Q_{10} of 3.3 would seem justified since the oxygen consumption of the carotid body is at least as high as that of the brain (5,9,15). Thus reduction in temperature from 38°C to 27.3°C (in Fig. 3b) should reduce the metabolic activity to about 28% of normal. In addition, lowering the temperature to 27.3°C will increase the percent saturation of the blood because the shift in the oxygen dissociation curve. At low levels of P_aO_2 such as 15-20 mmHg the increase would amount to about 1.7 times the saturation at 37°C (7). This increase in oxygen content at 27.3°C can be reckoned in terms of metabolic activity because increase in oxygen content will tend to raise the local PO_2 in the same way as it would be affected by reduction in metabolic activity. Therefore, increasing the oxygen content by 1.7 times can be represented as a reduction in the level of metabolism from 28% to about 16% of normal. Thus the curve at 27.3°C in Fig. 3b represents the state of affairs corresponding to a net metabolic activity of about 16% of normal. The shift in the curves from normal body temperature to 27.3°C in Fig. 3b thus represents a shift that is equivalent to a reduction of metabolic activity of about 84%. It is now possible to conjecture what a further reduction of metabolism amounting to 16% (thereby reducing the metabolism to zero) should do to this curve. The result of such an exercise is shown by the dashed line in Fig. 3b. No extrapolation is required at PO_2 of 0 mmHg. The value at 10% activity was obtained by using simple proportionality. The corresponding curve in Fig. 3a was arrived at in the same manner. These curves are an initial attempt to determine the relation of the actual PO_2 at the PO_2 sensor to the activity of the ending. It should be noted that, as is to be expected, the shift is very small at the lowest levels of P_aO_2.

The predicted curves of Fig. 3 suggest the following conclusions: {1} that at 10% activity in chemoreceptors, the local PO_2 will be about 5-9 mmHg in these two endings; {2} therefore, at normal body temperature the local PO_2 at the sensory ending will be about 5-9 mmHg when the P_aO_2 is about 40-50 mmHg (Fig. 3); {3} that at threshold activity the local PO_2 will be in the vicinity of 15-20 mmHg; {4} that when the P_aO_2 is about 20 mmHg at normal body temperature, the local PO_2 at the sensor will be about 1-2 mmHg. In this connection it is noteworthy that *Cater* et al., *Hill, Lindop, Nunn and Silver* (4) recorded a PO_2 of about 2.5 mmHg in the cortex of the brain when the P_aO_2 was about 20 mmHg (see Tables 2 and 3 in (4)).

Another conclusion that follows from the low values of local PO_2 predicted above is that a major involvement of the low-affinity cytochrome a_3 found by Mills and Jöbsis (11) in the chemoreceptors is unlikely. This is particularly true because, as shown in an earlier paper (13) as well as here (Fig. 3), the activity of aortic chemoreceptors at a P_aO_2 of about 9 mmHg, when the cytochrome a_3 is maximally reduced (11), is of the order of 20% of the activity at temperatures of about 27°C. This fact is contrary to the expectation of Mills and Jöbsis (see Fig. 7 in (11)).

References

1. Acker, H., Lübbers, D.W., Purves, M.J.: Pfluegers Arch. *329*, 136-155 (1971)
2. Acker, H., Weigelt, H., Lübbers, D.W., Bingmann, D., Caspers, H.: in: Morphology and Mechanisms of Chemoreceptors. Paintal, A.S. (ed.). New Delhi: Navchetan Press 1976, pp. 103-110
3. Angell James, J.E.: J. Physiol. *214*, 201-223 (1971)

4. Cater, D.B., Hill, D.W., Lindop, P.J., Nunn, J.F., Silver, I.A.: J. Appl. Physiol. *18*, 888-894 (1963)
5. Daly, M. de Burgh, Lambertson, C.J., Schweitzer, A.: J. Physiol. *125*, 67-89 (1954)
6. Diamond, J.: J. Physiol. *130*, 513-532 (1955)
7. Dill, D.B., Forbes, W.F.: Am. J. Physiol. *132*, 685-697 (1941)
8. Fitzgerald, R.S.: in: Morphology and Mechanisms of Chemoreceptors. Paintal, A.S. (ed.). New Delhi: Navchetan Press 1976, pp. 27-33
9. Leitner, L.M., Liaubet, M.J.: Pfluegers Arch. *323*, 315-322 (1971)
10. Matthews, B.H.C.: J. Physiol. *71*, 64-110 (1931)
11. Mills, E., Jöbsis, F.F.: J. Neurophysiol. *35*, 405-428 (1972)
12. Paintal, A.S.: in: Arterial Chemoreceptors. Torrance, R.W. (ed.). Oxford: Blackwell 1968, pp. 149-151
13. Paintal, A.S.: J. Physiol. *217*, 1-18 (1971)
14. Paintal, A.S., Riley, R.L.: J. Appl. Physiol. *21*, 543-548 (1966)
15. Purves, M.J.: J. Physiol. *209*, 395-416 (1970)
16. Rosomoff, H.L., Holaday, D.A.: Am. J. Physiol. *179*, 85-88 (1954)
17. Sampson, S.R., Hainsworth, R.: Am. J. Physiol. *222*, 953-958 (1972)
18. Whalen, W.J., Nair, P.: J. Appl. Physiol. *39*, 562-566 (1975)
19. Whalen, W.J., Nair, P.: in: Morphology and Mechanisms of Chemoreceptors. Paintal, A.S. (ed.). New Delhi: Navchetan Press 1976, pp. 91-100

DISCUSSION

Bingmann: We never observed that the highest activity of chemoreceptors was at a PO_2 of zero. When PO_2 was decreased, the maximum activity was reached when P_aO_2 was about 20-30 mmHg. With a further lowering of P_aO_2, receptor discharge tended to decline, and we often observed a wide hysteresis in the relation between receptor activity and PO_2 when the P_aO_2 reincreased. Also, we did not find that lowering temperature to about 20°C shifted the threshold of responses of chemoreceptors in a way you have shown. It is very difficult to detect changes in the response of activity of chemoreceptors at this temperature because of the randomness of firing at a low rate. You can obtain clear responses also at this temperature when the P_aO_2 is shifted between 100 and 60 mmHg. It is not necessary to reduce the P_aO_2 at this temperature to 20 mmHg to obtain a first response.

Paintal: I think one of the important differences is the preparation. I have used the aortic chemoreceptors in my preparation. They are in the intact chest, and when I cooled the aortic chemoreceptors I actually cooled the whole animal. Thus the temperature of the chemoreceptors is actually the temperature of the animal. We could find that the response curves of the chemoreceptors look identical at normal and at lower temperatures.

Eyzaguirre: We have just finished a long experimental series in vitro varying the temperature at different PO_2 and recording from single chemoreceptor fibers. It is interesting that, if you use the Arrhenius equation to plot the phenomena, when varying the PO_2 the curves occur at different levels but they are parallel.

Wiemer: I have similar observations as Bingmann regarding the chemoreceptor activity at PO_2 values close to zero. In the rabbit at very low oxygen tensions the activity after an initial overshoot decreases to frequencies very much lower than the maximum value at slightly higher PO_2.

Paintal: I think this is essentially a difference in preparation. We agree on the transient and on the steady state.

Lübbers: I find it confusing to talk about a single PO_2 value in the tissue and then to try to compare the single PO_2 with the P_aO_2 because all that is known is that there is an oxygen tension field. The absolute value of a single PO_2 value does not mean much for tissue if it is above a certain limiting value.

Paintal: I would agree with this. But if you shift the response curve, I am certain that for different cells, i.e., for different receptors, you will get different PO_2 thresholds.

Lübbers: We have also shown that in the brain, e.g., there is no definite single PO_2 but rather an oxygen tension field. I am sure that you can measure in one place about 10 mmHg and in another place, something close to zero.

Paintal: My figures were the mean values. I did not include a standard deviation.

Lübbers: In the cortex there is such a big difference between the lowest and the highest PO_2 values that it is somewhat risky to use the mean value.

Role of Calcium Ions in the Mechanism of Arterial Chemoreceptor Excitation

M. Roumy and L. M. Leitner

The dependence of carotid body oxygen consumption upon the oxygen partial pressure (PO_2) (3,10) provides a basic mechanism for PO_2 detection, and the low-affinity cytochrome a_3 found by Mills and Jöbsis (11) and probably located in type I cells accounts for the dependence of the carotid body oxygen consumption upon PO_2 in an unusually wide range. The excitatory effect of mitochondrial electron transport inhibitors (CO, HCN, etc.) on chemoreceptors implies that mitochondrial oxygen consumption is involved in PO_2 sensing. However, the excitation of chemoreceptors by uncouplers of oxidative phosphorylation means that the cessation of a process energetically coupled to the electron transport (ATP synthesis, calcium uptake by mitochondria) could represent a necessary step in this excitation mechanism.

The excitation mechanism proposed by Anichkov and Belen'Kii (1) relies on the assumption that the ATP content of type I cells decreased during their excitation and is strongly supported by the inefficacy of cyanide to excite arterial chemoreceptors when ATP (10^{-10} to $10^{-5}M$) is added to the solution perfusing the carotid body (8).

We propose an alternative hypothesis, which relates the chemoreceptor excitation to the impairment of calcium uptake by mitochondria during hypoxia and which will be called the calcium hypothesis. It relies on two assumptions: {1} The rate of transmitter released by type I cells during hypoxia is an increasing function of the cytoplasmic Ca^{2+} concentration, $(Ca^{2+})_c$ in the receptor; and {2} $(Ca^{2+})_c$ is controlled by mitochondrial calcium uptake, which depends on PO_2 since the chemoreceptor cells have a low-affinity cytochrome oxidase.

The experiments were performed on 15 cats anesthetized by an intraperitoneal injection of sodium pentobarbital (40 mg/kg), supplemented as needed through a venous catheter. A tracheal cannula was set low in the neck. The animal was then paralized with gallamine triethiodide and artificially ventilated. The carotid bifurcation was perfused through a catheter inserted in the external carotid artery. The common carotid artery was tied, but the perfusion of the carotid body with arterial blood was possible via the occipital pharyngeal trunk when artificial perfusion was stopped. The perfusing solutions were contained in 20-ml syringe bodies. A catheter, which could be switched from one solution to another, was connected to a constant-flow perfusion pump (1 ml/min). The output of the pump was connected to the catheter inserted into the external carotid artery through a stainless steel tube. The temperature of the perfusing solution was maintained at 38°C with a thermostated water jacket. The dead space of the system was 1.2 ml, and there war a time lag of 80 s when changing the perfusion solution.

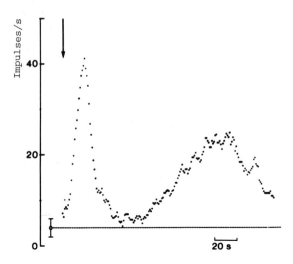

Fig. 1. Response of a few chemoreceptor units to 30-s perfusion with HCN (0.4mM). (*arrow*) End of 1-min period following change in perfusing solution. Ordinate: second-order moving average of discharge frequency during 1-s periods. (*open circle and dashed line*) Mean frequency of discharge (±SD) during control period

Two perfusing media were used. {1} Tris-buffered medium: tris, 49.4mM; NaCl, 110 mM; $CaCl_2$, 2mM. The pH was adjusted to 7.4-7.5 by adding HCl. {2} Krebs-Henseleit (pH = 7.4-7.5): NaCl, 118mM; KCl, 5.7mM; $NaHCO_3$, 25mM, $CaCl_2$, 2.5mM; $MgSO_4$, 1.2mM; NaH_2PO_4, 1.2mM. All the solutions were equilibrated with ambient air and contained glucose (1 g/liter) and Dextran (20 g/liter) (mol wt. 60,000-90,000).

Action potentials from one or a few chemoreceptor afferent fibers were recorded with platinum wire electrodes on a magnetic tape together with an event marker and displayed on an oscilloscope. Nerve activity was fed into a frequency meter whose output was connected to a computer to calculate the mean value (±SD) of the reference peroid activity and the second-order moving average of the discharge frequency for 1 s periods. The chemoreceptor response to 30-s cyanide perfusions showed two phases (Fig. 1): a rapid and large increase in discharge frequency, which decayed rapidly, followed by a second increase in activity of smaller magnitude, decreasing more slowly than the first.

Perfusions with $2 \times 10^{-5} M$ ATP, lasting from 15 to 60 min, did not modify either the resting chemoreceptor discharge or the response to cyanide. When the concentration of ATP was increased to 2mM, the resting frequency of discharge increased rapidly (Fig. 2A and B), but the magnitude of the cyanide response was little affected, and most importantly, its time course was unchanged. When the calcium concentration was raised to 5mM, the resting discharge decreased rapidly. Oligomycin (10 g/liter) added to the ATP perfusion did not modify the response to ATP.

Large concentrations of ruthenium red (RuR) (0.5mM) increased the resting firing frequency of discharge (Fig. 3B and D), decreased the magnitude of the response to cyanide (Fig. 3B), or even suppressed it when perfusion lasted long enough (Fig. 3D). The slope of the response to cyanide decreased (Fig. 3B). When a long-lasting RuR perfusion was changed for a normal Krebs-Henseleit solution, the resting frequency increased rapidly, but if at any time the cyanide was added, it was ineffective (Fig. 3C). After RuR was re-admitted in the perfusing

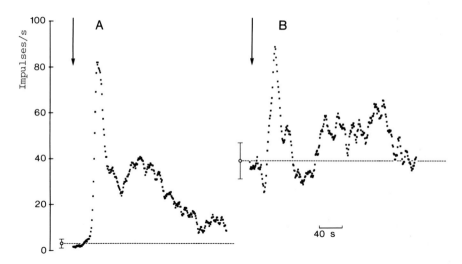

Fig. 2. Effect of ATP (2mM) on chemoreceptor response to cyanide. (*Arrow*) End of 1-min period following change in perfusing solution. (A) Control response to 30-s HCN (0.4mM) perfusion. (B) Response after 5-min perfusion with 2mM ATP. Ordinate: second-order moving average of discharge frequency. (*open circles and dashed lines*) Mean frequency of discharge (± SD) during reference period

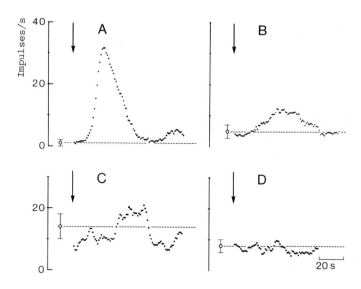

Fig. 3. Effect of RuR (0.5mM) perfusion on chemoreceptor response to cyanide. (A) Control response to 15-s HCN (0.4mM) perfusion. (B) Response after 10-min RuR perfusion. (C) Response 17 min after RuR perfusion. Medium has been replaced by normal Krebs-Henseleit medium. (D) Response after 15-min RuR perfusion. Ordinate: second-order moving average of discharge frequency. (*open circles and dashed lines*) Mean frequency of discharge (± SD) during reference period

fluid, the resting activity of the chemoreceptors decreased rapidly. Small RuR and La^{3+} concentrations (0.1mM) elicited the same response as larger doses except that, when RuR perfusion was interrupted, the chemoreceptor activity remained unchanged. The baroreceptor activity was not modified by La^{3+} perfusion.

Contrary to earlier findings (8), ATP perfusions ($2 \times 10^{-5}M$) lasting even as long as 1 h did not change the resting activity of the chemoreceptors and their response to cyanide. The difference in experimental perfusion technique (constant flow instead of constant pressure) could explain the difference in experimental results. Forrester et al. (6) reported a 40% increase in cerebral blood flow with ATP ($10^{-6}M$) perfusions. If such a vasodilatation occurred also in the carotid body under constant pressure perfusion, the blood flow through the organ would increase, the cytochrome a_3 would be more oxidized, and the response to cyanide would be lessened, since only the reaction of cyanide with reduced cytochrome a_3 is fast enough to be involved in the receptor response. Under constant flow perfusion the chemoreceptors would probably be much less sensitive to changes in vasomotor tone. Besides, if the suppression of the cyanide response by ATP has to be interpreted according to Anichkov's and Belen'Kii's hypotheses (1), ATP should act intracellularly, but it is well known that ATP does not cross membranes. It is likely, therefore, that the results obtained by Joels and Neil (8) resulted from a vasodilator effect of ATP on the carotid body vasculature. The excitation of chemoreceptors found by Krylow and Anichkov (9) with 1mM ATP solutions is confirmed by the present experiments. However, ATP excites the chemoreceptors by an extracellular mechanism in chelating a certain amount of extracellular calcium, as suggested by the fact that low calcium concentration decreased nerve fiber threshold and increased firing frequency when the stimulus was held constant (4). This effect of ATP was consistent with the following observations: {1} the discharge was still sensitive to an increase in flow rate; {2} the response to cyanide was not modified; {3} increasing calcium concentration to 5mM rapidly decreased the firing frequency; {4} addition of oligomycin (10 g/ml) to the perfusate did not change the chemoreceptor activity.

It is very unlikely that the excitation of chemoreceptors by low concentrations (0.1mM) of RuR and La^{3+} occurs through a direct effect on the afferent nerve fibers (i.e., the postsynaptic element). Indeed, La^{3+} (0.5mM) increased the excitation threshold of lobster axons (14) and of *Xenopus ronvier* nodes (13), whereas RuR depressed and even blocked the postsynaptic element of the neuromuscular junction by interfering with calcium movements (1). Moreover, in the present experiments La^{3+} did not affect baroreceptor activity, and a postsynaptic RuR-induced depression similar to that of the neuromuscular junction was found for large concentrations (0.5mM).

It is known that La^{3+} and RuR cross the cell membrane of nerve and Schwann cells (12) as well as of smooth muscle cells (7) where La^{3+} binds preferentially to mitochondria. Therefore, it does not seem unreasonable to assume that RuR and La^{3+} act mainly on type I cell mitochondria in inhibiting calcium uptake. As a consequence, the intracytoplasmic free calcium concentration would increase and so would the activity in chemoreceptor afferent fibers. When the inhibitor concentration was large enough, no more calcium would be released from the mitochondria and the chemoreceptor response to cyanide would disappear (Fig. 3D), whereas with low doses of inhibitors the magnitude and the slope of the response to cyanide would diminish only as was actually observed. Finally, the experimental results were consistent with the proposed calcium hypothesis.

References

1. Alnaes, E., Rahaminoff, H.: J. Physiol. 248, 285-306 (1975)
2. Anichkov, S.V., Belen'Kii, M.L.: Pharmacology of the Carotid Body Chemoreceptors. Oxford: Pergamon 1963
3. Daly, M. de Burgh, Lambertsen, C.J., Schweitzer, A.: J. Physiol. 125, 67-89 (1954)
4. Eyzaguirre, C., Zapata, P.: in: Arterial Chemoreceptors. Torrance, R.W. (ed.). Oxford: Blackwell 1968, pp. 213-247
5. Eyzaguirre, C., Leitner, L.M., Nishi, K., Fidone, S.: J. Neurophysiol. 33, 685-696 (1970)
6. Forrester, T., Harper, A.M., McKenzie, E.T.: J. Physiol. 250, 38-39 (1975)
7. Hodgson, B.J., Kidwat, A.M., Daniel, E.E.: Can. J. Physiol. Pharmacol. 50, 730-733 (1972)
8. Joels, N., Neil, E.: in: Arterial Chemoreceptors. Torrance, R.W. (ed.). Oxford: Blackwell 1968, pp. 153-176
9. Krylow, S.S., Anichkov, S.V.: in: Arterial Chemoreceptors. Torrance, R.W. (ed.). Oxford: Blackwell 1968, pp. 103-109
10. Leitner, L.M., Liaubet, N.J.: Pfluegers Arch. 323, 315-322 (1971)
11. Mills, F., Jöbsis, F.F.: J. Neurophysiol. 5, 405-428 (1972)
12. Singer, M., Krishnan, N., Fyfe, D.A.: Anat. Rec. 173, 375-390 (1972)
13. Vogel, W.: Pfluegers Arch. 350, 25-40 (1974)
14. Takata, M., Pickard, W.F., Lettvin, J.Y., Moore, J.W.: J. Gen. Physiol. 50, 461-471 (1966)

DISCUSSION

Ji: It is not necessary to tie your hypothesis to Mills and Jöbsis's of a high- and low-affinity cytochrome a_3. All you need is some oxygen-sensing mechanism that affects the mitochondrial calcium uptake activity. Your hypothesis can be easily linked to that of Lübbers and Acker about plasma skimming, which lowers the tissue PO_2 well enough to affect normal mitochondrial a_3.

Roumy: The real question is whether hypoxia is detected through a decrease of mitochondrial O_2 consumption. The observation by Mills and Jöbsis of a cytochrome a_3 undergoing reduction in the range of PO_2 where the receptor increased its discharge gave a positive answer to that question. Whether this reduction results from a different cytochrome or from a mechanism powering PO_2 enough to affect a normal a_3 is another question.

Eyzaguirre: Would it be possible that some of your results could be located in the mitochondria of nerve endings and not of type I cells?

Roumy: It is difficult to understand how nerve endings and fibers could have their maximum activity when their oxidative metabolism should be depressed. In addition, our experiments with frozen carotid bodies suggest that this cytochrome is not located in nerve endings.

Eyzaguirre: Have you checked the pH of the La^{3+} solution? That solution can be very acid.

Roumy: La^{3+} concentration was very low (less than 0.1mM), and the pH was in the normal range. The latency of 4-8 min for the effect of La^{3+} is not consistent with a pH effect.

Purves: Why not test this on neuroma?

Roumy: Because we should know before if the chemoreceptive properties shown by some neuroma do not result from remaining ectopic islet of type I cells.

Pallot: Do you know to what level the PO_2 has to fall in the region of mitochondria to affect Ca^{2+} movement?

Roumy: The values of critical PO_2 are unknown in the carotid body.

Pallot: In that case, is it true to say that the mechanism you are proposing does rely on there being this cytochrome a_3 with different oxygen affinities?

Roumy: Yes, in the sense I defined before, i.e. a cytochrome a_3 undergoing reduction at unusually high P_aO_2.

O'Regan: Our experience with repeated injection of cyanide is that the response gets less and in a lot of cases disappears. Have you tried other stimuli or other drugs?

Roumy: When the La^{3+} or RuR perfusion was turned off, the response to cyanide reappears progressively.

Wiemer: My question also relates to the cyanide. What was the concentration in your perfusate?

Roumy: Cyanide was 0.4mM. It was infused at a rate of 1 ml/min during 15 or 30 s so that the total amount for each test was 5 or 10 µg.

Wiemer: Did you keep the PO_2 and pH constant? I am asking because we know that the effect of cyanide depends on the PO_2 and pH. Such variations would surely not explain your findings, but they could modify results quantitatively.

Roumy: PO_2 and pH were kept constant.

Torrance: Surely the external cell membrane controls the total amount of Ca^{2+} within the cell. If there is something in the cell membrane that controls the Ca^{2+} level in the cell cytoplasm to some particular level and there is a leak of Ca^{2+} from the mitochondria into the cytoplasm, then this membrane mechanism will get rid of that Ca^{2+} from the cell. So I would have thought it is necessary to have a change also in the properties of the Ca^{2+} pump in the cell membrane or else the discharge would not be long sustained.

Roumy: The problem of the possible changes in the Ca^{2+} concentration in the long term needs to be considered.

Acker: We also did experiments with Ca^{2+} and agree that there is a depression of the nervous activity of the carotid body. But concomitantly with this depression, the oxygen consumption increases. This pointed to the special mechanism related to the oxygen consumption of the carotid body. Its dependence on flow, PO_2, and PCO_2 shows clearly that besides the normal mitochondrial oxygen consumption a special oxidase is present.

Roumy: Mitochondrial oxygen consumption is stimulated by increased Ca^{2+} concentration. However, I would suggest that this takes place in nerve endings and type II cells and not in the PO_2 sensor, where the rate of electron transport is limited by the reduction of cytochrome a_3 during hypoxia.

Lübbers: We did not see any effect of Ca^{2+} on the oxygen uptake of the excised carotid body under room temperature conditions (Starlinger, Lübbers: Pfluegers Arch. 366, 61-66, 1976). In this special case, the O_2 consumption of the carotid body was also independent in oxygen tension, so things are somewhat confusing. I really cannot understand how from your experiments on the Ca^{2+} effect you make assumptions as to whether mitochondria or membrane or what kind of cells are involved. There is a Ca^{2+} effect, but where will it be located?

Roumy: The effect of inhibitors of the respiratory chain on the discharge of chemoreceptors is established. This means that the mitochondrial respiration is combined with a mechanism of chemoreceptor excitation. The results of Mills and Jöbsis are sufficient evidence that the PO_2 sensing occurs by this mechanism through the mitochondrial respiratory chain.

Tissue PO₂ in the Cat Carotid Body During Respiratory Arrest After Breathing Pure Oxygen

D. Bingmann, H. Schulze, H. Caspers, H. Acker,
H. P. Keller, and D. W. Lübbers

Purves (11), as well as Keller and Lübbers (10), observed an increase in carotid body flow when PCO_2 was raised. Furthermore, Purves noted that oxygen consumption decreased at rising PCO_2 (11), a finding that is in line with results obtained in the brain cortex (4). Both hypercapnic effects should alter tissue PO_2 in the carotid body. Thus the question arose as to the relation between P_aO_2 and tissue PO_2 at rising P_aCO_2.

This relation was studied in anesthetized, paralized, artificially ventilated cats using the apnea technique, as described by Caspers and Speckmann (6,7). After ventilating the animal with pure oxygen for at least 20 min, the respiration pump was switched off with the O_2 container left connected to the trachea of the animal. This apnea caused a steady increase in P_aCO_2 up to more than 200 mmHg. However, at these extreme hypercapnic conditions the continuously decreasing P_aO_2 was still above 100 mmHg. In the present experiments both gas pressures were recorded in a bypass system inserted into a femoral arteriovenous loop. Blood pressure was measured in the aorta abdominalis using a statham transducer. In order to enable a continuous comparison between P_aO_2 tissue and PO_2, tissue PO_2 was recorded polarographically using needle electrodes, according to Baumgärtl and Lübbers (3). These electrodes were elastically suspended, thus being able to follow movements of the organ. This arrangement minimized the effects of the microelectrode upon microcirculation. In parallel with tissue PO_2 local flow was recorded in the carotid body by means of hydrogen clearance curves. Hydrogen clearances were recorded after bolus injections of H_2-equilibrated Ringer's solution into the muscular branch of the carotid artery and evaluated according to Ingvar and Lassen (9) on a PDP 12 computer (5). Finally, chemoreceptor fiber potentials were used as indicators of changes in gas tensions in the tissue of the carotid body. From the cut sinus nerve a maximum of five active afferent fibers was prepared. Thus baroreceptor activity could be excluded in the neurograms, which were recorded bipolarly from the prepared microstrands.

Using this experimental arrangement most of the analyzed parameters showed characteristic changes during apnea. A typical example is presented in Fig. 1. After artificial ventilation had been stopped, arterial blood pressure and chemoreceptor activity rose within a few seconds. Whereas blood pressure usually reached a steady level, the discharge frequency of chemoreceptor fiber units showed a gradual further increase. This slow rise in chemoreceptor activity was approximately linearly related to the increase in P_aCO_2. P_aO_2, ranging between 500 and 600 mmHg during oxygen application, declined slowly when mechanical ventilation had ceased. In contrast to this behavior, tissue PO_2 in the glomus remained rather constant for a long time. Within this first 7-8 min following respiratory arrest no significant changes in tissue PO_2 were observed. In subsequent phases of apnea the oxygen

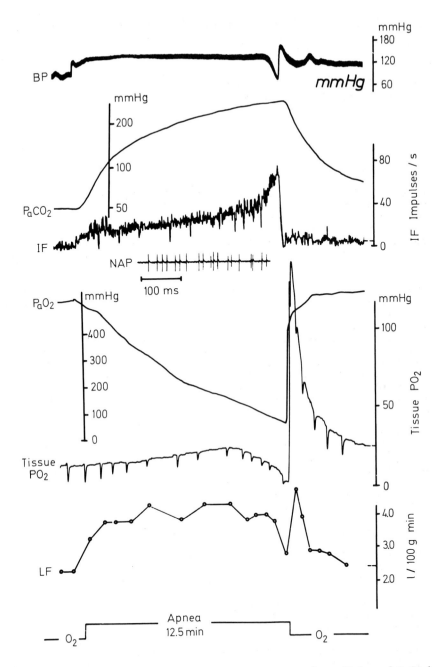

Fig. 1. Changes of blood pressure (BP) and of arterial gas tensions (P_aO_2 and P_aCO_2) during apnea and its effects on the course of tissue PO_2 in carotid body as well as on chemoreceptor activity (IF). IF was obtained by counting electronically in 1-s periods the nerve action potentials displayed in inset (NAP). Reactions of flow (LF) were recorded in glomus tissue by evaluating H_2-clearance curves produced by bolus injections of H_2-equilibrated Ringer's solution into blood stream of common carotid artery. Dips in tissue PO_2 curve were elicited by these bolus injections

pressure in the glomus tended to increase rather than to decay. Only when the P_aO_2 has reached a level between 100 and 200 mmHg did the tissue PO_2 decline in parallel with the further decrease in P_aO_2. At the same time arterial blood pressure fell rapidly, and the discharge rate of chemoreceptor fiber units showed a steep incline. After reventilation P_aO_2, on the one hand, returned immediately to the level encountered before the onset of ventilatory arrest. Tracings of the tissue PO_2, on the other hand, developed a considerable transient overshoot.

In all experiments chemoreceptor activity was strongly reduced often to zero level within a few seconds after reventilation. Such silent periods were found to last up to 1 min. The stability of tissue PO_2 during apnea raised the question as to the behavior of the flow. Simultaneous measurements of H_2-clearance curves showed that flow increased as soon as mechanical ventilation was interrupted. In Fig. 1 the rise occurred in parallel with the increase in P_aCO_2. This parallel course was interrupted only at the end of apnea when arterial blood pressure fell rapidly. In this final phase local flow always followed the decreasing blood pressure. These reactions point to the considerable influence of both parameters, P_aCO_2 and blood pressure, upon flow. In the experiment shown in Fig. 1 the stability of tissue PO_2, at least in the first minutes of apnea, seems to result mainly from the increasing flow, which compensates the falling P_aO_2. However, in other experiments the observed changes in local flow could not explain the course of tissue PO_2. Fig. 2 describes such a discrepancy between P_aO_2, tissue PO_2, and flow. After the onset of respiratory arrest a steep increase of flow up to 75% of the preceding level occurred simultaneously with a marked initial rise in blood pressure within the first minute. However, tissue PO_2 remained unchanged, as demonstrated in Fig. 1, although PO_2 had diminished not more than 5% below its initial level. In the same experiment the further course of apnea was characterized by a progressing decrease in P_aO_2 and by a clear reduction of flow. Nonetheless, tissue PO_2 tended to rise rather than to decline in this phase. In 12 cats the tissue PO_2 during apnea exhibited only a small variability, which might contribute to the constant linear relationship between receptor activity and the rising PCO_2 under these experimental conditions, up to PCO_2 values of more than 200 mmHg.

The small variabilities in tissue PO_2 and P_aO_2 are in contrast to the different responses of flow, which might reflect inhomogeneities in microcirculation. Therefore, we decided to find the mean reactions of flow during apnea. To this end the curves of flow and of the other parameters were averaged, and the results are presented in Fig. 3. The curves show the typical changes in chemoreceptor discharge, gas tensions, and blood pressure, as already demonstrated in Figs. 1 and 2. From Fig. 3 the mean flow seems to run parallel to the rising P_aCO_2 up to a maximum value. Only at the end of apnea, at P_aO_2 values of about 100 mmHg, did the mean flow decrease in parallel with the fall in blood pressure.

The averaged curves might lead to the conclusion that, besides a possible hypercapnic effect upon oxygen consumption during the first minutes of apnea, the rise in flow widely compensates the decreasing P_aO_2 down to 300 mmHg and thus considerably contributes to the stable course of tissue PO_2. However, the question is whether mean reactions of flow determined by H_2-clearance curves represent microflow at the site of the receptor and at the tip of the PO_2 microelectrode. The question can not be answered as yet. In any case, when P_aO_2 approaches normoxic values during apnea and after reventilation, the course of tissue PO_2 indicates that further factors attribute to the relation between arterial and tissue PO_2. Among these factors are {1} PO_2-depen-

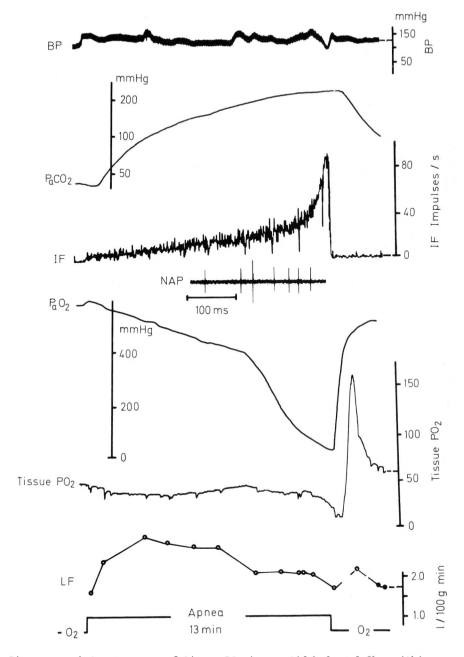

Fig. 2. Discrepancy between course of tissue PO_2 in carotid body and flow within glomus (LF) in relation to changes in P_aO_2 at changing P_aCO_2. BP, arterial blood pressure; IF, curve of chemoreceptor activity obtained by counting sinus nerve action potentials (NAP)

dent changes in oxygen consumption (Purves 11, Fay 8, Acker and Lübbers 1,2), and {2} an increase in the oxygen transport capacity by admixture of erythrocytes to plasma when PO_2 drcreases from hyperoxic to normoxic

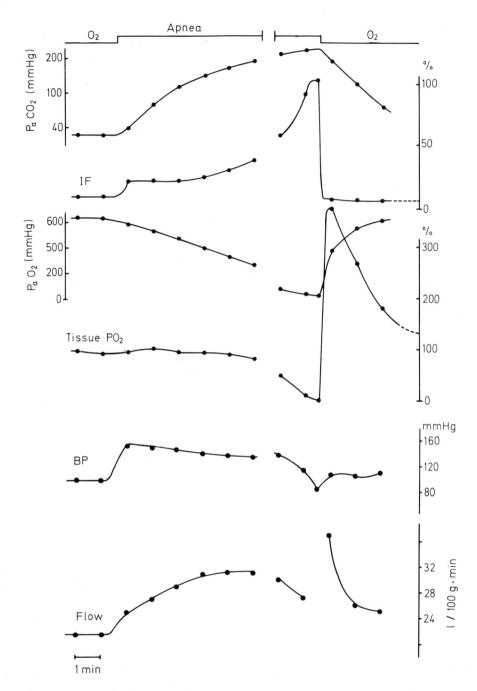

Fig. 3. Mean reactions of P_aCO_2, discharge rate of chemoreceptor fiber potentials (IF), P_aO_2 and tissue PO_2, and time course of arterial blood pressure (BP) and flow during apnea. Points of these curves were obtained by averaging each minute corresponding values of 6 experiments with a duration of about 12 min of ventilatory arrest. To normalize changes in tissue PO_2, difference between initial level of tissue PO_2 2 min before apnea began and lowest value was considered 100%. In chemoreceptor activity curve, maximum difference of each response was set equal to 100%

values (Acker and Lübbers 1). These mechanisms might explain the tendency of tissue PO_2 to rise when P_aO_2 decreased below 200 mmHg as demonstrated in Fig. 1. Furthermore, these mechanisms might contribute to the considerable overshoot in tissue PO_2 after reventilation. An analysis of the contribution of the various factors to the relation between P_aO_2 and tissue PO_2 in the course of apnea has to be done in further experiments.

References

1. Acker, H., Lübbers, D.W.: in: The Peripheral Arterial Chemoreceptors. Purves, M.J. (ed.). New York: Cambridge U. Pr. 1975, pp. 325-343
2. Acker, H., Lübbers, D.W.: Fed. Proc. 35, 527 (1976)
3. Baumgärtl, H., Lübbers, D.W.: in: Oxygen Supply. Kessler, M. et al. (eds.). München: Urban and Schwarzenberg 1973, pp. 130-136
4. Betz, E.: in: Der Hirntod. Penin, H., Käufer, C. (eds.). Stuttgart: Thieme 1969, pp. 1-9
5. Bingmann, D., Lehmenkühler, A.: Proc. of the Decus. The Hague: 1975, Vol. II, pp. 151-156
6. Caspers, H., Speckmann, E.-J.: Aerztl. Forsch. 25, 241-255 (1971)
7. Caspers, H., Speckmann, E.-J.: in: Handbook of Electrocenphalography and Clinical Neurophysiology. Remond, A. (ed.). Amsterdam: Elsevier Publishing 1973, Vol. XA, pp. 41-65
8. Fay, F.S.: J. Physiol. 218, 518-523 (1970)
9. Ingvar, D.H., Lassen, N.A.: Acta Physiol. Scand. 54, 325-338 (1962)
10. Keller, H.-P., Lübbers, D.W.: Pfluegers Arch. 336, 217-244 (1972)
11. Purves, M.J.: J. Physiol. 209, 395-416 (1970)

DISCUSSION

Wiemer: My question relates to the chemoreceptor activity you recorded. First, did you regularly observe under the initial conditions of your experiment, i.e., very high P_aO_2 and normal P_aCO_2, a resting activity? Second, you recorded an increase of activity that you correlated with the change in P_aO_2, and you found a constant local tissue PO_2. If the local PO_2 remained constant, what happened to the local PCO_2, which should be expected to change similarly to the recorded P_aCO_2?

Bingmann: In answer to the first question, at a PO_2 of about 600 mmHg or even higher we did not observe that recorded fiber potentials were silent. We always observed a low rate of discharge. As to the second question, up to now nobody has measured PCO_2 or pH in the carotid body. On the other hand, tissue PCO_2 is determined by other factors than tissue PO_2. I assume that, because of a higher diffusion coefficient, tissue PCO_2 is more running in parallel to the P_aCO_2 than tissue PO_2 is to the P_aO_2.

Purves: It seems to me that these experiments are really unnecessarily complicated to answer the question how flow is related to changes in oxygen tension.

Bingmann: We began with the observation of Purves et al. that flow increased in parallel with a rise in PCO_2. Also, in the literature the oxygen consumption in the brain was related to PCO_2, with points in the same direction. The question was whether apnea can be a useful model to test the relation between these parameters and a rising PCO_2.

Relationship Between Local Flow, Tissue PO$_2$, and Total Flow of the Cat Carotid Body

H. Acker and D. W. Lübbers

Since the local PO$_2$ and the distribution of local flow in the carotid body may play an important role in the chemoreceptive process, we compared the behavior of local flow and local tissue PO$_2$ with that of the total flow under different chemoreceptive stimuli.

These experiments were carried out on 10 spontaneously breathing cats, anesthetized with pentobarbital (40-60 mg/kg). The local PO$_2$ was measured polarographically with a PO$_2$ needle electrode (1). The local flow was measured with a new method described by Lübbers and Stosseck (4). The PO$_2$ needle electrode and the local flow electrode are elastically suspended with copper wire so that they are able to follow each movement of the carotid body. The PO$_2$ needle electrode has a tip diameter of 1-2 µm. For the measurement of the local flow two platinum microelectrodes with a tip diameter of 1-2 µm are glued together. The principle of the local flow measurement is shown in Fig. 1. With one microelectrode hydrogen is generated electrolytically, and with the other palladinized electrode the hydrogen pressure is measured (2). The generating current is about 6-10 nA. In a distance of less than 5 µm the current induces a hydrogen pressure within the perfused carotid body of about 35 mmHg. When hydrogen is generated continuously, an increase of the hydrogen pressure in the tissue corresponds to a decrease of the local flow and vice versa. For evaluation the initial value before the reaction is always taken as 100%. The catching volume is about 5×10^{-3} mm^3, i.e., the volume measured locally compared to the vascular arrangement of the carotid body. Histologically detectable damage of the carotid body tissue from the generating current or the hydrogen could not be seen (7). The total flow was measured by weighing the blood from the venous outflow, which had been carefully isolated. Changes of the tidal volume were continuously registered by a pneumotachogram to control the excitation of the peripheral chemoreceptors. Also the blood pressure was controlled.

The behavior of local PO$_2$, local flow, and total flow were investigated under the following conditions: arterial hypoxia and hyperoxia, arterial hypercapnia, and changing blood pressure. The effect of arterial hypoxia during ventilation of the animals by respiration of 5% O$_2$ in N$_2$ shows a distinct dissociation between local and total flow (Fig. 2). The local flow decreases by about 50%, whereas the total flow - as described by Purves - increases markedly (5). The local PO$_2$ decreases. The increased tidal volume shows that the peripheral chemoreceptors are excited by hypoxia. The blood pressure remains nearly unchanged. After reventilation with air the local PO$_2$ and the local flow as well as the total flow return undirectionally to their initial values. The tidal volume typically at first shows an undershoot before it returns to its initial value.

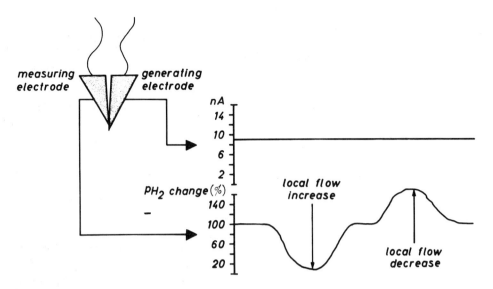

Fig. 1. Principle of local flow measurement. With one electrode hydrogen is generated electrolytically with a current of about 10 nA, while the other electrode measures changes of hydrogen pressure. A decrease of hydrogen pressure means an increase of local flow and vice versa

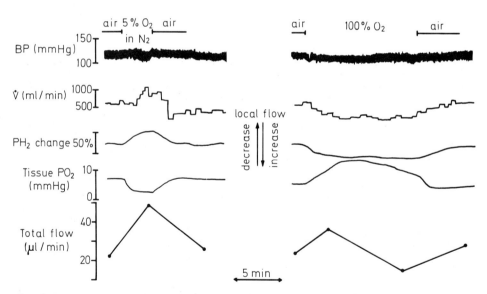

Fig. 2. Relationship between blood pressure (BP) and ventilation rate ($\dot{V}$) of animal, and local flow (PH_2 change), tissue PO_2, and total flow of carotid body during arterial hypoxia by respiration of 5% O_2 in N_2 and during arterial hyperoxia by respiration of 100% O_2

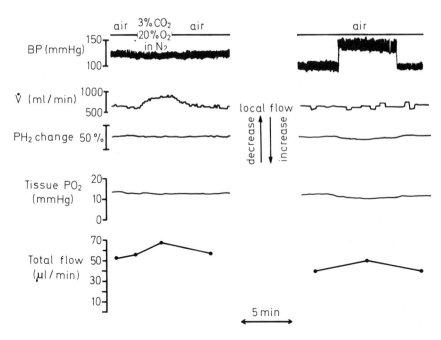

Fig. 3. Relationship between blood pressure (BP) and ventilation rate ($\dot{V}$) of animal, and local flow (PH_2 change), tissue PO_2, and total flow of carotid body during arterial hypercapnia by respiration of 3% CO_2, 20% O_2 in N_2 and during an increase of blood pressure

Under arterial hyperoxia caused by 100% O_2 ventilation of the animal, the local flow increases to about 50-70%, whereas the total flow at first increases and then decreases. The local PO_2 shows the typical overshoot, i.e., at first an increase and then a decrease. Under these conditions the blood pressure remains mostly unchanged. The tidal volume, however, shows a decrease. After reventilation with air the local flow immediately returns to its initial value, whereas the total flow remains in its initial range. The tissue PO_2 at first shows an undershoot and then increases to its initial value.

With arterial hypercapnia by ventilation of 3% CO_2 in air, in 40% of the experiments no reaction of the local flow is seen, whereas the total flow always increases, as shown in Fig. 3. In the remaining 60% the local flow increases or decreases by about 10-15%. The behavior of the local PO_2 mirrors that of the local flow. The tidal volume always increases. The blood pressure is not typically influenced by these conditions. After reventilation with air, tissue PO_2 and local flow and total return to their normal values.

When the blood pressure is increased by occlusion of the external carotid artery, the local flow remains nearly unchanged in 60% of the experiments, whereas the total flow always increases. Also the tissue PO_2 remains stable in these cases. In the remaining 40% of the cases the local flow at first increases by about 20% but then mostly decreases in spite of a high blood pressure. The tissue PO_2 decreases by 10-15% in spite of a higher local flow. After releasing the clamp blood pressure and total flow return to their initial values, while the local flow and the tissue PO_2 remain unchanged in 60% of the cases.

In 40% the local flow at first decreases and then increases to the initial values, while the tissue PO_2 returns to its initial value without an overshoot. The tidal volume is not markedly influenced by changes of the blood pressure.

In these experiments it is striking that the local flow and the total flow can show different responses. This means that besides the mechanisms to regulate the total flow there must be mechanisms to regulate the local flow separately. This could occur since the carotid body has a specific vascular structure (6). The specific carotid body tissue is supplied by a dense capillary network, while larger venous vessels have direct or short capillary connections with the arterial supply. The reaction of tissue PO_2 and local and total flow to changes of blood pressure and P_aCO_2 shows clearly these regulatory properties of the vessels of the carotid body. However, also the oxygen consumption of the carotid body with its dependence on PO_2 and flow must be taken into account. During arterial hypoxia in our experiments it was observed that the tissue PO_2 as well as the local flow decreases. This could mean that the hypoxic signal in the carotid body is augmented by lowering the flow but is also influenced by decreasing the flow-dependent oxygen consumption. Purves has shown (5) that by lowering the oxygen consumption the chemoreceptor activity increases and vice versa. During hyperoxia the local flow increases, which means that the flow-dependent oxygen consumption is also increased. The later reduction of the tissue PO_2 during hyperoxia can probably be explained by the increased flow, since we could show that the PO_2 dependence of the oxygen consumption did not change under these conditions.

Until now it has not been possible to determine which mechanisms are responsible for the different behavior of local and total flow during hypoxia and hyperoxia. The autoregulation of the local flow found in 60% of the experiments could explain the independence from blood pressure changes of the chemoreceptive activity, as described by Biscoe (3). The concomitant PO_2 decrease during flow increase with changes of blood pressure demonstrates the flow-dependent oxygen consumption. The predominant constancy of local flow and local tissue PO_2 in the carotid body during hypercapnia points to the direct influence of the CO_2 on the chemoreceptor.

References

1. Acker, H., Lübbers, D.W., Purves, M.J.: Pfluegers Arch. *329*, 136-155 (1971)
2. Acker, H., Lübbers, D.W., Durst, H.: in: Microcirculation. Lewis, D. (ed.). Basel: Karger 1977 (in press)
3. Biscoe, T.J.: J. Physiol. *208*, 121-131 (1970)
4. Lübbers, D.W., Stosseck, K.: Naturwissenschaften *67*, 311-312 (1970)
5. Purves, M.J.: J. Physiol. *209*, 395-416 (1970)
6. Schäfer, D., Seidl, E., Acker, H., Keller, H.P., Lübbers, D.W.: Z. Zellforsch. *142*, 515-524 (1973)
7. Seidl, E.: Unpublished results

DISCUSSION

Wiemer: You showed what happened when you raised blood pressure. What happened when you lowered it?

Acker: When you decrease blood pressure from about 100 to 60 mmHg, then local flow and total flow decrease. Here we are dealing with the normal range, i.e., with a normal blood pressure of 100 mmHg. When it increases, we can expect that the PO_2 of the carotid body would increase, which is avoided partly by the flow autoregulation.

Purves: By administering papaverine could you abolish the dissociation between local and total flow?

Acker: We did not do such experiments.

McDonald: With a simplified model could you describe where the dissociation between total and local flow occurs and where plasma skimming takes place?

Acker: As you can see from our casting models, the vascular structure of the carotid body consists of a dense capillary network and large shunt vessels. If we have any mechanism that regulates at the arterial inflow the amount of blood flowing through the shunt vessels, then we have the possibility of regulating the flow through the capillary network. Also by regulating the resistance of the capillary network by means of contractile proteins or swelling cells, the flow distribution can be controlled. Not only dissociation between total and local flow can be regulated by these mechanisms but also the degree of plasma skimming in the specific tissue of the carotid body.

O'Regan: Your results seem to me to confirm what de Castro has said, that under hypoxic conditions, shunt flow increases and local flow decreases. The mechanism, however, depends on an intact sinus nerve. Have you shown what happens when the nerve is cut?

Acker: No, the sinus nerve was always intact.

Lübbers: We really do not know at which site the plasma skimming takes place in the carotid body. But from rheology we know that you need a relatively high flow rate so that the red cells are swimming in the center of the stream. Also the branching angle of the vessels plays an important role.

Wiemer: In your concept plasma skimming is an important factor, which also depends on the degree of oxygenation and other parameters. So we have to deal with the regulation not only of flow but also of hemoglobin content in the carotid body.

Acker: Presumably our results show that more red cells flow into the carotid body under hypoxia with decreasing local flow. But this improvement of the oxygen supply by hemoglobin is abolished because the oxygen consumption is diminished by the decreased local flow and local PO_2.

Böck: The arteriovenous anastomoses, the supplying arteries, and the branching arteries of the carotid body are well provided by a dense adrenergic network. If you repeat your experiments with sympathectomized animals, the dissociation between local flow and total flow should disappear.

Acker: You are right, we have to do this kind of experiments.

Lübbers: It can be that the arteriovenous anastomoses, the branching arteries, or short connections between the arterial inflow and the venous outflow play an important role in the flow regulation of the carotid body. However, as Seidl has shown, the arteriovenous anastomoses are not regularly found in the carotid body; thus the carotid body must have different possibilities for flow regulation.

Ji: I thought that plasma skimming aided the induction of very low PO_2 values around the glomus cells in hypoxia, but you just mentioned that in hypoxia there are more red cells around the glomus cells so that the oxygen supply is improved.

Acker: As Purves has shown, the oxygen consumption of the carotid body depends on flow and PO_2 reciprocally, i.e., that not the hemoglobin content but the decreased oxygen consumption determines the chemoreceptor activity.

Paintal: We have to go back to 1939 when Comroe said that the oxygen consumption of the carotid body is so low that it can be satisfied merely by the oxygen dissolved in the plasma, and that is what you are saying now.

Acker: When the local flow is high enough, the oxygen dissolved in plasma is sufficient to supply the tissue in spite of a high oxygen consumption.

Whalen: In regard to the evidence that the oxygen disappearance curves in the blood and Ringer's-perfused carotid body point to plasma skimming, it is very possible that the high tonus in the carotid body during perfusion stop essentially squeezes out all the red cells.

Acker: Immediately with the perfusion stop the venous outflow of the carotid body stops. Therefore, we can say from our experiments that no additional flow occurs after the perfusion stop.

Willshaw: Have you got any findings on lymphatic drainage in the carotid body during the different stimuli? I think the movement of the lymphatic fluid can also influence the flow measurements.

Acker: That is a very interesting question. But as yet I have not investigated the lymphatic flow in the carotid body.

Effects of Temperature on Steady-State Activity and Dynamic Responses of Carotid Baro- and Chemoreceptors

D. Bingmann

It is well known that carotid baroreceptors are highly responsive to the rate of rise and fall in arterial blood pressure. Like other mechanoreceptors, such as muscle spindles (12), crayfish stretch receptors (11), and Pacinian corpuscles (6), they adapt markedly to applied stimuli (2,4,7,8). McCloskey (9), Gray (5), and Black et al. (1) have demonstrated that chemoreceptors of the carotid body also adapt if they are stimulated by hypercapnia.

In both baro- and chemoreceptors the mechanisms underlying the process of adaptation have not been understood up to now. Findings in the literature indicate that an electrogenic sodium pump could be responsible for adaptation processes. Thus, in the properly controlled crayfish stretch receptor, Nakajima and Onodera (11) suggested that activation of an electrogenic sodium pump evoked adaptation to constant stimuli. These observations lead to the question whether electrogenic pumps or other energy-consuming mechanisms cause adaptation in baro- and chemoreceptors. In energy-consuming processes the activation energy ranges above 10 kcal/M. The temperature coefficient Q_{10} is, correspondingly, about 2. If adaptation in baroreceptors or chemoreceptors results mainly from active mechanisms, the time course of adaptation should be markedly affected by changes of temperature. Therefore, we decided to examine here the relation between temperature and time course of adaptation in both receptor types of the carotid sinus region. The results of these tests were compared with the effects of temperature on the steady-state activity of the receptors.

Fig. 1 demonstrates the experimental arrangement used to test chemoreceptor reactions to hypercapnic and hypoxic stimuli at different temperature levels. Blood of the carotid communis artery was pumped through one of two gas exchangers where blood contacted different gas mixtures. The blood stream could be switched alternatively between both gas exchangers. After passing a module where P_aO_2 and P_aCO_2 were monitored continuously, blood was pumped through a heat exchanger and then flowed into the carotid sinus. Blood temperature and blood pressure were recorded at the carotid sinus with the pressure being adjusted to at least 20 mmHg above the pressure in the aorta abdominalis. Thus, a rapid blood stream perfused the carotid sinus, and blood temperature in the external carotid artery differed by less than 0.5°C from the temperature at the input of the carotid sinus.

Neurograms were recorded bipolarly from few-fiber preparations of the cut carotid sinus nerve. In these recordings individual fiber potentials could be clearly separated. The nerve action potentials were counted electronically in 1-s periods. Such counting periods did not allow a detailed analysis of peak activities in the dynamic responses. However, shorter counting periods often masked the course of adaptation in chemoreceptors because of the randomness in chemoreceptor discharge.

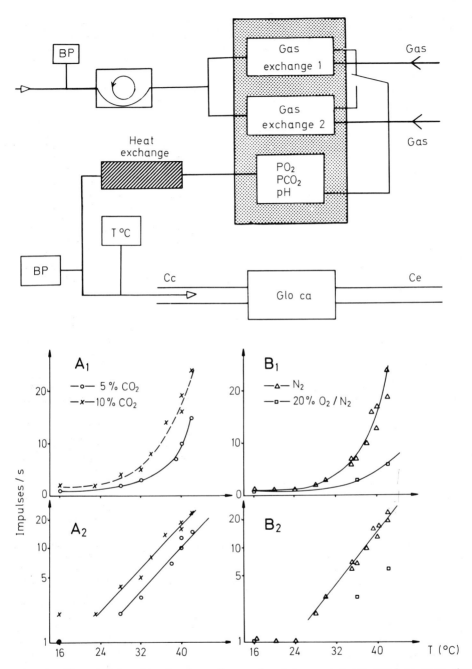

Fig. 1. (*Upper*) Schematic drawing of experimental arrangement used to test chemoreceptor reactions at different levels of temperature. BP, transducer elements for recording blood pressure in aorta and common carotid artery (Cc). T°C, thermistor; Glo ca, glomus caroticum; Ce, external carotid artery. (*Lower*) In graphs A_1-B_2, adapted discharge rates of chemoreceptor fiber units are plotted vs blood temperature (T), (A_1 and A_2) Receptor activity recorded at different levels of PCO_2, which resulted from extracorporal gas exchanges with 5% CO_2 in O_2 and 10% CO_2 in O_2. (B_1 and B_2) Blood contacted gas mixtures of 20% O_2 in N_2 and 100% N_2 with the latter gas causing a P_aO_2 of about 30 mmHg

From the literature, moreover, it is known that half-times of adaptation in chemoreceptors range between 5 and 10 s (1,9). Therefore, the frequency of chemoreceptor fiber potentials was measured in 1-s periods, and the evaluation of the time course of adaptation started about 5 s after the maximum of the frequency curve. To compare the thermal effects on the dynamic responses in baro- and chemoreceptors, the frequency of baroreceptors was counted and evaluated in the same way. Using this experimental arrangement, chemoreceptor responses were analyzed at different levels of blood temperature, which was varied between 42 and 25°C. Parts A_1-B_2 in Fig. 1 demonstrate that only within this range was the thermal sensitivity of the adapted receptor activity rather constant. At various levels of P_aO_2 and P_aCO_2 the logarithmic plot of the discharge rate vs temperature showed a linear relationship between about 25 and 42°C. Below this range the thermal sensitivity of chemoreceptor discharge changed. Furthermore, at about 20°C the decay time constants of the dynamic reactions to hypercapnic stimuli could not be determined since the receptor response hardly appeared through the randomness of the discharge. If blood temperature exceeded 42°C by about 1°C, sensitivity of chemoreceptors to changes in P_aCO_2 and P_aO_2 decreased after a latency of some minutes, and normal responses did not reappear when blood temperature was lowered. These phenomena limited the range of temperature in which the thermal sensitivity of chemoreceptors was tested.

Fig. 2A demonstrates how chemoreceptors reacted to hypercapnic stimuli at different levels of temperature. With a rise in temperature from 36 to 42°C, receptor discharge increased phasically. Then a stable rate of discharge was recorded, which ranged about 100% above the initial level at 36°C. After a rapid rise of the P_aCO_2 from 55 to 65 mmHg, chemoreceptor fiber units behaved as would be expected from adapting receptors. A steep increase in the discharge rate was followed by a decline in receptor activity. This decline, in a rough approximation, followed an e-function with a mean time constant of 12-13 s, which proved to be independent of the starting level of PCO_2. Correspondingly, the decay of P_aCO_2 induced an off-effect in the receptor discharge, which widely reflected the on-reactions. In contrast to the obvious sensitivity of the adapted chemoreceptor discharge to changes in temperature, the time of decay seemed to be rather unaffected when temperature of the blood was lowered from 42 to 26°C. The evaluation of the experiment shown in Fig. 2A is demonstrated in Fig. 2B. The steady-state activity of the fiber preparation is plotted vs P_aCO_2 at 38, 35, 29, and 26°C. The fiber activity recorded at a P_aCO_2 of 55 mmHg and at a temperature of 37°C was set equal to 100%. The graph shows that an increase in temperature causes an increase in the slope of the response curves, but changes in the shape of these curves cannot be detected. In the next step of evaluation the activity found at a P_aCO_2 of 55 mmHg was plotted against temperature. This plot yields a temperature coefficient of 2.7-3.0, which corresponds to the data of Paintal (13) and McQueen and Eyzaguirre (10). In the experiments shown in Fig. 1 the Q_{10} value ranged between 3 and 4. In contrast to this sensitivity the effect of temperature on the time course of adaptation is small, as shown in the plot of the decay time vs temperature. In order to compare the effects of temperature on the adapted frequency and on the time course of adaptation, the mean time constant observed at 37°C was set equal to 100%. The plot shows that the time constant was prolonged by only 20% when temperature was lowered by 10°C. This finding did not depend on the level of the P_aO_2. In Fig. 2B the squares represent mean values of time constants observed at a P_aO_2 between 30 and 80 mmHg. Time constants indicated by circles were determined at a P_aO_2 of 600 mmHg. The thermal insensitivity of the time course of adaptation in chemoreceptors raises the question as to the effects of temperature on the course of adaptation in carotid baroreceptors.

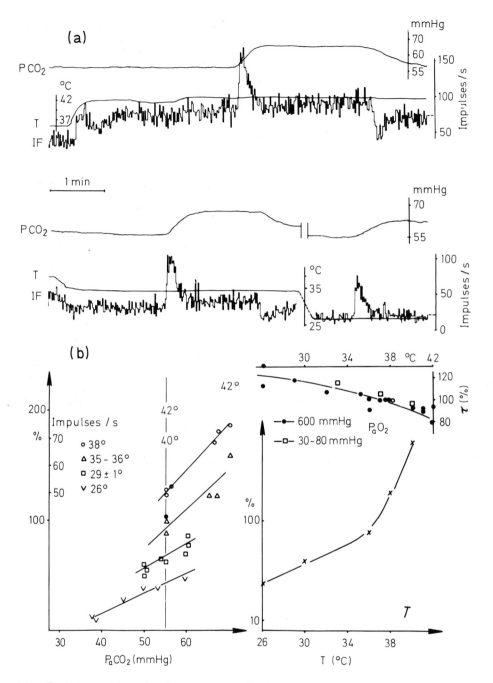

Fig. 2. (A) Reactions in chemoreceptor discharge rate (IF) to changes in arterial PCO_2 at various levels of temperature (T). (B) (*left*) Adapted frequency of chemoreceptor discharge is plotted vs P_aCO_2 at different temperatures. (*right*) Relation between temperature and adapted discharge rate recorded at a P_aCO_2 of 55 mmHg. Also decay time constants of adaptation (τ) are plotted against temperature. Adapted discharge rate found at P_aCO_2 of 55 mmHg and 37°C as well as mean decay time constant observed at same temperature were set equal to 100%

Almost the same arrangement as was described in Fig. 1 was used to test baroreceptor units. Stepwise changes in blood pressure were achieved by opening alternatively different shunts between the carotid sinus and the femoral vein. Vasoreceptor activity and adaptation were evaluated in the same way as the data for chemoreceptors. This mode of evaluation could be used because 95% of the final adaptation in baroreceptors is reached in about 30 s (4,7,8). This equals a mean time constant of approximately 10 s within the late phase of adaptation, i.e., in both the baro- and chemoreceptors the time constants are the same order of magnitude. Fig. 3A shows the reaction of carotid baroreceptors to rapid changes in blood pressure at 40, 35, and 26°C. Similar to the reactions in chemoreceptors the steady-state activity of baroreceptors was widely affected by changes in temperature, and also, striking effects of temperature on the time course of adaptation were missing. Fig. 3B shows the evaluation of this experiment. The response curves of the analyzed baroreceptor units were flattened markedly when blood temperature was reduced from 37 to 20°C. The plot of temperature vs the adapted discharge rate at 150 mmHg and 200 mmHg demonstrates Q_{10} values of about 2.5. This result is in line with temperature coefficients of carotid baroreceptors found by Diamond (3) and corresponds with the Q_{10} values determined in the stretch receptors of the lungs (13) and muscle spindles (12), and with the generator potentials of Pacinian corpuscles (6). Furthermore, Fig. 3B demonstrates that the time constants of adaptation were prolonged only 7-9 s when blood temperature was lowered from 40 to 30°C. Thus, the temperature coefficient resulting from these data does not differ significantly from the Q_{10} values found in the decay time constants of other mechanoreceptors and chemoreceptors. Summarizing in the range from 26 to 40°C dynamic and nondynamic reactions to temperature of carotid body chemoreceptors and baroreceptors were determined similarly. The Q_{10} of the adapted frequency in both receptor types ranged above 2, whereas the decay time constants reflected a Q_{10} of 1.2-1.3. The small effects of temperature on the time course of adaptation in carotid chemoreceptors cannot support the hypothesis that the phenomenon of adaptation results mainly from active transport mechanisms. It is more likely that passive processes in the membrane of the receptor units contribute considerably to the adaptive behavior of both baro- and chemoreceptors.

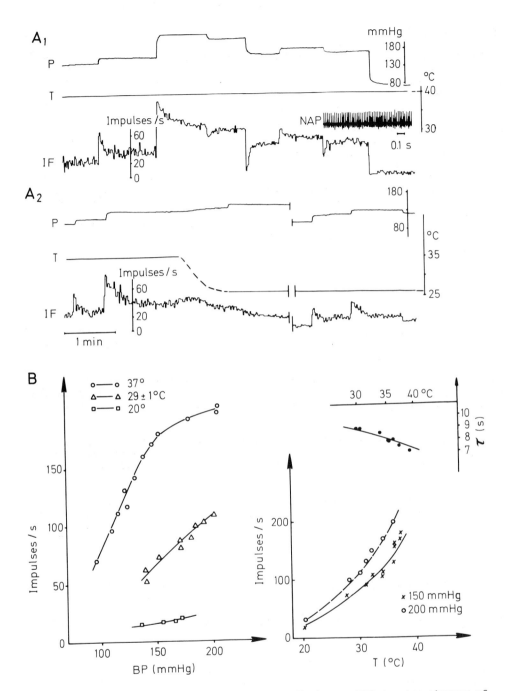

Fig. 3. (A_1 and A_2) Reactions in baroreceptor discharge (IF) to step changes of blood pressure (P) in carotid sinus shown at 3 different temperatures (T). Impulse frequency curve (IF) was obtained by counting electronically nerve action potentials (NAP). (B) Evaluation of experiment demonstrated partly in A_1 and A_2. (*left*) Blood pressure (BP) vs impulse frequency. Adapted baroreceptor response curves are displayed at different temperatures. (*right*) Temperature (T) is plotted vs adapted discharge rate recorded at 150 and 200 mmHg . Also, relation between decay time constants of adaptation in discharge rate (τ) and temperature are shown

References

1. Black, A.M.S., McCloskey, D.I., Torrance, R.W.: Respir. Physiol. *13*, 36-49 (1971)
2. Bronk, D.W., Stella, G.: Am. J. Physiol. *110*, 708-714 (1934)
3. Diamond, J.: J. Physiol. *130*, 513-532 (1955)
4. Franz, G.N., Scher, A.M., Ito, C.S.: J. Appl. Physiol. *30*, 527-535 (1971)
5. Gray, B.A.: Respir. Physiol. *4*, 229-245 (1968)
6. Ishiko, N., Loewenstein, W.R.: J. Gen. Physiol. *45*, 105-124 (1961)
7. Kirchheim, H.R.: Physiol. Rev. *56*, 100-175 (1976)
8. Landgren, S.: Acta Physiol. Scand. *26*, 1-34 (1952)
9. McCloskey, D.I.: in: Arterial Chemoreceptors. Torrance, R.W. (ed.). Oxford-Edinburgh: Blackwell 1968, pp. 279-294
10. McQueen, D.S., Eyzaguirre, C.: J. Neurophysiol. *37*, 1287-1296 (1974)
11. Makajima, S., Onodera, K.: J. Physiol. *200*, 161-185 (1969)
12. Ottoson, D.: J. Physiol. *180*, 636-648 (1965)
13. Paintal, A.S.: J. Physiol. *217*, 1-18 (1971)

DISCUSSION

Wiemer: You showed that the steady-state activity after adaptation depended on temperature. My question is, was the peak frequency also changed by temperature?

Bingmann: In a few-fiber preparation it is difficult to evaluate changes in the peak activity of chemoreceptors because sharp peaks in the activity curves typically were missing. We have not tried such an evaluation as yet.

Zapata: In one of your slides the relationship between temperature and chemosensory activity was not linear. In that case can you give a single Q_{10}? Also, I want to know if these values for the Q_{10} are determined from a single-fiber or a multifiber preparation.

Bingmann: In the range analyzed the half logarithmic relationship between discharge rate and temperature was linear down to 25°C. In our multifiber preparation we had a maximum of about 10 active fibers. The results with multifiber preparations and few-fiber preparations did not differ.

Purves: Why did you decide to take PCO_2 values of 55 mmHg to calculate the Q_{10}? From your slide it seems that you would get different results for Q_{10} if you take, e.g., a normal PCO_2 of about 30 mmHg.

Bingmann: If we use a PCO_2 of 30 mmHg, we get the same Q_{10} as with a PCO_2 of 55 mmHg.

Purves: Another point: you said that the half-time of adaptation of baroreceptors is 10 s. That seems to me to be rather large because in many people's experience, the rate of adaptation is much faster. What was the time constant of your averager?

Bingmann: From the literature it is known that after a stepwise rise in blood pressure 95% of the final adaptation in baroreceptors occurs in about 30 s. This equals a mean time constant of about 10 s within the late phase of adaptation. Our averager has had a count rate of 1 s.

Purves: That will not be adequate for things that have time constants in milliseconds, which is more likely in mechanoreceptors.

Bingmann: In baroreceptors there are time constants in the range I have discussed here.

Paintal: It should be noted that the temperature coefficient of the chemoreceptors has to be determined at the same local PO_2. This cannot be achieved by using a hypoxic mixture of a known concentration, because when you lower the temperature the local PO_2 is higher for the same gas mixture or the same P_aO_2. With your method the Q_{10} was about 16, a similar value to what Eyzaguirre found. Did you determine your Q_{10} by administering nitrogen?

Bingmann: In most of our experiments the thermal sensitivity of chemoreceptors was determined at hyperoxic values. Thus, the effects of temperature on local PO_2 which result from shifts of the oxygen binding curve, should be minimized. However, if nitrogen was blown through the oxygenator, the Q_{10} value at a P_aO_2 of about 20 mmHg did not differ significantly from the Q_{10} found at hyperoxic values.

Trzebski: Would it be possible to give an exact time course of adaptation of the baroreceptor by recording the whole sinus nerve activity, if we take into account that we have quite a large spectrum of baroreceptor sensitivity? If you analyze single fibers, you can see that there is a big variation in the rate of adaptation from fiber to fiber. I wonder if it is possible to get one exact number to characterize the variable population of baroreceptors. Is it justified, therefore, to speak of adaptation when you use a multifiber preparation?

Bingmann: In the literature there are many data about adaptation of such receptors. The time course of adaptation of the baroreceptors surely differs from one baroreceptor to the other. So only a mean course of adaptation can result from a test of the baroreceptor population. The question here was, does the time course of adaptation change markedly when temperature is changed? By testing the influence of temperature on the course of adaptation in baroreceptors and in chemoreceptors, we wanted to find out if active mechanisms contribute markedly to the phenomenon of adaptation in one of these receptors.

Eyzaguirre: Just a few comments. We have abandoned Q_{10} measurements because the curves are not linear. Therefore, depending on the way you measure, you obtain different Q_{10} values. If you average at very low values, you get a low Q_{10}; but if you use temperatures from 32 to 40°C, the Q_{10} will be very high because the curve is very steep. Also, with single-fiber preparations we found in recent in vitro experiments also high Q_{10} values.

Nishi: The time course of adaptation of the frequency of the baroreceptor discharge does not necessarily mean the real characteristics of the baroreceptor nerve terminals.

Bingmann: The time course of adaptation of baroreceptors is, of course, determined first by the receptor element and second by the elastic elements of the surrounding tissue. So we have an overall response of the baroreceptors, which includes all these compartments.

Zapata: I found in the carotid body in vitro that the Q_{10} for a single-fiber preparation differs completely from that for a multifiber preparation. In this last condition, recruitment will result in very high Q_{10} values.

Bingmann: In our in vivo experiments we found a homogeneous course in the half logarithmic plot of nervous activity and temperature. The curves in the above range of temperature were linear in single- and and multifiber preparations, and the Q_{10} values did not change systematically when the number of active fibers led from varied between 1 and 10.

Manipulation of Bicarbonate in the Carotid Body

R. W. Torrance

A hypothesis has already been presented that attempts to account for the convergence of hypoxia and hypercapnia to give impulses in the same afferent arterial chemoreceptor fiber. The nerve endings are regarded as acid receptors. Carbon dioxide, by becoming carbonic acid, naturally affects the environment, and hypoxia does so also because the mechanism that tends to hold the acidity of this environment constant fails in hypoxia. Here I shall briefly outline the behavior of chemoreceptors and the explanation offered by the hypothesis. Then I shall discuss particularly the way in which the hypothesis deals with adaptation of the response of chemoreceptors to a step change in P_aCO_2, a subject about which Bingmann has made some important observations (p. 277).

Behavior

The behavior of chemoreceptors that has to be expained by any hypothesis is illustrated in Fig. 1. Fig. 1A shows the hyperbolic relation between discharge and P_aO_2 at any fixed P_aCO_2. Fig. 1B shows that there is a linear between discharge and P_aCO_2 at any fixed P_aO_2. As hypoxia is made more intense, the slope of this relation becomes steeper and its intercept on the P_aCO_2 axis becomes less.

The time course of the response to a sudden change of stimulus depends on which stimulus is changed. If P_aO_2 is reduced, the response develops over a few seconds and adapts little (Fig. 1C). If arterial pH is lowered by lowering $\{HCO_3^-\}_a$, the response develops slowly and does not adapt (Fig. 1D), but if arterial pH is lowered by raising P_aCO_2, the response reaches a peak over about 1 s and adapts with a half-time of about 10 s (Fig. 1E). Thus at any level of hypoxia, there is a single steady-state response curve to CO_2 (Fig. 1F), but if P_aCO_2 suddenly rises to a new steady level, the immediate response rises to a point on a steeper transient response curve and then adapts down to lie again on the steady-state curve (1 → 2 → 3). At a reduction of P_aCO_2, the sequence is 1 → 4 → 5, the off-effect of an adapting receptor. There is not a single transient response curve to CO_2 for any level of hypoxia; rather one should conceive of a transient response curve passing through each point on a steady-state response curve.

Structure

In chemoreceptors groups of type I cells are surrounded by type II cells. Peroxidase, and presumably small ions also, can pass from the capillary vessels into the cleft between the type I and type II cells. The sensory nerve endings lie in this cleft. Fig. 2 shows this with the type I cells drawn as a sphere for the sake of mathematical simplicity (see later).

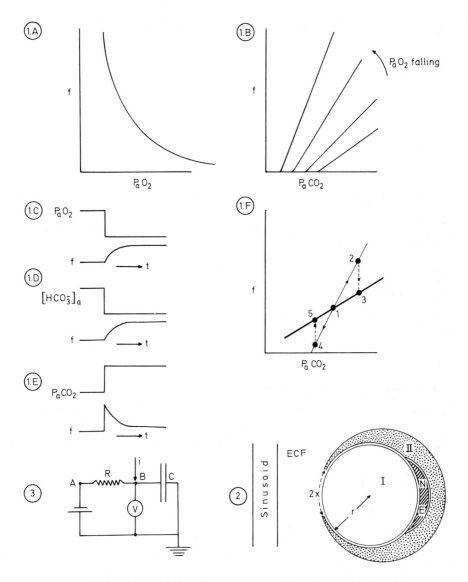

Fig. 1A-F. Frequency of impulses (f) in chemoreceptor fiber in steady-state as a function of P_aO_2 at a constant level of P_aCO_2 (A) and as a function of P_aCO_2 at various constant levels of P_aO_2 (B). Time course of development of response to a step charge of P_aO_2 (C), $\{HCO_3^-\}_a$ (D), and P_aCO_2 (E). (F) Steady-state (5, 1, 3) and transient (4, 1, 2) response curves to P_aCO_2 at constant P_aO_2. When P_aCO_2 is suddenly raised or lowered, response follows transient curve, but when adaptation is complete, response again lies on steady-state curve, i.e., 1, 2, 3 and 1, 4, 5

Fig. 2. Type I cell (I) of radius r with nerve ending (NE) against it and surrounded by type II cell (II) with opening from cleft between I and II to extracellular fluid (ECF). 2×, diameter of opening

Fig. 3. Analogue of Fig. 2. Ground represents zero bicarbonate concentration. A, in ECF, is held at a potential above ground corresponding to $\{HCO_3^-\}_a$. Bicarbonate pumping is represented by a current (i) into point B. Capacity (C) represents volume of cleft, and resistance (R), leak from cleft to extracellular space at A. Nerve ending is represented by voltmeter (V) measuring potential at B

The Model (3,4,5,6)

In the model it is assumed {1} that the nerve ending responds instantly and without adaptation to acidity in the cleft in which it lies and {2} that the frequency of response of the afferent fiber is linearly related to the pH function of acidity in the cleft. PCO_2 in the cleft is taken to be equal to P_aCO_2 since blood flow is huge and CO_2 diffusese easily through cell membranes. Bicarbonate can diffuse into and out of the cleft along the extracellular pathway followed by peroxidase, and its concentration in the cleft can be manipulated to be greater than $\{HCO_3^-\}_a$, so that local pH is given by

$$pH_L = pK + \log \frac{\{HCO_3^-\}_a + \Delta\{HCO_3^-\}}{\alpha\ PCO_2}$$

Since peroxidase can enter the cleft, it must be a leaky space, and so $\Delta\{HCO_3^-\}$ can be maintained only by mainting pumping into the cleft. Also, if the rate of pumping changes suddenly, $\Delta\{HCO_3^-\}$ will take some time to reach a new steady level, for the space has a finite volume in which the concentration changes. The rate of pumping should be linearly related to PCO_2 (cf. cerebrospinal fluid and kidney), and at any value of PCO_2 it tends to a maximum in hyperoxia and to zero in hypoxia. Thus in the steady state

$$pH_L = pK + \log \frac{\{HCO_3^-\}_a + \beta\ PCO_2}{\alpha\ PCO_2}$$

where β is affected by PO_2 in the same way as the rate of pumping.

If pH_L is plotted against PCO_2 for various values of β, a fan of curves results rather like the CO_2 response curves of chemoreceptors at various levels of hypoxia. Put another way, the slope of the curve is given by

$$\frac{d(pH_L)}{d(PCO_2)} = - \frac{\{HCO_3^-\}}{\{HCO_3^-\} + \beta\ PCO_2} \cdot \frac{1}{PCO_2}$$

so for any value of PCO_2 the slope is greater when β is less.

The transient curve is given by the change that occurs in pH_L when PCO_2 has altered locally but the change in the rate of pumping that it immediately produces has not yet significantly altered local $\{HCO_3^-\}$. The slope of the transient curve is given by

$$\frac{d(pH_L)}{d(PCO_2)} = - \frac{1}{PCO_2}$$

and so is inversely proportional to PCO_2 but is independent of β and so of the level of hypoxia.

The ratio of the initial shift in pH_L to the final shift at a small change in P_aCO_2 is given by the ratio of the slopes of the transient and the steady-state curves, i.e.,

$$\frac{\{HCO_3^-\} + \beta\ PCO_2}{\{HCO_3^-\}}$$

Thus the degree of adaptation is greater in hyperoxia and at a high PCO_2 and is less in hypoxia. There is evidence that adaptation of the response of chemoreceptors to CO_2 has these characteristics.

Now consider diffusion of bicarbonate in the cleft. Suppose that the cleft has a constant width and that type II cells pump bicarbonate into the cleft at a uniform rate per surface. Since the cleft is narrow, this is equivalent to bicarbonate being generated in the cleft at a constant rate per volume and diffusing out at the aperture to extra-cellular fluid (ECF). Problems arise similar to those considered for oxygen consumption by Krogh and Hill. Using the same terminology that Hill (2) used for oxygen and the geometrical simplification of Fig. 2, the difference between the concentrations at the point most remote from the aperture and at the aperture (i.e., in ECF) is

$$\frac{a}{K} \cdot 2r^2 \log\left[\sec\frac{\pi r - x}{2r}\right]$$

where $2x$ equals the diameter of the aperture measured as an arc around the sphere, r is the radius of the sphere, a is oxygen uptake, and K is the diffusion constant.

The expression $\log\left[\sec\frac{(\pi r - x)}{2r}\right]$ increases rapidly as $x \to 0$, and so the principal resistance to diffusion is near the aperture if the aperture is small. Thus one can use an oversimplification in which the part of the cleft near the aperture acts as a resistance R to the escape of bicarbonate by diffusion and the part of the cleft away from the aperture acts as a capacity C with little resistance to diffusion in it. (This is not a necessary simplification (1)). Such a setup has a time constant RC. The capacity is set simply by the geometry of the system, but the resistance is set by the geometry near the aperture and the reciprocal of the diffusion constant for the substance considered. The electrical analogue is shown in Fig. 3. Bicarbonate pumping corresponds to a current i passed into point B and blood bicarbonate to the potential at A. The bicarbonate at the nerve ending is given by the potential at B.

A change in blood bicarbonate corresponds to a change in the potential at A, and this eventually causes an equal change in potential at B, which develops with a time constant RC. A change in bicarbonate pumping corresponds to a change in current i, and this also causes a change in potential at B, which develops with the same time constant RC. The extent of the change in potential at B depends on the size of the potential change at A or of the current change into B, but the time constant of each of the changes of potential at B is the same, RC, and is dependent of the size of the change at B.

Thus the time constant of adaptation to CO_2 should be indepedent of PO_2 and PCO_2, which Bingmann has shown. If a change of temperature the geometry of the system does not change, its capacity C should not change, but the resistance R should change because it is dependent on the diffusion constant for bicarbonate. Diffusion constants increase by about 25% for a rise in temperature of 10°C. Thus the time constant of the system should be reduced by about 20% by a temperature rise of 10°C. Bingmann has found that the time constant of adaptation to CO_2 is affected in this way.

It is to be expected that a change in temperature will also alter the relation between local PO_2, PCO_2, and the rate of pumping, and such a change could well alter the degree of adaptation.

Peto has suggested that hypoxia might lead to an opening of the aperture through which bicarbonate escapes to the ECF and hyperoxia, to a closing, so that a constant rate of pumping would lead to a PO_2-dependent concentration of bicarbonate near the ending because PO_2 affects R of the system. If this were so, the time constant of the system, RC, would be reduced in hypoxia and increased in hyperoxia. Bingmann's finding that the time constant of adaptation to CO_2 is independent of PO_2 contradicts this chemomechanochemoreceptor variant of the hypothesis. The hypothesis does, however, predict, as has already been stated, that the degree of adaptation is dependent on the level of hypoxia.

References

1. Carslaw, R.S., Jaeger, J.C.: Conduction of Heat in Solids. Oxford: Clarendon Press 1959
2. Hill, A.V.: Trails and Trials in Physiology. London: Arnold 1965, pp. 208-214
3. Torrance, R.W.: in: Respiratory Physiology, M.T.P. International Review of Science Physiology. Widdicombe, J.G. (ed.). London: Butterworth 1974, Ser. 1, Vol. II, pp. 247-271
4. Torrance, R.W.: J. Physiol. 244, 64-66 (1975)
5. Torrance, R.W.: in: Morphology and Mechanisms of Chemoreceptors. Paintal, A.S. (ed.). Delhi: Vallabhbhai Patel Chest Institute 1976 pp. 131-137
6. Torrance, R.W.: in: Acid Base Homeostasis of the Brain Extracellular Fluid and the Respiratory Control System. Loeschcke, H.H. (ed.). Stuttgart: Georg Thieme 1976, pp. 95-103

DISCUSSION

Kiwull-Schöne: Do you know if there is a high content of carbonic anhydrase in the type I cell?

Torrance: There is carbonic anhydrase in the carotid body. You would expect it to be in the type II cell for that is the cell rather like the glial cell, which contains the carbonic anhydrase of the nervous system. If carbonic anhydrase is on the cleft surface of the type II cell, one could most easily account for the effects of acetazolamide in slowing the response to CO_2 and reducing the steady-state responses.

Willshaw: There is, in fact, evidence that a change in oxygen tension does not influence the time course of adaptation. Is it necessary to suppose that CO_2 itself changes the rate of pumping?

Torrance: I introduced that into the hypothesis originally to account for the fan of CO_2 response curves, because if HCO_3^- is independent of PCO_2 at any PO_2 the slope of the CO_2 response curves would be independent of oxygen tension and the response curves would be parallel. The abstact I submitted attempts to justifiy the idea that pumping is dependent on PCO_2 by a bit of mathematical ingenuity, but it seemed to me preferable to discuss Bingmann's interesting observation here.

Willshaw: Does the recent work of Delany and Lahiri on single-fiber responses in fact show much of a fan?

Torrance: They have some very striking fans, and we find a four- or fivefold range of slopes. But it was Fitzgerald and Parkes's observation of a fan I was trying to account for originally.

O'Regan: Would there be a sufficient amount of bicarbonate diffusing in from the capillary blood?

Torrance: I am not supposing that bicarbonate diffuses into or out of the capillaries. I suppose that bicarbonate is pumped into the cleft and then it leaks out of it and is then pumped back in, so that it can be regarded as going around in a circle. The ion that goes with it or moves in the opposite direction to exchange with it may be sodium or chloride, but it follows such a pathway. Of course, if you have carbonic anhydrase and CO_2, it is rather tricky to make any definite statement about bicarbonate.

O'Regan: It is a relatively small ion so that the resistance to diffusion through cell membranes may be slight.

Torrance: I suppose that ions diffuse along the extracellular peroxidase pathway but that fat-soluble particles go straight through cell membranes. That is one of the basic assumptions of the hypothesis.

Eyzaguirre: If you are postulating a bicarbonate pump, would you not expect it intuitively to be more dependent on temperature unless you have evidence from other systems that this type of pump is not temperature-dependent?

Torrance: I do not have any evidence that the pump is not dependent on the temperature. I have evidence only about one of the consequences of the pump. The time constant of the space into which it is pumping is only slightly dependent upon temperature. This is not a matter of the temperature coefficient of the pump itself; it is rather the temperature coefficient of the space into which it is pumping, and this is set simply by the geometry of the space and the diffusion constant within it. It is the temperature coefficient of the diffusion constant that sets the temperature coefficient of adaptation.

Eyzaguirre: So you do not know where the pump is located?

Torrance: I do not know. I am postulating there is pumping into the cleft. By analogy with the nervous system, the type II cell should do the pumping. But you could conceive of one pole of the type I cell pumping out bicarbonate that leaks within the cleft to be absorbed by the opposite pole of the type I cell, as has been suggested for glial cells in the central nervous system. I think the type II cell is the most likely cell to do in the carotid body what glial cells do in the central nervous system.

McDonald: Why do you postulate that the hydrogen ions act on the type I cells rather than the sensory endings themselves?

Torrance: There is a nerve ending that the hydrogen ions act on. If the leak to ECF acts as the principal resistance to escape of bicarbonate from the cleft, you have a fairly uniform concentration in the rest of the cleft. I could be that the glomus cell is sensitive to pH and releases a transmitter that acts upon the nerve endings. Basically I am considering how oxygen and CO_2 converge, and I suggest that they do so before any nervous activity is concerned.

Acker: What has the type I cell to do in your model?

Torrance: It creates a space. At this symposium the type I cell has been supposed to control the ending. It is geometrically convenient if there is a cell, enclosed by a type II cell, that controls the ending. The basic problem is whether it is a phasic control, which varies with PO_2 and PCO_2, or whether it is a tonic control, which sets the sensitivity and is affected by efferent activity.

Loeschcke: You know I do not like bicarbonate pumps, and we have had this dicussion previously. The metabolism of the type I cell depends on the PO_2 for oxidative metabolism, so you have the influence of oxygen on bicarbonate production and thus of the pH. Would that be too much of a simplification?

Torrance: I should better explain what I mean by a bicarbonate pump. I simply mean something that leads to a change in bicarbonate concentration. As for its mechanism, if you poison the carotid body with Diamox, the discharge at any PO_2 becomes less. This is a real effect in the carotid body, and I think it is caused by something analogous to the disequilibrium pH in the kidney tubule. The thing responds less, so that within our hypothesis the observation suggests that hydroxyl ions are pushed into the cleft, there combine with CO_2 at a rate depending on whether carbonic anhydrase is active, and then escape through the leak to ECF.

Wiemer: How does your model explain the nonlinearity of the oxygen response curve?

Torrance: That waits for Lübbers and Whalen to agree. Basically I suggest that the rate of pumping is a function of local PO_2. We have a curve relating P_aO_2 to response, but we need to know the relation of P_aO_2 to tissue PO_2. Then the simplest assumption to make is that the rate of pumping obeys Michaelis-Mente kinetics; thus the rate of pumping would bear the standard relation to a K_m and the local PO_2.

Kiwull-Schöne: What about anaerobic glycolysis and a possible exchange of lactate ions for bicarbonate ions and in such way as to change buffer capacity?

Torrance: I think there are two points there - the original hypothesis of Winder was that in hypoxia lactic acid was formed. The chemoreceptors simply are acid receptors and so are excited by CO_2. If there is not enough oxygen, lactic acid is formed and contributes to excitation. This does not, however, fit with observations on poisons of metabolism such as fluoride and iodoacetate, which are supposed to stop the formation of lactic acid but do excite chemoreceptors. However, a problem with my hypothesis is that to account for reactions in hyperoxia the concentration of bicarbonate has to be fairly high. If the response of the afferent fiber is the same at a high PO_2 as at a low PO_2, the PCO_2 is several times greater at the high PO_2 than at the low PO_2. But the local bicarbonate must be in the same ratio to PCO_2 in each case because the identity of the response is attributed to identity of local pH. This local bicarbonate must reach very high levels in hyperoxia. If, however, lactate were formed at a steady rate and leaked into the cleft, the problem would not be so intense.

Willshaw: I know that Torrance likes the idea that the type I cell may, in fact, be manufacturing some type of protein. If it has to push large quantities of bicarbonate into the space, it will probably form equal quantities of hydrogen ions, which have to be buffered somewhere. Perhaps they are buffered by the protein being made in the same process.

Torrance: You are making two distinct points. So far as quantity of bicarbonate pumped is concerned, there are possibilities of working this out. You need to know what the volume of the cleft is and also its time constant. You would get the latter from the response to changing the bicarbonate in the perfusate. If you know the volume and the time constant, you can work out the rate of pumping and then perhaps the oxygen cost.

Paintal: What tests have you made of the hypothesis?

Torrance: I first used the hypothesis simply to account for the steady-state curves, and then it becomes clear that it accounted for adaptation, which you may regard as a test. It was for me, but it may not be regarded so publicly. So far as the carbonic anhydrase is concerned, inhibition of carbonic anhydrase should have some effect. If it had had no effect, I would have been very worried, but it did have quite a marked effect, so the hypothesis was not refuted. The hypothesis suggests that discharge to any degree of hypoxia can be abolished by severe enough hypocapnia, which is true. It makes predictions about the effect of PO_2 and PCO_2 on the degree of adaptation to CO_2, but these have not been fully tested yet. Finally, as I pointed out in my paper, it predicts many of the observations that Bingmann has just reported.

Subject Index

Acetylcholine, carotid body
   tissue   107
Adaptation   289
Alpha-adrenoceptor, carotid
   body   152, 161
Aortic body   250
Apnea, technique   36, 264
Autoradiography, nerve endings
   11, 19, 109
Autoregulation, local flow   273
AVO_2-difference, oxygen consumption   236
Axon, sprout   29
Axoplasmic flow   9

Baroreceptor   252, 277
-, carotid body   55
Beta-adrenoceptor, carotid
   body   152
Brain-stem, neurons, carotid
   body   182
Büngner's cord   30

Calcium, chemoreceptor discharge   258
-, hypoxia   117
-, type I cell   72
-, vesicle   202
cAMP   116
Capillaries, reconstruction   65
Carotid labyrinth   246
- nerve fiber, regeneration   45
- sinus nerve, crush   46
Carotidin   107
Catecholamine, flourescence   130
-, -, tyrosine hydroxylase   131
Chemoreceptor activity, fluctuation   175
- -, regeneration   36, 38, 45
Chemoreflex, mutant animal   51
Chief cell, see type I cell
Chloride, type I cell   72
CO_2, type I cell   73

Diameter, vessels   2
Differentiation, carotid body
   cells, hypoxia, hypercapnia
   89

Diffusion calculation   240
-, bicarbonate   289
- distances   62
Dopamine   99, 114, 123, 132
-, chemoreceptor function   145, 152
-, hypoxia   123
- beta-hydroxylase activity
   100, 139
- beta-hydroxylase, inhibitor
   138
Dopaminergic receptor, stimulating agent   147, 153
Dorsal medulla, see brain-stem

Ehrenritter ganglion   17
Electrogenic ion transport,
   receptor system   79

Glomoid   1
Glossopharyngeal nerve   17

Horseradish peroxidase   21
Hyperneurotization   31
Hypoxia, type I cell   73

Impulse generation   71
Input resistance, type I cell
   71, 92

Local flow   266, 271

Magnesium, type I cell   72
Mechanoreceptor, see baroreceptor
Membrane capacity, type I cell
   71
- potential, type I cell   71, 92
Mitochondria, concentration   5
-, oxygen consumption   257
Mitosis, type I cell   218
Monoaminooxidase, inhibitor   136
Mouse, wobbler   51
Muscle gracilis   234

NaCN, type I cell   73
Nerve, Jacobson   17
- endings   11, 21, 22, 207, 281

- endings, degeneration 27
- endings, mutant animal 52
- endings, type 34
- suturing 30
Neuroma 29
Nodose ganglion 13
Norepinephrine 123, 132
Nucleus ambiguus 187

Osmolarity, type I cell 73
Ouabain, type I cell 72
Oxygen, diffusion 62
- consumption, chemoreceptor mechanism 233, 257
- consumption, PO_2 240, 257
- consumption, qualitative morphology 1

Paraneuron 221
Petrosal ganglion 9, 11, 21
PH, type I cell 73
Plasmaskimming 242, 247
PO_2, diffusion distances 62
-, tissue, carotid body 234, 240, 244, 250, 264, 271
-, tissue, carotid body, comparative 246
-, type I cell 93
Potassium conductance, temperature 81
-, type I cell 72
Pressoreceptor, see baroreceptor

Receptor, dopaminergic blocking agents 147
Reconstruction, three dimensional 65
Regeneration, carotid sinus nerve 25, 30
-, nerve endings 30
Respiratory frequency, chemoreceptor discharge 176, 190
RNA 116

Serotonin 132
SIF cell 133
Sinus nerve, activity, efferent 168
Stratum nervosum 1
Sympathetic nervous system, carotid body 160

Temperature, chemoreceptor activity 252
- coefficient 250, 277
- coefficient, baroreceptor activity 281
- coefficient, chemoreceptor activity 279
-, type I cell 71
Tissue culture, type I cell 86
- -, type II cell 86
Total flow 271
Type I cell, characteristics 223
Tyrosine hydroxylase 114

Vagus, chemoreflex response 193
Vascular structure 221, 244
Vesicle 201
-, catecholamine, calcium 202
-, content 209
-, diameter 216
-, large 123
-, small 123
Vessel, shunt 245
Vessels, efferent stimulation 169
Volume measurement, carotid body tissue 1

Wallerian degeneration 30

Handbuch der allgemeinen Pathologie

Herausgeber: H.-W. Altmann, F. Büchner, H. Cottier, E. Grundmann, G. Holle, E. Letterer, W. Masshoff, H. Meessen, F. Roulet, G. Seifert, G. Siebert
Band 3
Zwischensubstanzen, Gewebe, Organe
Teil 7
Mikrozirkulation/Microcirculation
Redigiert von H. Meessen
Etwa 260 Abbildungen. Etwa 1090 Seiten. (Etwa 360 Seiten in Englisch). 1977
ISBN 3-540-07750-2

A. LABHART
Clinical Endocrinology
Theory and Practice
With a Foreword by G. W. Thorn. In collaboration with numerous experts.
Translators: A. Trachsler, J. Dodsworth-Philips
400 figures. XXXII, 1092 pages. 1974
ISBN 3-540-06307-2
Distribution rights for Japan: Igaku Shoin Ltd., Tokyo

W. A. McALPIN
Heart and Coronary Arteries
An Anatomical Atlas for Clinical Diagnosis, Radiological Investigation and Surgical Treatment
1098 figures, mostly in color. XVI, 224 pages. 1975
ISBN 3-540-06985-2
Distribution rights for Japan: Igaku Shoin Ltd., Tokyo

P. OTTO, K. EWE
Atlas der Rectoskopie und Coloskopie
31 Schwarzweißabbildungen, 115 farbige Abbildungen auf 21 Tafeln, 1 Tabelle. IX, 96 Seiten. 1976
ISBN 3-540-07489-9

M. R. PARWARESCH
The Human Blood Basophil
Morphology, Orgin, Kinetics, Function, and Pathology
With a Foreword by K. Lennert.
58 figures, some in color. XI, 235 pages. 1976
ISBN 3-540-07649-2

K. SIGG
Varizen
Ulcus cruris und Thrombose
Mit Beiträgen von C. C. Arnoldi, E. Imhoff, R. Kressig, H. J. Leu, C. Montigel, T. Wuppermann
4., neubearbeitete und erweiterte Auflage.
130 farbige und 411 Schwarzweiß-Abbildungen. XV, 403 Seiten. 1976
ISBN 3-540-07373-6

Atherosclerosis 3
Proceedings of the Third International Symposium, 24-28 Oct., 1973
Editors: G. Schettler, A. Weizel
349 figures, 222 tables. XXXV, 1034 pages. 1974
ISBN 3-540-06909-7
Distribution rights for Japan:
Maruzen Co. Ltd., Tokyo

Springer-Verlag
Berlin
Heidelberg
New York

M. BESSIS
Corpuscles
Atlas of Red Blood Cell Shapes
121 figures, 147 pages. 1974
ISBN 3-540-06375-7
Distribution rights for Japan:
Maruzen Co. Ltd., Tokyo

M. BESSIS
Blood Smears Reinterpreted
Translated from the French by G. Brecher
342 figures mostly in color. Approx. 290 pages.
1977
ISBN 3-540-07206-3

M. BESSIS
Living Blood Cells and Their Ultrastructure
Translated by Robert I. Weed
521 figures, 2 color-plates. XXI, 767 pages. 1973
ISBN 3-540-05981-4
Distribution rights for Japan:
Maruzen Co. Ltd., Tokyo

Red Cell Shape
Proceedings of a Symposium held June 20 and 21, 1972 at the Institute of Cell Pathology, Hôpital de Bicêtre.
Physiology, Pathology, Ultrastructure
Edited by M. Bessis, R. I. Weed, P. F. Leblond
147 figures. VIII, 180 pages. 1973
ISBN 3-540-06257-2
Distribution rights for Japan:
Maruzen Co. Ltd., Tokyo

Unclassifiable Leukemias
Proceedings of a Symposium, held October 11-13, 1974, at the Institute of Cell Pathology, Hôpital de Bicêtre, Paris, France.
Editors: M. Bessis, G. Brecher
81 figures, 1 color-plate. 38 tables. VI, 270 pages.
1975
ISBN 3-540-07242-X

Springer-Verlag
Berlin
Heidelberg
New York